ARTICULATING DINOSAURS

A Political Anthropology

In this ambitious interdisciplinary study, anthropologist Brian Noble traces how dinosaurs and their natural worlds are articulated into being by the action of specimens and humans together. Following the complex exchanges of palaeontologists, museum specialists, film- and media-makers, science fiction writers, and their diverse publics, he witnesses how fossil remains are taken from their partial state and recomposed into astonishingly precise, animated presences within the modern world, with profound political consequences.

Articulating Dinosaurs examines the resurrecting of two of the most iconic and gendered of dinosaurs. First Noble traces the emergence of *Tyrannosaurus rex* (the "king of the tyrant lizards") in the early twentieth-century scientific, literary, and filmic cross-currents associated with the American Museum of Natural History under the direction of palaeontologist and eugenicist Henry Fairfield Osborn. Then he offers his detailed ethnographic study of the multimedia, model-making, curatorial, and laboratory preparation work behind the Royal Ontario Museum's ground-breaking 1990s exhibit of *Maiasaura peeblesorum* (the "good mother lizard"). Setting the exhibits at the AMNH and the ROM against each other, Noble is able to place the political natures of *T. rex* and *Maiasaura* into high relief and to raise vital questions about how our choices make a difference in what comes to count as "nature." An original and illuminating study of science, culture, and museums, *Articulating Dinosaurs* is a remarkable look at not just how we visualize the prehistoric past, but how we make it palpable in our everyday lives.

BRIAN NOBLE is an associate professor in the Department of Sociology and Social Anthropology at Dalhousie University.

BRIAN NOBLE

Articulating Dinosaurs

A Political Anthropology

UNIVERSITY OF TORONTO PRESS
Toronto Buffalo London

Toronto Buffalo London
www.utppublishing.com
Printed in Canada

ISBN 978-0-8020-9696-8 (cloth)
ISBN 978-1-4426-2705-0 (paper)

♾ Printed on acid-free, 100% post-consumer recycled paper with vegetable-based inks.

Library and Archives Canada Cataloguing in Publication

Noble, Brian, author
Articulating dinosaurs : a political anthropology / Brian Noble.

Includes bibliographical references and index.
ISBN 978-0-8020-9696-8 (cloth). – ISBN 978-1-4426-2705-5 (paper)

1. American Museum of Natural History. 2. Royal Ontario Museum.
3. Museum exhibits – Political aspects – New York (State) – New York.
4. Museum exhibits – Political aspects – Ontario. 5. Political anthropology.
6. Dinosaurs in popular culture. 7. Tyrannosaurus rex. 8. Maiasaura.
I. Title.

QE861.4.N62 2016 567.9074 C2016-900134-2

University of Toronto Press acknowledges the financial assistance to its publishing program of the Canada Council for the Arts and the Ontario Arts Council, an agency of the Government of Ontario.

Canada Council for the Arts
Conseil des Arts du Canada

Funded by the Government of Canada
Financé par le gouvernement du Canada

Contents

Figures

Acknowledgments

Being anthropology, this work is necessarily a philosophy and practice of the many, even if written only by one. Illuminating my thinking, writing, craft, and my commitments are so many individuals whom I wish to honour. Even if I cannot name each and every one, my gratitude goes out to all.

I must first express my deepest thanks to palaeontologist and friend Philip Currie, a lifelong interlocutor on things dinosaurian, whose combined generosity, guidance, frank conversation, and critical reading of my texts has left its mark indelibly on this book and on how I think and write in this area of research. I also thank the many other palaeobiologists who, over the years, have candidly shared their knowledge and work with me, most notably Hans Dieter Sues, Eva Koppelhus, Jack Horner, Dale Russell, Meeman Zhang, Michael Caldwell, Paul Sereno, David Norman, Mike Nowacek, and Lowell Dingus. I also acknowledge the many anthropological, science studies, and cultural studies scholars whom I have been delighted to "think with," among them Stefan Helmreich, Isabelle Stengers, Chris Kelty, Hannah Landecker, Donna Haraway, Bruno Latour, Kregg Hetherington, Annelise Riles, Marilyn Strathern, Joe Dumit, Marianne DeLaet, Tom Mitchell, Emily Martin, Ken Little, Julie Cruikshank, Bruce Miller, Janice Graham, Petra Rethman, and Gordon McOuat. I extend special gratitude to Sarah Franklin, with whom I have found the warmest and smartest of intellectual kinship. I am grateful to the anonymous reviewers of this work, who provided excellent comments helping me to reflect on aspects of my arguments that could benefit from some reworking, and where possible I did make key changes in response. I also acknowledge the support of my publishers, the University of Toronto Press, and their various editorial leaders who worked steadfastly to see the book to completion: the late Virgil Duff, Stephen Kotowych, and Douglas Hildebrand. Judith Williams and Anne Laughlin were such helpful copy and proof editors. I also thank Nina Hoeschele for her able and efficient text editing and indexing, and Amy Donovan, my former graduate student, for additional editorial support.

During my doctoral program, when I was immersed in ethnographic components of the research, I enjoyed the finest blend of mentorship and freedom to think, given by my supervising committee at the University of Alberta – Milton Freeman, Linda Fedigan, Eric Higgs, and Jean Debernardi – and by Sarah Franklin, now at Cambridge University, who was external examiner. My research could not have succeeded without the generous help of the staff and curators in many departments at both Toronto's Royal Ontario Museum (ROM) and New York's American Museum of Natural History (AMNH). For those at the ROM, you will recognize your contributions in these pages, if under the pseudonyms I have used. I am also grateful to the Royal Ontario Museum for providing me with visiting researcher status and a place to sit and work within the Vertebrate Palaeontology department, when I was conducting the crucial intensive ethnographic research at the museum that would come to form the basis of the second part of this volume. Several individuals were key in arranging for permissions and licensing of many of the images appearing in this work – I especially thank Nicola Wood at the ROM, Mai Reitmeyer and Greg Rami at the AMNH, and Mary Rose MacLachlan of MRM Associates. My thanks go as well to several archives and their staff who aided my historical research, including those at the AMNH, the British Museum of Natural History, Columbia University, University of Toronto, and the late William Sarjeant's Archive, when it was housed at the University of Saskatchewan. I must also heartily acknowledge the early inspiration to "think with" dinosaurs that I gained in the communities of wonderful folks I worked with at the Royal Tyrrell Museum of Palaeontology, and all those associated with the Canada-China Dinosaur Project, in which I was honoured to play a key role.

I also extend my appreciation to all those in my home institution who have encouraged me in this research, including colleagues in my own Department of Sociology and Social Anthropology, Deans of my Faculty, and so many in diverse disciplines across campus at Dalhousie University who appreciate the inescapable enmeshing of the natural, the human, the political. I have benefited greatly from interchanges with many fine scholars and students when presenting research and ideas from this work in various universities, programs, and meetings: McMaster University, Harvard/MIT, Dalhousie University, University of King's College, University of Toronto, York University, University of Alberta, University of British Columbia, the Situating Science Knowledge Cluster, and in the meetings of the American Anthropological Association, the Canadian Anthropological Society, and the Society for the Social Studies of Science. Several organizations and programs have provided generous funding support for the PhD research, fieldwork, and ongoing research and writing that allowed this book to be realized. These include the Canada Council; the SSHRC and Izaak Walton Killam Memorial Post Doctoral Fellowship at UBC; the Doctoral Izaak Walton Killam Memorial Scholarship at the University of Alberta; the Wenner Gren Foundation for Anthropological Research (New York), Predoctoral Research Fellowship; the Social Sciences and Humanities Research Council Doctoral Fellowship; and a Dalhousie University Faculty of Arts and Social Sciences Publications Grant.

I also must acknowledge all those who walked alongside me and offered encouragement and support in thinking, as friends and colleagues along the lengthy path of this endeavour, including Michael and Margaret Asch, Seth Asch, Regna Darnell, Joshua Smith, Reg and Rose Crowshoe, Michael Ross, Marc Pinkoski, Rob Hancock, Justin Kenrick, Mario Blaser, and Harvey Feit, all of my wonderful graduate students, and, without question, my always-believing parents, Paul and Ollie Noble.

My wonderful children Brenna and Cole have lived pretty much their whole lives excited to know their father was, odd as it may sound, an "anthropologist of dinosaurs." They continually affirm so much of the novel and playful ways that young people come to engage these performative beasts. Finally, there is one person, my partner in life Constance MacIntosh, who has traced the complex steps of the coming-into-being of this work. She has supported each and every one of those steps through her incisive reading, suggestions, idea sharing, patience, laughter, and her endless encouragement. I save my deepest gratitude for Constance, who, along with our two children, has been my fellow traveller in the terrains of dinosaurs, palaeontologies, anthropologies, badlands – all good lands really. In this we have found a world chock-full of inspiring people and of living and bygone stuff – a world that has taught both of us why we must care for the living places in which our lives unfold. I dedicate this book to her.

ARTICULATING DINOSAURS

A Political Anthropology

1 Can There Really Be an Anthropology of Dinosaurs?

I realized this was all based on these very fragmentary remains and that intrigued me … this was something where you had these rare and precious objects that then helped you in your mind to reconstruct these ancient worlds …

Andreas Henson, curator for "The Maiasaur Project," speaking of his childhood fascination with illustrated dinosaur books

BN: Why do you suppose people are so fixated on dinosaurs as something to support and talk about in natural science and in museums?

MARTIN: It's just like – they are not around anymore, and what happened to them, they were so huge. They are just so – like nothing we have ever seen. That's why.

AMY: I kind of think, the more you learn about them, and where they came from and what they lived like, it tells the story about us.

MARTIN: Yeah, the story about us.

AMY: Where we came from. It is our history too.

From interview with twenty-one-year-old Amy and twenty-one-year-old Martin, visitors to the Maiasaur Project Exhibition at the Royal Ontario Museum, 1998[1]

It is palaeontologists, not anthropologists, who consider dinosaurs as their research object – so how can there really be an anthropology of dinosaurs? The question that the title of this introduction poses is, of course, rhetorical. This book *is* an anthropology of dinosaurs, or at least a small contribution to that anthropology. It is a book about dinosaurs, but more particularly a book about *the action that brings them about*, back to life, so to speak. It is a book of dinosaurs as things and beings with which *we as humans are very much involved*. It is about the action of sciences, of sciences and natures as public cultures – that is, of the decidedly political natures of dinosaurs. Importantly, it is also about the always-changing relation between sciences and publics and the very real effects that those changing relations have upon contemporary life, for humans in all sorts of collective groupings, and for what count as natures – the geological past, prospects for possible futures.

Young Amy, with whose words I opened this chapter, really put it in the simplest of terms: *dinosaurs tell the story of us, where we came from; they are our history too*. Could there be a more anthropological claim than this one? Amy is not alone in such speculative anthropological thinking about dinosaurs. American palaeontologist and sauropod specialist Matthew Bonnan commented on his public webpage, entitled "Dead dinosaurs and reasons for hope," saying:

> Dinosaurs put our place in the world into perspective – this is not a world meant for us, but one we have had the happy fortune to inherit from previous generations of life.[2]

Bonnan's provocative statement provides a partial answer to the question implied by Amy's statement: *How do our stories participate in and with theirs, and theirs with ours? Do we share a singular story with them, or are there many stories?* There are several responses that come to mind fairly easily. The first, that of Bonnan, is arguably the most conventional one, that palaeobiological accounts reconstructing dinosaurs can be understood as "stories" of dinosaurs, stories that we come to accept as part of our story of earth and biological history. A second, more disengaged version of this is that dinosaurs and humans are simply narrated temporally, sequenced in history, with dinosaurs in the past and humans nowadays. They are players from different moments in the story. Another more responsive possibility might suggest something of a causal relation: that dinosaurs came before us, and then as a consequence of their dying off and thus becoming the erstwhile "most successful vertebrates" in earth history (as they are often described), they left open the evolutionary possibility for humans to come along eventually. Yet another possibility is that dinosaurs are analogically informative, sometimes in the most gripping fashion, where understanding what happened to them gives guidance to what might happen to us. There is also is the more fully anthropocentric possibility that dinosaurs are entailed entirely by human signification, that is, by human storying, imaginative construction, and nothing else – they are total, unmitigated fabrications. In that rather godlike and equally problematic version, we are the authors of their becoming. They are quite literally "our" story, the story from us, made by us, and so necessarily a part of us, rather than autonomous beings in their own right.

The problem is that there are elements of all of these possibilities that are interesting, just as there are strong arguments that can be marshalled against each proposition. One challenge for me is accepting that they might have value in some manner, underscoring the speculative potency inhering in dinosaurs around the possibility of relating dinosaurs and humans. So, in this book, rather than casting out any of the possibilities, I allow each of them and others to come into play. Working from my stance as an anthropologist, I consider how dinosaurs act as cultured and political beings of multiple natural possibilities in contemporary social history. It is through practical action that they become the story about us – our history too.

I attend to two particularly well known and complexly gendered dinosaurs, *Tyrannosaurus rex* and *Maiasaura peeblesorum* – through their remains, through the Mesozoic worlds in which they are *known* to have lived, and through the situated and complex actions of science and public practices in concert with dinosaur remains. In this sense, this book is also a study in *how dinosaurs act in ways that produce us* as we recognize our story and theirs to be so mutually entangled. As will be seen, that entanglement is anything but simple, even though the imaginings we have of *T. rex* and *Maiasaura* may to us seem clear, uncomplicated, and fairly well established.[3]

Such clarity is, I contend, ephemeral. The book is also about how dinosaurs and the Mesozoic natures they inhabit have never quite stayed the same. Both are in flux, both are multiple, in response to the collisions of new fossil specimens, new scientists, technical practices, new fictions, and public concerns that continually envelop them, relate to each other, and are so necessary for dinosaurs and Mesozoic natures coming into reliable, physical being. In studying this flux in the coming-into-being of dinosaurs, however, many recurrent ideas and practices of modern human/natural relations are brought to the fore, linking the most mundane to the most technical.

Consider some of the following puzzles about dinosaurs. Is it possible that scientific practice itself – the knowledge-fashioning technique meant to give us facts derived from material evidence – only contributes a small fraction to inform what we come to know about dinosaurs in museums and schools? Or, more brazenly, in palaeobiology itself? How is it that so many millions of European, Japanese, and North American children under the age of eight – if not children from across a media- and science-saturated late capitalist "world" – make sense of myriad polysyllabic names of dinosaurs, going further to sort them into complex groupings of different kinds, and then to know the time period in which they lived? Why might they even care? How much do fossil specimens tell us about how dinosaurs behaved and interacted as living creatures? How, and why, do the fossils tell us such things differently at different moments? Why is it that the first popular exhibition of dinosaurs in 1854 presented them on an island? Why, beyond coincidence, were Michael Crichton's book and Steven Spielberg's subsequent 1993 film, *Jurassic Park*, similarly set on an island? What is the historical flaw in the statement from a major international museum, "We discovered them before Hollywood"? Could it be that film and television dinosaur reconstructions are more accurate than those that the best natural science museums in the world have assembled? What does it mean *to reconstruct* a dinosaur? Or *to animate* a dinosaur?

Cutting to the specific topical creatures considered in this book, what were the actions in palaeontology, politics, and natures such that what would become the world's most influential dinosaur could emerge from museum-situated, palaeobiological research at the start of the twentieth century – that is, of course, *Tyrannosaurus rex*, "King of the Tyrant Saurians." I ask a similar question about the dinosaur *Maiasaura*, the "Good Mother Lizard," and its emergence at the end of the twentieth century. How do palaeontological experts secure their authority, and their own fascinations with dinosaur specimens, against the massive commodification witnessed in publicly and privately marketed dinosaurian spectacles? Put another way, how do scientists manage incursions of what is only supposed to be fictional into what is only supposed to be factual? Or do they at all? Do they – or, just as importantly, why should they – care about this?

The larger puzzle this book grapples with is the character of the social-material, natural-cultural relations that potently hook all of these more limited puzzles and multiple stories together. In total, this work is an extended array of historical and contemporary case studies addressed to this puzzle. At the same time, it offers a philosophical experiment in the form of a practical anthropology addressing the composing and recomposing of natures garnered from the partial cases of some particular dinosaurs, particular material specimens of them, and all this situated and in action. Two central formulations are put forward: one is on *Mesozoic Performativity*, where both science-authorized and more imaginative, fictional manifestations of dinosaurian nature have depended upon locating and materializing ever-refined and altered forms of dinosaurs in *time and space;* the second is that the mutiplicity of actions taking place across technoscientific and everyday public practices of dinosaurian reconstruction and performativity articulate in intricate, mutual ways. These two action formulations, as such, work together in bringing dinosaurs into humanly apprehendable being, while providing the arrangement for the alteration of that being over time.

So, many forces and actions work together in *articulating dinosaurs*, sometimes in rather predictable ways, other times in very surprising ways, and often in ways that substantiate particular if questionable relations of power, gender, and other normalized practices of difference. These relations can (and eventually should and will) be undone by new and alternative articulations. The ultimate proposition of this book, then, is that articulating dinosaurs is a mode of articulating power in different ways – dinosaurs as politics. The book is, quite literally, a political anthropology of dinosaurs. It is, at the same time, an extended case study, tracing the public emergences and transformations of two dinosaurs in particular, following how politics and natures of dinosaurs, as with any creatures of public concern, are always and already intimately entwined.

My task now is to situate this extended study, as historical and ethnographic encounter with the *political nature of nature*, first in relation to anthropological approaches and then in relation to aspects of the larger field of science studies.

First Milieu: Anthropologies, Political Natures

> Although the impetus for contextualizing what one studies as deeply as possible has an old and venerable tradition in anthropology, the contexts in which this fieldwork on science is being done are not like we imagined the field sites of our forebears to be ...
>
> Emily Martin[4]

William Dawe, the historically modelled but fictional dinosaur collector in Robert Kroetsch's novel *Badlands*, remarked aptly how in natural historical study, "There is nothing that does not leave its effect ... we study the accumulated remains."[5] Anthropology – like palaeobiology and nature and society – is anything but ready-made, even though it is sometimes presented that way. As a tradition that continually shifts, anthropology nonetheless carries forward some of the *remains* of its forebears.

In harmony with Dawe's sedimentological thinking, elements of two anthropologies layer up to constitute this project, a political anthropology of Mesozoic natures: one anthropology from older disciplinary conventions and a second that follows newer ones. There are good reasons for allowing both of these traditions to guide this project. Older anthropologies – such as those of Malinowski, Boas, Evans-Pritchard – worked very well in a late colonial travel mode of ethnography, looking for local "culture," often without directly suggesting the privilege and domination in knowledge/power that were entailed in practising such a "natural history" mode of fieldwork. Nonetheless, and while noting the coloniality of their approach, which even these authors struggled against, these older anthropologies developed a remarkable body of method, theory, and analysis – which may be borrowed from, if in a rigorous, resituated, and cautious manner. On the other hand, anthropologies of the last three decades have increasingly stayed home to study the highly distributed institutions and practices of the "home world" of the anthropologist (including for example sciences, markets, legal struggles, medical practices, insurance and risk, food security, bureaucracy, intellectual property, work life, state policy, and more) – or if their students travelled, they attempted to trace how these practices and institutions, from "home" and elsewhere, reached far and wide. They also increasingly borrowed from outside anthropology: from sociology, philosophy, gender and race studies, cultural studies, political and legal thought, media studies, economics, and so on. Two decades ago Akhil Gupta and James Ferguson pointed out that this decentring helped to reformulate the locus for anthropology:

> The idea that anthropology's distinctive trademark might not be found in its commitment to "the local" but in its attentiveness to epistemological and political issues of location, surely takes us far from the classical history model of fieldwork as "the detailed study of a limited area."[6]

Sarah Franklin, writing about the same time, was then at the forefront of the newly emerging anthropology of science, and commented, "Were Western science to be reassessed as a cultural practice, in the narrowest and widest senses, it arguably stands to gain, in both resources and on its own terms, as an effective, predictive, useful and interested account of its objects."[7]

Now, and extending on the senses of culture that Franklin spoke to, in this volume I am seeking to make yet another move – one that reassesses science not just as culture, but also as a practice immersed in the collective composition of natures, as Bruno Latour has phrased it.[8] That is to say, dinosaurs are not only cultural beings; they are also political beings, as a consequence of their emergence as things that capture collective human concerns, and also as fossil-born creatures that, once engaged by us, impose new concerns upon us. We belong to a common collective with dinosaurs. More than merely being influenced by political interests, their materialized outcomes actually embody, mobilize, and at every turn instantiate politics in the public life of the natural. I will refer to this and discuss it throughout this volume as *dinosaur politics/natures, or political natures*.

In this way, I am taking seriously what John Law and John Urry have noted, which is to say that what we take as the natural is, in the most dynamic sense possible, "a relational effect ... produced and stabilized in interaction that is simultaneously material and social."[9]

The technical practices of dinosaur reconstruction and public reanimation likewise can be understood as mutually immersive practices caught in perpetual trade. We know that older anthropologies studied and understood technology and material practices as products and actions of social collectives bringing the human and non-human into organized relations. Clearly, we are able to approach palaeontology as the technological practice it is. It has "raw" matter, instruments, techniques, and relations, and it produces artefacts – from diagrams to laboratories to systematically ordered collections – and these in turn are ever folded back into the circuits of palaeontological action, affecting and organizing further relations. Dinosaurs, as we also know, have come to be known through the technical practices of museums and media, composed into art, films, books, web-based media, educational curricula, exhibitions, fictional texts, advertising, toys – transformed into humanly fashioned matter – and in some manner these two technologies, palaeontological and public mediation practices, can trace relations to each other. Moreover, none of these relations, practices, and technologies are ahistorical – they each have shifted with each other across histories. In order to study this very distributed matter of the political natures of dinosaurs – across time and space – the newer anthropology has to be highly mobile, following the action wherever it leads.

Studying the political natures of dinosaurs presses us to recognize how, as Emily Martin suggested, such relations are to be found in multiple and diverse locations and encounters and all that flows around, through, and among them. Consequently, the very separation of what we thought was *context* from what we thought was *content* becomes moot in these newer anthropologies: interests and materializations are shown to merge indissolubly into one another.

A second contrast between older and newer anthropologies is that the older approaches tended to align with the dividing up of scientism and humanism. Scientistic and humanistic anthropologies opposed each other across a modern divide that saw the natural and non-human on one side, and the cultural and human on the other.[10] This divide was felt across the academy, beyond anthropology, as famously expressed in C.P. Snow's consideration of what he called the "Two Cultures," for lack of a better expression.[11] It reared itself up again in the form of the 1990s "Science Wars." In effect, the older anthropology, operating so resolutely in relation to the nature/culture opposition, offered a critical point of departure for development of the newer anthropology.

In the course of things, the very oppositional categories which had for so long organized the older anthropological project came into question – most notably from feminist anthropologists, postcolonial anthropologists, and anthropologists of science who noticed that things like "sex" and "nature" were highly cultured matters, to the point that many of the distinctions between nature and culture, or sex and gender, scientism and humanism, came undone. Marilyn Strathern anticipated what is now the current situation: "The old double model for the production of culture – society improves nature, society reflects nature – no longer works."[12]

A singular, universal "Nature" and a singular, universal history came to be understood themselves as a contingent effect of complex histories which followed and allowed only a particular conception of the working of humans and non-human things: nature and the non-human over there, culture and humans over here. Now, the newer anthropologies take the nature/culture opposition itself as but one more dimension of public practice to be considered. This may very well be why science studies scholars Donna Haraway and Bruno Latour – two notoriously anthropological non-anthropologists whose thinking I draw upon substantially – have often claimed anthropology as a chosen disciplinary location: it had always dealt with the intricate relations and intermeshing of nature *and* culture, even amidst its internal battles over whether to apply a more humanistic or scientistic approach.[13]

The shift and extension in anthropology from an intellectual ethos engaging oppositional nature vs. culture to one that would collapse or dissolve natures/cultures has been in many ways an unsurprising and consequential move, if not a wholly fluid or predictable one.[14] So, today, we see the continuing development of anthropological projects that turn their attention to the intersections and collapse of what was formerly held apart by the nature/culture divide, as exampled in the rise of "multi-species ethnography"[15] and the rising anthropological and sociological engagement with epigenetics, neuroscience, and microbiomics.[16] A particular anthropological and ethnographic engagement with macro-fauna (i.e., large animals) in the wealthy "West," so to speak, as opposed to engagement with the microbiota and genetics or with animals of "other" (i.e., usually "non-Western") societies, has been slower to develop. Franklin has addressed the historical, domestic, and biotechnological production of sheep and the collapsing of human and sheep genealogy; Raffles has considered insects in diverse and ubiquitous human relations; Haraway the complexity of companion animal relations and histories – though all of these have taken on the question of contingency of inter-species boundedness, which is something they share in common with other inter-species and multi-species studies.[17] Suffice to say that there still remain only limited rigorous anthropological ethnographic studies engaging with the larger-scale animals among us, and this current work stands as an augmentation to the literature in this regard.

Marilyn Strathern was also prescient on the importance of the shift from the older to the newer anthropology, noting "the potential consequences of the present ecological necessity – namely that we make explicit the participation of nature and culture in each other."[18] Here – and whether we are speaking about climate change, desertification, biodiversity collapse, the juggernaut of capitalist resource exploitation, or the failure of synthetic antibacterial vaccines – the question of the reciprocal ecopolitical impositions of non-human forces and human ones becomes palpable.

Bruno Latour, in conversation with philosopher Isabelle Stengers, has responded to our most recent moment of ecological necessity, though he cautions us not to assume that the interplay of nature and politics is only a recent matter:

> [N]o matter how novel this situation appears to be to us, it is certainly not the case that we have suddenly moved from a situation where nature was kept away from politics to a situation where it is now entangled into political arenas. In a very deep sense, politics has always been about things

> and matter. It has always been, to take up again the old and beautiful term rejuvenated by the Belgian philosopher Isabelle Stengers, a cosmopolitics, by which she means not an appeal to universality or to life in big metropolis [*sic*], but a politics of the cosmos.[19]

When taken together, the otherwise contrasting older and newer ethnographic anthropologies turn out to be consonant with the cosmopolitics that Stengers and Latour rejuvenate. Hearkening a rather Boasian mode of practice, Latour commented how "even the most rationalist ethnographer is perfectly capable of bringing together in a single monograph the myths, ethnosciences, genealogies, political forms, techniques, religions, epics and rites of the people she is studying," adding how in this "you will not find a single trait that is not simultaneously real, social, and narrated."[20] Such a proposition applies readily to the political anthropology of dinosaurs I set out here, one that certainly considers dinosaurs as "simultaneously real, social, and narrated," and as such, "the story of us."

The perennial problem faced by anthropologists, whether in the older or newer praxes, remains one of *articulation*, that is, the question of *how what is real (being), social (collective), and narrated (storied) are assembled together substantively*, and conversely how they are taken apart, disarticulated. Articulating collective, storied being is resolutely political work – as it is in this anthropology of dinosaurs, and in understanding why articulation has become a central term and practice in the work that follows. It is also a clearly fitting term for thinking of and with dinosaurs, given its use in palaeontology and vertebrate anatomy to describe how skeletal and other bodily elements are assembled together into a corporeal, structural, or biofunctional relation. I use the verb form "articulating" in titling this book, since much of what I am considering is the pragmatic action in the relation between scientific workings and public workings as dinosaurs come into being. I am also keenly aware of the multiple meanings of articulation (which I expand upon in chapter 8), from the vernacular sense to the sense suggested in Gramscian theory (as taken up by Stuart Hall, Ernesto Laclau, and Chantal Mouffe, among others).[21] However, the approach to dinosaur articulation I offer has particular affinities (which arise and are considered in differing degrees, at different places in the volume) with the ideas and practices from Deleuze and Guattari, Stengers, Latour, Mol, and Fujimura, about which I will say more in later chapters – chapters 2, and 5, and especially in chapter 8, which introduces the ethnographic section of the book.

A Second Milieu: Monsters, Specimens, Spectacles, Beings

Apart from anthropology, the second milieu I work within and take up in thinking about dinosaurs is the intersection of scientific and public practices. Specifically, I consider dinosaurs as political forms of life emerging within and across what I call the *specimen-spectacle complex* – most recognizable in, though not exclusive to, the action of museums.

The ideas and accounts here extend on a longer involvement with the question of the culturing of nature through the relations of dinosaurs, monstrosity, and the geographical and historical imagination.[22] That research began with a consideration of propositions from science historian Martin Rudwick on the notion of "scenes from deep time" – the seventeenth- to nineteenth-century genre of pictorial rendering of ancient life-worlds produced under the guidance of natural theologians, natural historians, geologists, and palaeontologists, but available as well for public circulation.[23]

In these earlier explorations, I tracked the Western history of monsters (and "monstrous races") and the manner in which they have come to be narrated, depicted, and made intelligible by being bounded in remote locales in time and space. In medieval Europe, monstrous races were thought variously to live in Egypt, Ethiopia, India – a decidedly orientalist tradition of othering.[24] Through visualizing stories and street-criers' accounts, monsters became a matter of common street or village *spectacle*. Paralleling but diverging some from this history, in seventeenth-century Enlightenment natural philosophy, monsters and anomalous things also became objects of wonders, *specimens* for philosophical investigation and ordering as in the case of Baconian "teratology."[25] As geology and palaeontology came into being in the eighteenth and nineteenth centuries, one of the most useful means of disseminating the sense of ancient life was through "scenes from deep time," to use Rudwick's terms. Dinosaurs and their ancient kin had the character of monsters – being anomalous, hybrids, giants – and they were also amenable to both scientific investigation and public wondering. Effectively, as examined in the more recent work of historians Bernard Lightman and Ralph O'Connor, they could be taken as specimens and the object of scientific investigation, or as spectacles and the object of public marvelling.[26]

Moving between these two domains – of fossil specimens for science and spectacular visions for society – were the pictorial visions of which Rudwick wrote. In 1854, with the otherwise bizarre, new fossil forms of ancient terrestrial creatures having been ordered through comparative study of fossil specimens, the pictorialization of dinosaurian time-space was recomposed in a materialized, three-dimensional scene – an artificial saurian island display in Crystal Palace Park (see figure 1.1). The south London pleasure garden and architectural space displayed a progressive natural and human history in an overall project of civil improvement, becoming a model for the newly emerging public natural history museums of the day.[27] Authorized by Sir Richard Owen, dean of British natural history, the work of transposing the scientific procedure of specimen study into a visible naturalistic spectacle had been achieved.[28] Moreover, the display of dinosaurian nature was, from the outset, as much a social practice as it was a scientific one. Dinosaurs had become a part of public human history as well as natural history.

At the same moment, the great public museums of the world were emerging. As cultural theorist Tony Bennett noted, this combination of objects of study with spectacles of display could now be coordinated in a larger governmental project of public improvement and civil disciplining through the public museum:

> [T]he museum might be regarded as a machinery for producing "progressive subjects." Its routines served to induct the visitor into an improving relationship to the self ... the space of the museum was also an emulative one; it was envisaged as a place in which the working classes would acquire more civilized habits by imitating their betters ... In these respects, the museum provided its visitors with a set of resources through which they might actively insert themselves within a particular vision of history by fashioning themselves to contribute to its development.[29]

The museum became an apparatus for enrolling the people into particular histories of nature with a motive to improve – progress. It is the *vision* of history, or, more pointedly, *each of the envisioned time-spaces from history*, which focused my initial interests, for those assembled visions stand as elements of what counted as "nature" – the bounded picture or time-space was what connected the scientific and the public in whatever class- or self-improving project was deemed appropriate. Following Krzysztof Pomian, Bennett also remarked on the "ocular-centric" dimensions of natural historical collections and displays since the eighteenth century:

> What can be seen on display is viewed as valuable and meaningful because of the access it offers to a realm of significance which cannot itself

Figure 1.1. Benjamin Waterhouse Hawkins's rendering of the Secondary Island, ancient inhabitants, Crystal Palace Park, London (1854). Source: Phillips 1859, 169.

> be seen. The visible is significant not for its own sake but because it affords a glimpse of something beyond itself: the order of nature, say, in the case of eighteenth-century natural history collections.[30]

Natural historians, then, had a special purchase on hidden ordering that was accessible through vision, and museums became the institutional locale for gaining that access and regulating it socially.

Bennett referred to the wider array of commonly available public spectacles, arcades, department stores, museums, country fairs, pleasure gardens, and shows in the nineteenth century with the phrase "the exhibitionary complex," which skews his work more to the action of spectacle practices, from Leicester Square spectacles to museums to proto-cinematic spectacle.[31] Modifying the terms of the "exhibitionary complex," in my discussion I have attended to the more specific and yet broader relation of specimens with spectacles, and the actions of the trading between them – the "specimen-spectacle complex." Such a complex, and all that it organizes and generates within it, can be thought of as an "apparatus" or a "dispositif," after the propositions of Michel Foucault and Giorgio Agamben, whose work I contextualize in chapter 5, "Politics/Natures, All the Way Down."[32] Arguably, this is a major entry point for conjoining the story of humans and dinosaurs, the common story mentioned by young Amy. Dinosaurs were an entity both constituted as specimen and, in turn, reconstitutable in public form as spectacle, and indeed as characters in science fiction literature and film. The ordering and regulating of that form was in part achieved by incorporating it into a bounded time-space. Incorporating dinosaurs and other ancient creatures into a stabilized natural order and natural world picture, a time and a space, made them, to some extent, more normal – that is, natural rather than monstrous or preternatural beings. The project of incorporating more and more creatures through systematic naturalistic study and consequent pictorialization and world-making will without doubt continue in contemporary palaeobiology, as the practice is foundational in both the technical and imaginative action of the discipline.

As I discuss in the first section of this book, two prominent (and, indeed, intergrading) times-spaces for such bounding are highlighted: from geology and palaeontology, the "Mesozoic era"; and from literature and natural historical exploration, "the lost world." These fields of containment form a nexus for concentrating meanings, fossils, techniques, and practices in an intelligible, visible form. Since the later nineteenth century, with dinosaurs well entailed by palaeontological description, they have become, along one axis, normalized matters of natural fact known from the ongoing studying of specimens. Along a second axis, they are still widely used as spectacles and as figures in science fiction, continuing to borrow on their figurative effect as monsters, giants, and "terrifying" creatures – witness Hollywood's long tradition of giant reptilian monster movies. The dinosaur in spectacle and publicly consumable form is mobilized across an amazing array of sites and media, from popular books and toys to cartoons, kitsch, the internet, and school programs.

To this day, however, it remains that the dinosaur specimen-spectacle complex is most clearly expressed and enacted in museums.[33] One of the important effects of museums through exhibitions is that they do connect the action of scientists with the action of the public – and hence of what counts as nature *and* society. They are "zones of contact" between the networks of science and society more widely considered. Other such zones or crossover locales/productions are popular science films, semi-popular books by scientists, public lectures, and so on. As I point out, it is in these locales that scientists often relax their technical performativity enough to allow the ever-present culturing of their practices to be revealed.

A starting proposition in the first part of this book, then, is that bounded time-spaces – chronotopes – have continued to connect the action of science with the action of public culture, effecting a rich trading and fusing of interested practices in the mix. So I start the book by tracing a genealogy of the dinosaur specimen-spectacle relation, with large meat-eating dinosaurs, and *Tyrannosaurus rex* in particular, as a key locus, spring-boarding from the early twentieth century by considering the palaeontological activities of the American Museum of Natural History and Arthur Conan Doyle's novel *The Lost World*.

I must also situate the work in this book in relation to American literary scholar and cultural theorist W.J.T. Mitchell's *The Last Dinosaur Book*, which stands as a major effort to consider dinosaurs as/in culture. Mitchell's wide-ranging, at times tongue-in-cheek analysis accounts for dinosaurs as "cultural icons" – or, as he put it, as the "totem animal of modernity," delving into popular, scientific, museological, and pictorial representations. It stands more or less as a representational history of transatlantic and American iconology of dinosaurs.

Reckoned as totemic figure, the dinosaur (which Mitchell writes of generically) "has an uncanny capacity for working both symptomatically and diagnostically ... [expressing] the political unconscious of each era of modern life."[34] While it is hard to argue with Mitchell on the point of the correspondence between modernity and dinosaurs – a point extending on earlier propositions of Theodor Adorno, who referred to the "dinosaur as symbol of the 'monstrous total state'" and specifically the modern capitalist American state – the matter Mitchell largely eludes is that dinosaurs are more than icons, symbols, totems, or projections of the political unconscious. They are also *natural beings*.[35] Mitchell comes closest to the question of the natural-cultural *positioning* of dinosaurs within modernity – and how they are practically definitive of modernity therefore – in his allusion to Lévi-Strauss on totemism, quoting Lévi-Strauss's classic work on the topic: "The term totemism covers relations, posed ideologically, between two series, one natural, the other cultural."[36] To Lévi-Strauss's point, Mitchell adds, "the whole 'world of living things' – at least for human beings – is actually constituted as a world of acts of naming, image making, and classification ... We don't ever 'see nature' in the raw, but always coded in categories and clothed in the garments of language and representation."[37]

So, we still need to move beyond this sort of decidedly representational approach from Mitchell, knowing full well that dinosaurs are actual, once-living, scientifically verified, reconstructed, and always reconstructable beings. Indeed, one dinosaur palaeontologist, Phillip Currie, suggested to me that this was the unaddressed concern with Mitchell's book, and the reason why palaeontologists were likely to forget about it quickly. A second dinosaur palaeontologist, Hans Sues, put the question to me differently: "So, we know that dinosaurs are cultural icons. My question is, how did they actually get that way?" It is the challenge of engaging and merging both the actuality of dinosaurs as beings and their actuality as icons – that I take up in this volume. This is also a turn on the interesting proposition of geoscientist David Fastovsky when he notes, citing Mitchell's work, "The influence of dinosaurs on social climates has been well documented ... [i]ts converse – the effect of culture on dinosaur paleontology – has not been thoroughly investigated."[38] The turn I take is to reckon with "both/and" as it were, that is to consider the actions on their own and together, the outputs of each, the sourcing for each, the trade between, and the modifications made as this action unfolds.

Therefore, the question I ask is what sorts of actions and relations of humans and non-human matter have been mobilized to bring dinosaurs into particular reanimated being and notably iconic being, and how both iconic and non-iconic being then works its way into the mobilizing of non-human matter, including the supposedly "raw" matter, fossils. How has this been achieved so forcefully in the modernist milieu that we now share with dinosaurs? This means making careful and considered moves beyond natural historical, museological, fictional, cinematic, and popular engagements with dinosaurs that others have offered, interesting as these cases may be.[39] It also means moving beyond thinking excessively of dinosaurs as, for instance, a kind of psychoanalytic "transitional object." However, in sympathy with Mitchell, I do seek the readmission of *phantasies* and *imaginaries* into individual or collective fashioning of materialized dinosaur natures.[40] Such moves demand greater precision in relation to Mitchell's astute observation that "the dinosaur" is interesting not so much because it helps to distinguish fantasy from reality, "but because it frustrates and challenges that skill with a complex, intractable object."[41]

To put this all in rather pithy terms, this book investigates how dinosaurs and their worlds are materialized performatively, articulating much that has been separated otherwise. But I do not stop with that as though it were the end of the story. Once again, I relate this to the larger political and anthropological project of the book, which is *to take Amy's proposition seriously*, that dinosaurs *are* the story of us, that they are *our* history too. This is the project of following actions of people, allowing dinosaurs to be constituted by our participation, by our concern to activate and animate fossils well, by our concern for what we call nature – an unfolding together between humans and animals which Donna Haraway has smartly referred to as "living with" and "becoming with."[42]

Tyrannosaurus to *Maiasaura*, Performativity to Articulation: Two Complementary Accounts

What the scholarly literature on dinosaurs as public/scientific creatures most lacks are detailed case studies. This book offers a slow case study, presented in two complementary and more or less chronological sections. It is a political anthropology of dinosaurs in two movements.

The first part of the book, "Animating the Tyrant Kingdoms," principally addresses the genealogy and actions of twentieth-century Mesozoic performativity, paying notable attention to the appearance of the dinosaur *Tyrannosaurus rex*. It speaks more to the animating topos of the Mesozoic and lost world as that which resolves at the modern nexus of the specimen-spectacle complex, bringing past natures and creatures to life, so to speak. The second, "Articulating the Good Mother Lizard," is an ethnographic study of the public precipitations of the dinosaur *Maiasaura peeblesorum*, in a museum exhibit of that creature. This part of the book attends more to the practices of articulation at play bringing this dinosaur into scientific/public being. The first section also sets out historical conditions and terms that ground the articulation practices contoured in the second section's ethnographic accounts. This then moves the focus from the natures performed into being to the means by which such performativity is conducted.

At this juncture, I will provide a brief summary of several chapters for part 1 of the book. Part 2 has its own framing chapter (chapter 8).

The following six chapters track a particular history of this specimen-spectacle trade. I address how the Mesozoic and its counterpart, the lost world, set up a very material "performative" topos – or *chronotopos* – both within and produced by the specimen-spectacle apparatus. They both *result from* and *effect the trade between* fictional and scientific otherworld-making, infusing these otherworlds with physical, technical, political, social, masculinist, and racializing concerns. Out of this twentieth-century genealogy came a fixation on large meat-eating dinosaurs, most notably *Tyrannosaurus rex*. Ultimately, the first section foregrounds and offers some crucial terms regarding the precipitating nexus which generates Mesozoic and lost worlds and the generalized trade between them, and the emergence of *T. rex*.

Chapter 2 sets out some basic terms of Mesozoic and lost world performativity and considers how this allows the simultaneous articulation of particular practices of "human" and masculinist sociality and *phantasmatics*. Next, chapter 3 presents the particular case of the "The Doyle-Osborn Nexus" – conjoining the action in Arthur Conan Doyle's racially and gender-configured adventure novel, *The Lost World*, with the evolutionary exploits of the American Museum of Natural History in New York City under the presidency of eugenics advocate Henry Fairfield Osborn.

In chapter 4, something of a pivot point for the first section, the appearance in palaeontological and public life of *Tyrannosaurus rex* under the influence of Osborn's evolutionary modelling impulses towards *supreme* animal kinds is considered in more detail. The discussion also works to demonstrate how particular techniques of dinosaurian reanimation in palaeontological and cinematic practices emerged together in the work of museum display development for the first major exhibit of *T. rex*. The meeting of techniques also facilitated the embedding of Osborn's racial-apical theory formulations in this very influential museum locale, captured in the reanimated *Tyrannosaurus* skeletal constructions.

In chapter 5, I discuss how the Doyle/Osborn phantasies and practices, when set against the case of the American Museum of Natural History's Akeley Gorilla diorama as engaged in the work of Donna Haraway, demonstrate "kinship" between the dioramic and the Mesozoic, including the neocolonial effect of both of these "apparatuses." The insertion of simian figures sets up a more resolute public-scientific nexus around the phantasized progression and juxtaposition: dinosaur – ape – human. Here I provide a more particular discussion of the idea and praxis of "politics/natures."

Chapter 6 discusses the resilience and reuptake of the human – simian – saurian modelling practices that became routinely entrenched in the traffic of dinosaur performative nexuses, most familiar in popular culture. In this instance, I track the masculinist technical and phantasy trade between authenticated dinosaur palaeontology and the envisioning practices of popular Hollywood and fictional literary production.

Then, in chapter 7, I take up a question posed earlier on the actions scientists take to manage phantasmatic engagement. This question I address through a consideration of scenarization in dinosaur illustration practices, and then through the now paradigmatic approaches of phylogenetic systematics, also known as cladistics. While, arguably, the figure of the lost world is in certain ways slowly receding from Mesozoic performativity, a critical point here is to note the ongoing potency of phantasmatic action in the changing disciplinary work of palaeobiology.

The eight chapters of the second section of the book constitute an ethnographic account of the action of *articulating* work which instantiates, transforms, and complicates the performative nexus of the Mesozoic, with the dinosaur *Maiasaura*, "the Good Mother Lizard," as its locus. The ethnography focuses on the research, development, and marketing of – and public engagement with – a major high-tech museum exhibition on this dinosaur. The Maiasaur Project exhibit ran at Toronto's Royal Ontario Museum between 1996 and 2005. The ethnography tracks, among other things, the engagements of the dinosaur palaeontologist and ROM curator Andreas Henson, who oversaw the acquisition of the fine Maiasaur specimen ROM #44770 and who curated the exhibition. It follows the development and transformations of the specimen, an in-gallery laboratory, interactive animation displays, and the participation of those who assembled the exhibit, those who operated it, and those who engaged it as visitors, revealing the often unpredictable way that the Mesozoic and its constituent fauna are recomposed in the making and engaging with this exhibit. The account tracks *articulations* and *disarticulations* in public and technical engagements with *Maiasaura*, detailing how dinosaur reanimation is diverted along kinship lines when scientists and exhibit staff cooperated and struggled – always with attendant care – in the planning and coordinating of their efforts to activate specimen ROM #44770. My eventual proposition concerns how the specimen and its allied pewter model stand out as *factishes* – a neologism coined by Bruno Latour but one very befitting dinosaurs – that could articulate to understandings of "family relations," dynamic processes understood in palaeobiology, and presumed audience-markets for the exhibit in what are so often consumer capitalist circuits.

This two-part work *also* stands, therefore, as a slow case study giving shape to the idea of the "factish," after Latour (though also discussed by Stengers and Haraway), by tracing the emergence and action of dinosaurs as beings that symmetrically preserve, act, and fuse both fact and fetish in their constitution.

This juxtaposition and flow in the book – two dinosaurs, two curators, two confluent historical spans – are also, in certain ways, experimental. They allow me to propose the sort of political natures modernity offers up, and what comes into play through disciplined practices within mostly museum-situated palaeontology, achieved through the specimen-spectacle trade. *Maiasaura*, *T. rex*, and the Mesozoic, in many ways, *become* the method for understanding performativity and articulation. So, while on first blush *T. rex* and *Maiasaura* may appear as simple gendered binaries, and while gendering is certainly working here, the story turns out to be much more complex. The two sections lead us through this palaeontologically and publicly instrumental binary to its eventual collapse.[43]

However, the juxtaposition is also much *more* than methodological. It is empirical, drawn by direct engagement with experiences in human-material action. The accounts make it very evident, for example, that the action of performing the Mesozoic turns out to have been more orchestrated and authorial in the early part of the twentieth century, as exampled by the emergence and stabilization of *T. rex* at the hands of Osborn. In contrast, the action of articulating work in the latter twentieth century, in the emergence of *Maiasaura*, turns more on a democratic involvement of a "team" of people working together and drawing upon a more heterogeneous genealogy of antecedents. That transformation in itself is telling of the shift in museum milieus from a top-down, expert-privileging modality in the early twentieth century to a more democratically configured, if market-interested, modality in the late twentieth century. While Mesozoic performativity continues apace, the manner in which more complex sets of actors and active forces are involved in it has changed the possibility of what kinds of dinosaurs and conformations of the Mesozoic might appear. This of course portends futures of ongoing transformations both in dinosaur and Mesozoic kinds, and in the complexes that bring them about.

With that point in mind, the book ends with the chapter "Just Trying to Be a Scientist," the lament and wish uttered by the Maiasaur Project exhibit curator. The lament is recognizable as a plaintive reclaiming of curatorial care in the face of complex, market-trending milieus of contemporary museological practice, which I also take up in this closing chapter. Setting the ROM curator's action against that of Osborn in the early twentieth century, the book closes by considering the place of curatorial scientists and museums generally – as practices of care – and their potency in furthering the realization that *another Mesozoic is possible,* one that can be responsive to the ecological necessity of our current moment. It leaves open the question of how new public concerns merge both in research and in the political transformation of dinosaurs and other fossil creatures.

In arriving at the closing of this book, one contribution that I hope readers will have gained from its two sections, each with its own orientation and approach, is how this juxtaposition also points to two techniques of knowing and being. One technique is that of drawing out the performativity of worlds in time and space. The second is the technique of following the intricate action of articulating dinosaurs into humanly apprehensible beings. Of course, the book conveys the story of the congress between these two actions, these two knowledge techniques, something that came from simply engaging with the specimen-spectacle complex, with attention to dinosaurs. To reiterate, performative worlds are wrought by the action of articulating, but, as they are constituted, they also constrain how articulations may proceed.

This too is the contribution I hope to make to the emerging horizons of anthropology, and beyond anthropology. My hope is that readers from various areas of scholarship and public life who read this book may be able to draw upon these actions and techniques in fruitful ways, especially in thinking about politics/natures. Palaeontologists might reflect on the choices they make, knowing that world-making is such a powerful aspect of what they do. Museums and exhibit planners and designers may consider differently how they channel or redirect the knowledges that scientists and curators so carefully assemble. And those who seek to engage with the larger concerns of the politics of difference – notably anthropologists, sociologists, gender- and race-critical scholars, cultural studies scholars – will find these techniques helpful, especially when turned to considering the intimacy of human/non-human relations in shaping socio-natural worlds in which we may live well together.

These are some modest tools and practices that this anthropology of dinosaurs might also provide.

Some Remarks on Reading, Writing, and Materializing

When taking on such a wide-ranging matter as the articulation of dinosaurian political natures, which reaches into geoscience, museums, popular movie making, literary fiction, everyday experience, anthropology, science studies, history, gender studies, sociology, cultural studies, and political theory, I am clearly faced with the challenge of multiple potential audiences. To narrow matters, I have tried to write with four specialist audiences in mind: anthropologists, science studies specialists, diverse cultural studies specialists, and palaeobiologists.

Some years ago, Sarah Franklin wrote of the challenge in writing across the very mixed field of studies captured within "science studies," referring specifically to anthropology:

> One of the most important concerns facing anthropologists of science is how to enable their work to speak to the broadest audience of scientists, social scientists, and other scholars. It remains unclear what language is needed for this to occur.[44]

I anticipate, however, that the writing will challenge scientists the most. I have been fortunate to have maintained friendships and professional associations with many palaeobiologists over the last thirty years, and my hope is that they will take what follows as an extension of the discussions we have engaged in together over the years, and that they will respect any technical peculiarities in language and description presented, much as non-palaeontologists respect the technical languages of palaeobiologists in their scholarly work. A reading of this book and its scholarly sources will show that although there is a strong orienting to the languages of social sciences and humanities, I also write and source the work in scholarship from the biological and geological sciences, and of course from palaeobiology.

In the spirit of any scholarly engagement, I attempt at times to describe complex concepts and practices for which no particularly stablized lexicon exists, so a specific working lexicon has been developed. The terms and lexical constructions I introduce are tools for knowledge sharing across multiple intellectual approaches. That said, the languages and terms used are not entirely novel, often having been borrowed and modified from the wide disciplinary gamut informing this research. But through much of the text, I strive for plain language.

In those places where I intently explore certain terms, I provide definitions or glosses either in the notes or in the body of the text. I also use long notes to detail matters that may be more salient to a particular audience. I ask the reader to bear with the slower reading that may occasionally come at these points, in order to get a sense of the complex matters to which I am gesturing. I move back and forth between the cases and analytic commentaries, putting these various terms to work, with the aim of bringing the reader to a fuller imaginative engagement with the propositions developed over the many chapters of the book.

Readers should also anticipate a sort of ricochet of concepts and practices between the two sections, two time frames, two movements of the book – opening the possibility for speculating on future unfoldings of the Mesozoic. One example that emerges as a major, recurrent account in the book, the history of dinosaur scale-model techniques and techniques of cinematic animation, appears in chapter 4 on the American Museum of Natural History and Osborn's reconstruction of *T. rex*, in chapter 6, "Vestiges of the Lost World," and then again in the discussion of the Maiasaur Project's "technotheatrics," chapter 14.

Finally, given how highly circulated and recirculated dinosaurs are in public, I am also relying on the *imaginative capacity of readers* to make practical connections between their own experiences of dinosaurs and what is discussed in the book. That readers possess this straightforward, participatory capacity to connect their imagining through palaeontology, cinema, reading, gaming, museums, etc. serves to affirm how we readily inhabit the worlds of dinosaurs, just as they so readily inhabit ours.

PART ONE

Animating the Tyrant Kingdoms

2 Materializing Mesozoic Time-Space

Dinosaurs erupted into human consciousness in an age when rationality – in the invention and development of great Victorian disciplines like geology, glaciology, and evolutionary biology – contended with a legacy of Romantic melodramatic cosmology from Percy Bysshe Shelley, Mary Shelley, Emily Brontë, and the dream worlds of writers like Samuel Taylor Coleridge. A century and a half later we cannot place dinosaurs firmly in our world picture. Do they lurk in the glowering shadow of a primordial consciousness like that which summoned a Heathcliff or Frankenstein's monster? Or do they stand majestically in the bright sunlight of Charles Lyell's vision of an endlessly recreating earth?

John Whyte[1]

Where do we place dinosaurs in our world? Few would argue that they can be very dramatic, sensational, and at times unserious entities. Yet, they convey much that overflows the serious technical commitments from which they are also derived: the stakes in scientific truth, the determining of human/non-human relations, the possible anxieties over our own ultimate demise akin to dinosaur extinction, or the dream of excess embodied power. Tom Mitchell calls them the "totem animal of modernity," using "totem" in both the Freudian and Lévi-Straussian senses: they stand in for so much that is modern, and they bond people together (or separate them) in meaningfully modern ways.[2] When dealing with dinosaurs as unserious stand-ins, we are allowed to be exuberant, excessive, irreverent, parodic – all because they are both physically not us and materializations of much we may believe in. In a single stroke, we can embrace them and cast them away.

At the same time, certain representatives of their ilk – notably the giant, sharp-toothed kinds – have been so extremely *re*-presented as to become the regular inhabitants of contemporary, middle-class North American nightmares, threatening that which provides the greatest security – the American family dog, for instance. In a scene from the first of three sequels to the film *Jurassic Park*, the parents from a canonically white, heterosexual, middle-class San Diego family rush to their child's room to comfort him after his exclamation, "There's a dinosaur in the backyard." They turn to look out the window and with horror and disbelief meet the gaze of the rampant papa *T. rex*, chain and dog house of the devoured family pet hanging impotently from the beast's maw.

That was the summer of 1997, when the largely forgettable Steven Spielberg film – which unashamedly adopted the title of the Arthur Conan Doyle book, *The Lost World* – was released upon the world, earning more than $230 million in box office receipts.[3] What is far more disturbing about this scene than the threat portrayed to American-dream security is that Spielberg could consciously compose and direct it, knowing all too well that this could be a box-office selling feature. He knew how great its cultural purchase could be, and followed the formula: *First*, let a state-of-the-art animated *Tyrannosaurus rex*, the giant bloodthirsty alien, out of its bounded otherworld, wherever that might be: on an island; seventy million years in the past; in the psyche; in a foreign land. *Second*, set it upon middling American domesticity. *Result*: you stand to frighten people – most notably the domestic middling Americans who are Spielberg's primary target market – think of *E.T.*, *Jaws*, *Close Encounters*. *Finally*, in the socio-economic domains of Western industrial techno-spectacle, you also stand to make enormous profits by peddling such thrills.

The monster, released from its boundedness into the world of humans, is the operative apparatus here; in this instance, an apparatus that pits the exotic and terrifying against the domestically stable. It is the same sort of set-up that rationalizes the commercial viability of gated communities and homeland security policies as a buffer against spectacular fears of drive-by shootings, car-jackings, home invasions, and terrorist attacks on the most hallowed of American institutions.

In keeping with this apparatus, these next discussions introduce, by example and analysis, how the imagining of time and space has ensured and continues to ensure that humans, dinosaurs, and other non-humans alike are persistently *entangled* via the equally imagined geography of dinosaurs.[4] The geography I concentrate on here is known alternately as the "lost world" and the "Mesozoic": two intergrading terms that designate the quasi-fictional, quasi-factual time-space locales in which dinosaurs are resurrected into scientific and public being. Some, like Crichton, Arthur Conan Doyle, Spielberg, Edgar Rice Burroughs[5] – and indeed, on occasion, some scientists – refer us to the first term, the "lost world," as a locale in which "life continues." It keeps dinosaurs and otherwise bygone creatures safely bound until, as in the Crichton-Spielberg case, they escape into domestic backyards or break through electrified fences or, with human technical assistance, defy their otherwise normative destiny as extinct beings.

Of course, what I have mostly foregrounded in these introductory descriptions are filmic and literary figurations. A more challenging issue is to consider how mass media, public, *and* scientific practices all share in such technologies of saurian otherworld-making, grounded at times in the evidence of their existence: the fossil matter that speaks to us of the once-being of these astonishing creatures.

Time-Space and Performativity in the Lost World and Mesozoic

The title of this book, "Articulating Dinosaurs," is shorthand for a highly recurrent action in both expert and public engagements with dinosaurs.

Here is the two-part proposition I offer: *(1) In the work of fossil reconstruction, dinosaurs are simultaneously performed and articulated from out of a plausible past time, and from fragmentary bits of matter, into a more or less complete, recognizable, materially palpable, sometimes animated form in the here and now. (2) This procedure simultaneously effects a secondary articulation: that between the humans and the fossils and all that is put to work in generating the robust effect of a time-space in which dinosaurs may live, and (usually) outside of which humans live.*

In this first section of the book (up to and including chapter 7), I will be emphasizing the first part of the proposition. To give shape to these kinds of actions, I will spend the next few pages discussing three particularly helpful terms from social and literary theory: *chronotope*, *performativity*, and *phantasy* – pointing to their applicability when thinking of the figures of both "the lost world" and "the Mesozoic." This brief foray into key terms and theory is a prelude to some further examples of how these notions are manifest in everyday practices of palaeontologists, which in turn will help to orient the discussions in the following chapters – using "orient" here in the rather *apropos* verb sense "to find one's position in relation to new and strange surroundings."[6] In that spirit, I begin with "the lost world," followed by "the Mesozoic," and then move into the three social theory terms.

A featured lead character in many of the novels of Arthur Conan Doyle is the British professor of zoology, George Challenger. Challenger, the idealized male adventurer-scientist, was first introduced in Doyle's *The Lost World*. In that novel, while attempting to rationalize reports of yet-living prehistoric creatures, Challenger offered a succinct description of the lost world, outlining it as lucidly as any living cryptozoologist[7] might:

> [T]here can only be one explanation. South America is, as you may have heard, a granite continent. At this single point in the interior there has been, in some far distant age, a great, sudden volcanic upheaval. These cliffs, I may remark, are basaltic, and therefore plutonic. An area, as large perhaps as Sussex, has been lifted up en bloc with all its living contents, and cut off by perpendicular precipices of a hardness which defies erosion from all the rest of the continent. What is the result? Why, the ordinary laws of nature are suspended. The various checks which influence the struggle for existence in the world at large are all neutralized or altered. Creatures survive which would otherwise disappear. You will observe that both the *Pterodactyl* and the *Stegosaurus* are Jurassic, and therefore of a great age in the order of life. They have been artificially conserved by those strange accidental conditions.[8]

In effect, some rather *monstrous* forms of life are isolated from what are accepted as regular natural-historical processes that should otherwise produce familiar, *normal* forms of life. Notwithstanding the Darwinian struggle that it suggests, this technical-sounding description might just as readily have been used as an illustration for Francis Bacon's *Advancement of Learning* (1605). The logics operating here are remarkably parallel to Baconian *teratology* – that is, the study of monstrosities: *nature in course* is interrupted by *nature erring*.[9] Effectively, in Doyle's lost world, nature's artifice (nature for Bacon being decidedly feminine and reproductive) allows the monstrous to persist over time by bounding it in space. In the 1990s, Michael Crichton's *Jurassic Park* would, in effect, repeat the same basic conventions that had informed Bacon's teratological reasoning.[10]

In contrast, and complement, to the "lost world," the technical term for the geological and faunal time-space locale of dinosaurs is the "Mesozoic Era," said to span the time period from 248 to 65 million years ago and known through study of sedimentary geological localities. Both terms have come to signify journeying, especially masculine journeying – a central thematic of these discussions and those of later chapters. The Mesozoic and the lost world have become a unified and highly influential location around and through which the actions of science practitioners and public authorities have *cultivated* senses of nature, of certain forms of humanness, and of particular histories of life on earth. They are "formalized" entities that have significant generative force for creating a rational ordering of life – they are outcomes of modernity, and thus celebrations of modernity. Anxiety, fear, and expressions of courage and bravery have become a large part of the sense they entail – as has the numbing multiplication of menacing, meat-eating dinosaurs from *Tyrannosaurus* to *Velociraptor*, creatures made all the more menacing by their naturalized juxtaposition against placid, herbivorous dinosaurs.

Both the lost world and the Mesozoic are time-space figures. They are at once *locales* and *moments*, and they organize and mobilize certain sorts of action. Mikhail Bakhtin's figure of the *chronotope* is eminently applicable in setting the terms of consideration for how dinosaur geographies – palaeontologically authenticated or not – have continued to operate through the past and present. He remarks:

> We will give the name chronotope (literally, "time space") to the intrinsic connectedness of temporal and spatial relationships ... What counts for us is the fact that it expresses the inseparability of space and time (time as the fourth dimension of space). We understand the chronotope as a formally constitutive category of literature; we will not deal with the chronotope in other areas of culture.[11]

The Mesozoic/Lost World allows for a ready extension of the chronotope into another area of "culture": that of modern palaeontological science and its organizing ethos of travelling. Bakhtin dealt mostly with the chronotope of "the road" as an organizer of orderly narrative movement through time-space in "familiar territory." He also compares this genre to that of "wandering" narratives characterized by Greek Sophist or Baroque novels, where "a function analogous to the road is played by an 'alien world' separated from one's own narrative land by sea and distance."[12] In all such genres, Bakhtin suggests:

> The chronotope is the place where the knots of narrative are tied and untied. It can be said without qualification that to them belongs the meaning that shapes narrative.[13]

Bakhtin goes further:

> Without such temporal-spatial expression, even abstract thought is impossible. Consequently, every entry into the sphere of meanings is accomplished only through the gates of the chronotope.[14]

Another key effect of chronotopes is how they necessarily engage in and produce a reality-phantasy[15] exchange; a movement between what counts as concrete and what counts as abstract. This is salient in considering both the lost world and Mesozoic, which stand as struggles against incredulity and phantasy to establish a sense of the real. Again, while resonating with Bakhtin, this also complicates his point:

> However forcefully the real and the represented world resist fusion, however immutable the presence of that categorical boundary line between them, they are nevertheless indissolubly tied up with each other and find themselves in continual mutual interaction; uninterrupted exchange goes on between them, similar to the uninterrupted exchange of matter between living organisms and the environment that surrounds them ... this process of exchange is itself chronotopic: it occurs first and foremost in the historically developing social world, but without ever losing contact with changing historical space.[16]

Chronotopes, then, may organize worlds, times, and narratives.[17] However, as will be suggested in regard to the variety of practices that make dinosaurs intelligible, chronotopes can also be tacitly drawn upon, and so unify palaeontological and literary practices. In both instances, chronotopes aid in navigating the ever-uneasy relation between what is accepted as real or accepted as imagined. They are a means for creating intelligibility, bounding out spatially or temporally material worlds and their constituents. At the same time, they confer *materiality* upon things that are extraordinarily imaginary in their constitution. Considered side by side, the Mesozoic and the lost world upset any strict distinction between reality and fabrication, truth and lies, fact and phantasy. As such, I am suggesting that they also cross between and conjoin the actions of technical, scientific, and public knowledges.

Further, part of the action reinforcing such fusing of reality/fabrication takes place beyond the laboratory, the museum, the scientific collection. What are reckoned as sites for fictional accounts, as in the case of science fiction literature and cinema, are often engaged by scientists as locales to assert influence over scientific debates. This observation has been put forward forcefully by David Kirby, who followed science consultants, including American palaeontologist Jack Horner, a key figure in the second part of this book, working on Hollywood films. Kirby rightly notes:

> The ground over scientific ideas is not limited to scientific meetings and publications, or even to traditional popularizing realms such as documentaries and newspapers. Fiction provides an open, "free" space to put forward speculative conceptualizations.[18]

To press these points further into service, I want now to focus on the effects of Mesozoic or lost-world chronotopes in the mutual making of the scientific and public cultures of nature associated with dinosaurs. I work to advance the proposition, therefore, that what are taken as media and literary performances or phantasies[19] are also, quite properly, part of the *performativity* of palaeobiological practice. They are part of the reality-making work of science, not something outside it, nor do they simply influence or bias it; rather, chronotopes are actually engaged as part of the syntax of technical scientific practice, making the findings of palaeontology – and thereby the "Mesozoic period" and the "fossil record" – so much more palpable.[20]

Judith Butler points out that such sorts of phantasies, "when wielded within political discourse … posture as the real"; or, as Jean Laplanche and J.B. Pontalis put it, "phantasy constitutes a dimension of the real."[21] That crucial point – of the phantasmatic, imaginary *reality* of the Mesozoic chronotope, along with its constituents – is what I will be pressing and expanding upon throughout this first section of the book. In addressing palaeontology and dinosaurian "natures," I also seek to move from Butler's concentration on gender performativity to a more widely distributed sense of spatial-temporal-material performativity[22] and of human and non-human performativity, in order to recognize how the Mesozoic has acquired its remarkably unquestioned status as real.[23] Reality, at least in this form, is nothing short of materialized human phantasies in exchange with such non-human, yet equally contingent, entities as fossils, strata, and previous representations.[24] I should emphasize that in raising "phantasy" I am not privileging its ontological status, but rather putting it forward as a tool; a competency associated with abstraction, thinking, imagining, and desire.

Butler emphasized the importance of materialization as a means of avoiding the restrictive conceptions of "social" or "cultural" constructions: "What I would propose in place of these conceptions of construction is a return to the notion of matter, not as site or surface, but as *a process of materialization that stabilizes over time to produce the effect of boundary, fixity, and surface we call matter*" (italics in original).[25] That process, or proposition, is referred to as "performativity." To be sure, what I offer here is one contribution to a still-growing scholarship on how performativity, in Butler's sense, can be modified and extended to explain how "the economy" or "the market" is generated and comes to have such potent effects – on people, on what comes to count as "society," and on human or non-human "things" and environments more generally.[26]

Of course, I am pressing performativity in ways that considerably exceed its conventional origins in linguistics, most notably in J.L. Austin's speech act theory.[27] Butler writes:

> Within speech act theory, a performative is that discursive practice that enacts or produces that which it names. According to the biblical rendition of the performative, i.e., "Let there be light!," it appears that it is by virtue of the power of a subject or its will that a phenomenon is named into being.[28]

Citing Jacques Derrida, Butler notes that the "subject" does not have to be "God" – as the biblical example suggests – but can simply be those who are accorded agency to act, and so to enact performatives.[29] The most cited case from Austin is the illocutionary performative, the action of marriage ceremonies where the priest or adjudicator utters the performative, "I pronounce you …" and so produces the effective "reality" of marital union. As Butler points out, such kinds of performative acts "are forms of authoritative speech: most performatives, for instance, are statements that, in the uttering, also perform a certain action and exercise a binding power."[30] Honing this further, Gilles Deleuze and Félix Guattari might suggest that performatives operate as "order-words," insofar as they make an authoritative declaration of what the majority ought to or, even more, is obliged to accept.[31] At the same time – and this is something I wish to keep hold of in my discussion all through this book – for Deleuze and Guattari there is always the potential for escape, or in their words *lines of flight*, staged by and inherent to order-words, such lines being actions that allow alternative possibilities to be tried, and even to succeed in destabilizing the order enacted in, by, and through order-words.[32]

In the case of the Mesozoic and the lost world, some particularly authoritative subjects who act through both their utterances and practices are geological or biological scientists and fiction and popular literature writers; those who bring these respective hidden worlds into being, in part by pronouncing them, in part by mobilizing a complex of rationalizations along with an array of things (including fossils) and accounts offered as "evidence." Part of my project is to show how some of these other agents – whether human or non-human, beyond just authorized subjects – are brought into play in the performativity of the Mesozoic. As Butler suggests, this is considerably more than, if not wholly different from, social construction.

Butler does gesture to some of the larger complexities at work beyond the simple action of a single actor at one moment in time. She writes:

> Performativity is ... not a singular "act," for it is always a reiteration of a norm or set of norms, and to the extent that it acquires an act-like status in the present, it conceals or dissimulates the conventions of which it is a repetition. Moreover, this act is not primarily theatrical; indeed, its apparent theatricality is produced to the extent that its historicity remains dissimulated.[33]

Here, Butler suggests that that which is performed into being has force because it draws upon and mimics performances, or performatives, which have preceded it historically – that is, by a complicated history of citation and recitation. But she also points to the second sort of performative act after Austin, the perlocutionary, which is

> those utterances from which effects follow only when certain other kinds of conditions are in place. A politician may claim that "a new day has arrived" but that new day only has a chance of arriving if people take up the utterance and endeavor to make that happen. The utterance alone does not bring about the day, and yet it can set into motion a set of actions that can, under certain felicitous circumstances, bring the day around.[34]

The Mesozoic works in this manner, as it has precursors, relies on conditions that make it sharable and effective with others, and opens to new circumstances that will allow it to be reconstituted, at least partially. Yet, whether illocutionary or perlocutionary, these performances no longer appear to be performances, or even strategic actions, because they have become so accepted as to hide the history and particularity of the practices informing them. A plain, if rather simplistic, linguistic example is in how the repeated use of the term "xerox" to signify a photocopy ultimately came to be so stabilized in language by the 1980s that the word's corporate origin actually became lost, concealed in the utterance, dissimulated. Obviously, with the term "xerox," much more came into play along the way from marketing and trademarking practices, socio-economic justification for copying, technical developments of photocopying machinery, effective monopolization of workplaces in the distribution of photocopying equipment, securities trading, the history of photography and mechanical reproduction, and so on.[35] Thinking deeply about performativity allows dimensions of such lost actions to be recovered.

It must be noted that Judith Butler's fine-grained analyses on the politics of gender performativity are closely allied to Lacanian psychoanalysis.[36] My intention here is not to follow such an explicit psychoanalytic approach in these discussions – even though the terms used are significantly parallel. But neither am I suggesting that such an approach should be precluded, or that it might not be productive – indeed, this discussion welcomes the readmission of such alternate approaches to studies of science. My intention, rather, is to turn the terms "phantasy" and "performativity" in a different direction: towards modern techniques for materializing a physical nature, nature that is generated – or more accurately, co-generated – rather than given, that is, a nature that is articulated.[37]

My use of *phantasmatic* is aligned with the idea of phantasm as apparition, or with that of a projection as in the nineteenth-century spectacle apparatus, the "phantasmagoria" – a still-image precursor to the theatrical film projector.[38] As such, it is a use of "phantasy" which is not grounded restrictively in psychoanalytic discourse. Moreover, this particular sort of phantasy is grounded in the "ground," as geologists have come to study it. At the same time the Mesozoic, being a figure derived from the study of subterranean features, has a very literal *underworldly* character, such that discussion should no doubt resonate with the subconscious as discussed in psychoanalysis – or, indeed, in gothic literary genres.[39] This subterranean aspect of the Mesozoic, as a sort of buried imaginary realm, stands it as something that undergirds materialized natural being – in this instance, dinosaurs themselves. One could as readily argue that geology's sedimented earth is the model for the psyche, as well as the reverse: that psychic interests are the model for geology. It is through these blended "psychogeologies" that the Mesozoic/Lost World figure, at least in part, organizes dinosaurian intelligibility while hiding within its terms the personal phantasies by which human actors are guided. This will become clear over the next five chapters, which discuss the constitution of great meat-eaters (most notably *Tyrannosaurus rex*) in relation to the actions of Arthur Conan Doyle, Henry Fairfield Osborn, their allies, and those who followed them in practices of Mesozoic performativity and dinosaurian animation.

The performative proposition of a highly complex "set" of relations – between what happens in palaeontological reconstruction, in literary composition, in exhibit design and construction, and on and on – is meant to point to a series of human/non-human engagements that gather together to produce the reality-effect we call "dinosaurs," inhabiting equally effective time-space "worlds" such as the Mesozoic. To think of dinosaurs as these complex outcomes or phenomena is also to extend on and layer up physicist Karen Barad's formulation of performativity in Bohr's quantum physics where, as she puts it, "phenomena are the ontological inseparability of agentially intra-acting 'components.'"[40] As Barad explains, there is never an actual, but only a provisional separation between the observer and what is being observed. Rather, it is the complex varieties of moments of intra-action that generate *phenomena* that have the equally provisional effect of being bounded "things." Subatomic physicists regularly set up experimental situations with technical apparatuses that allow them to apprehend phenomena within the provisional constraints of their experimental techniques. Barad refers to such moments as "agential cuts," where the human, the technical, and the phenomena are brought into play together, and are necessarily dependent upon each other for the efficacy in generating "positive" results.[41] As the next several chapters unfold, it should become clear how dinosaur beings – as, for example, *Tyrannosaurus rex* – and their mutual inhabited apparatus of worlds come about positively in a similar way. Phantasies, parlayed by humans, come to play a significant role in this.

With all this set out, the propositions I make are *about the scientific performativity of the materialized phantasy of the Mesozoic* and *its constituent fauna*, and equally how this complex (noun) trades with public cultural practices, and so is effectively reliant on that trade for its very being. Nature is modelled, performed, animated – and, in short, enacted – into being, and that action brings people and things incessantly into exchange. This enacting also helps to regulate and produce palaeontology's authority along with many other entities distributed across the spectrum of public-scientific activity. It connects certain publics to the relevant science, and to scientists. The reckoning of life, the histories of biological relations, the replaying of masculinist phantasies, the revisiting of colonial logics, the performing of social relations, the anxieties over human and species vulnerability and environmental destruction – all are worked upon in this ever-revisable landscape of modern, naturalistic life and death. In larger, anthropological terms, while the historical reiteration of performatives gives them their force, my proposition is that performatives akin to the Mesozoic are also densely and widely represented in human and non-human action. The performances range from literature to science, to museum displays, to entertainment, to everyday discourse, to child's play. In this sense Mesozoic performativity is both historical (diachronically emergent and mobile) and heterogeneous (synchronically distributed and adjustable), and consequently a persistent, modern technique for generating normative yet revisable *natures*.

I will make many returns to these ideas throughout this book, and ultimately in its conclusions – which will bring to bear the second part of my proposition regarding the articulation work of dinosaur palaeontology as public practice, via animated film, literature, and museums practices, etc.[42] But to bring the performativity points home, I will complete my discussion with some specific examples of how it is that scientists address themselves to what is recognizably phantasmatic: the very notion of lost worlds.

Bracketing Out Phantasmatic Worlds

Since, as I have already suggested, dinosaurs can be taken quite unseriously, it is not surprising to find out that those who study them also have both serious and less-than-serious relations with these arguably monstrous creatures. Of the highly published dinosaur palaeontologists with whom I have worked closely over the last few decades, every one has been engaged in what is more often seen as "imaginary" or "speculative" work – sometimes spoken of as part of their extra-scholarly activity – but which nonetheless articulates to their serious palaeontological work.[43] One crucial consistency across both their technical and non-technical work, however, is the speculative and innocuous imagining of an otherworld inhabited by alter-beings.

Take Dr Philip Currie, for instance: formerly the senior curator of dinosaurs at Alberta's Royal Tyrrell Museum of Palaeontology; currently a Senior Research Chair holder at one of Canada's leading universities; and the editor of *ERBivore*, a journal of enthusiasts for Edgar Rice Burroughs's Tarzan. Notably for Currie, in the *Pellucidar* trilogy Tarzan encounters dinosaurian creatures called "Gryfs."[44] As Currie has remarked in many public lectures, the Burroughs books were a childhood key to launching his palaeontological career trajectory, as were his boyhood discoveries of plastic dinosaur figures in packages of Cheerios in the late 1950s.

Or consider vertebrate palaeontologist Dr Hans Dieter Sues, who has demonstrated to me in many conversations his encyclopaedic knowledge of lost land, giant monster movies, from the 1933 *King Kong* to the Toho Godzilla films. Sues, recently the associate director for collections and research at the Smithsonian National Museum of Natural History, regularly gets asked by the media to speak to the palaeontological veracity of the monsters in films from *Jurassic Park* to the American film remake of *Godzilla* (1998) (figure 2.1).

Then there is Dr Dale Russell – professor emeritus of earth sciences at North Carolina State University, former curator of vertebrate palaeobiology at the Canadian Museum of Nature, and one of the world's leading authorities on dinosaur extinction. Some of Russell's use of imaginative modelling has been explicit, becoming a flashpoint for both public and palaeobiological controversy. A faithful Roman Catholic, Russell has long admired the philosophy of Pierre Teilhard de Chardin, and consequently developed an intellectual frame (in accord with Teilhardian ideas of Platonic "shadows" which could be cast everywhere in the fossil record)[45] where divinely sourced design in organic forms might very likely recur convergently in evolutionary history.[46]

One of Russell's daring "thought experiments" was given considerable credence by NASA's SETI ("Search for Extraterrestrial Intelligence") program: in this case, Russell offered expert speculation of a humanlike "Dinosauroid" (see figure 2.2), a form that could plausibly, if not actually, have evolved on earth had there been no mass extinction at the end of the Mesozoic. The Dinosauroid is an imaginative figure converging upon and blending forms found in large-brained, small, carnivorous dinosaurs and in *Homo sapiens*. Russell's speculative creature was inspired in part by some notable UFO close-encounter reports.[47] SETI's interest is clear enough here: if we have theoretic models allowing for repeated evolution of the humanoid form on earth, then why should such models not be applicable elsewhere in the cosmos, and so lend some kind of credence to reports of encounters with humanoid extra-terrestrials?

Figure 2.1. Godzilla, and fellow inhabitants of Toho's "Monster Island." Source: Toho/The Kobal Collection

Both Currie and Russell have ended up having their tales re-presented in some popular press media: Currie himself has been a figure in Japanese Manga magazines, and Russell has found his Dinosauroid drawn into conspiracy-theory-mongering when appropriated by sensationalist tabloids – there, his creatures plot worldwide conquest from their hide-out at the centre of the earth.[48]

In a very thoughtful and rather tongue-in-cheek response to the public reaction to his Dinosauroid, having lauded one particularly "well-written article in *Omni*" magazine, Russell points out that the media were quite sensible in their engagement with his thought experiment: "The public owes a great debt to the intelligence, breadth of knowledge, and integrity that I have uniformly encountered among science reporters."[49] In contrast, Russell encountered very different sorts of responses from anonymous peer reviewers of his technical manuscript on the speculative cousins, *Tröodon* (*Stenonychosaurus* [sic]), which is a palaeonotologically described creature, and the Dinosauroid, his thought experiment. One wrote:

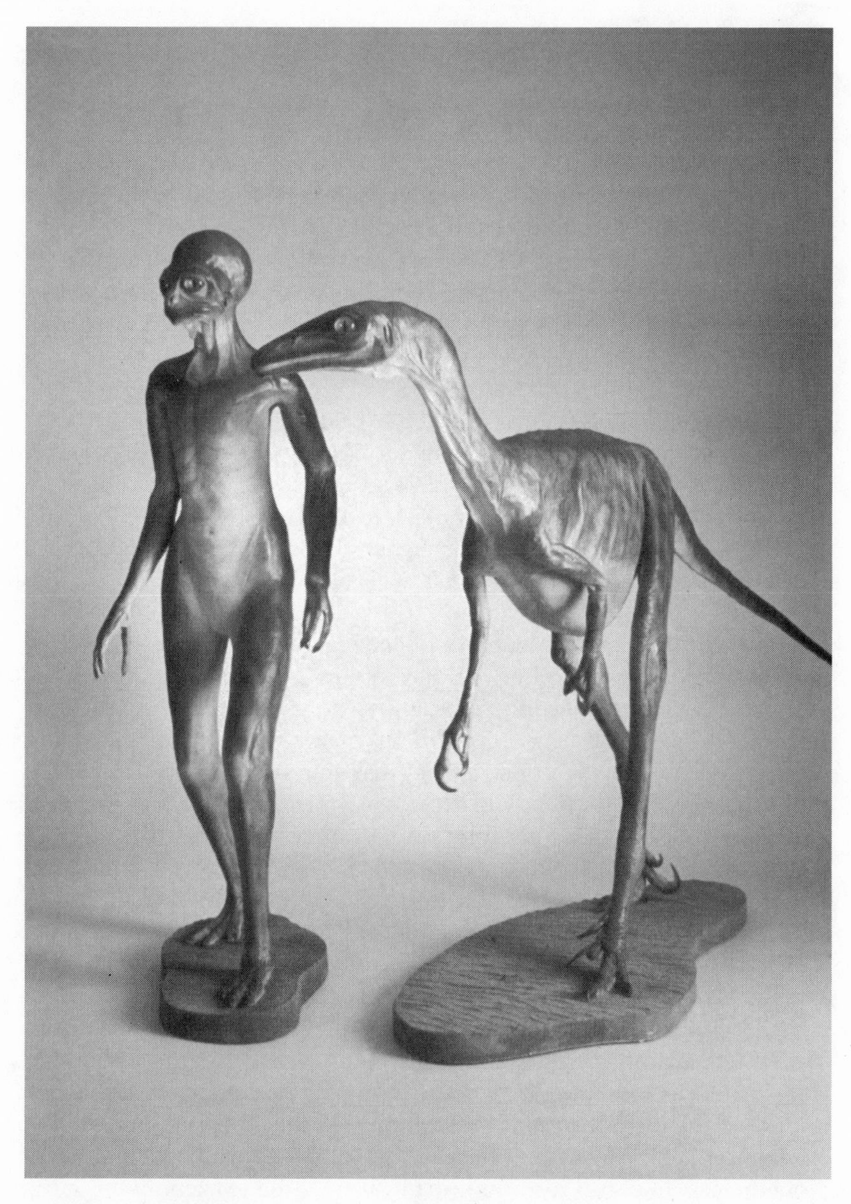

Figure 2.2. Models of *Tröodon* and a hypothetical large-brained descendant. Source: Photograph by Robert Fillion, Dale Russell, and Ron Séguin, © Canadian Museum of Nature, Ottawa. Cf. Russell 2009, 217; Russell and Séguin 1982.

> The ideas and methods used in the reconstruction are thorough and even elegant. However, I do not see much value in the extremely speculative "dinosauroid" discussion. Dinosaur studies today are already characterized by a prominent science fiction component.

It was such comments, no doubt, that prompted Russell to explain to his colleagues that in developing the Dinosauroid "there was no contact with the producers of 'ET.'" Expressing even greater concern about the potential erosion of scientists' credibility and practices – and with a concerned jab at Russell's convictions towards directional causality in evolution – a second reviewer cautioned:

> The [dinosauroid] model you caused into being with such surprising results would be difficult to display without encountering retorts you would probably not welcome. Among your peers you would stir much trouble, righteous ridicule and much tut-tutting. I would recommend that – after you have a sufficiency of good photographs – you keep it securely out of sight; or destroy it if the skilled artists will not be too hurt.[50]

Many, including most palaeontologists reviewing Russell's article, would say that all this otherworldly thinking, the drawing of curious affinities between dinosaurs, aliens, and fictional literature, is just extra-scientific noise, hobbyism, amusement, media excesses, or indulgent speculation, with little effect on the real action of science. But these same three scientists (i.e., Currie, Sues, and Russell) have also told me explicitly that they quite literally visualize Mesozoic dinosaurian worlds and imagine ecological scenarios in the act of conducting their research – even as they prospect for fossils or gaze down their microscopes at tiny surface morphologies. Despite this direct phantasizing, they usually claim to draw a clear separation between this imagining and the technical work (i.e., the translation of physical evidence to information to models) of observing, collecting, comparing, reconstructing, systematizing, and so on. In effect, they see themselves as making rigorous efforts to *bracket phantasies out*, giving their technical work a kind of purity, authority, and a more reliable truth-like character. They know all too well that they're not supposed to let such stuff in. Nonetheless, like any forbidden fruit, it must be extremely enticing, and I argue throughout this book that it does make its way in – indeed, it is a necessary feature of how scientific work actually functions, and it is precisely that which makes science a human endeavour in its exchanges with all that may be taken to constitute the non-human and the human cosmos, including dinosaurs.

Bracketing In and Performing Mesozoic Time-Space

> If I say "we assume that the earth has existed for many years past" (or something similar), then of course it may sound strange that we should assume such a thing. But in the entire system of our language games it belongs to the foundations. The assumption, one might say, forms the basis of action, and therefore, naturally, of thought.
>
> Ludwig Wittgenstein[51]

One has to wonder what truly does get bracketed out, and how palaeontologists resist the potential flood of supposedly phantasmatic incursions into their scientific practices. Well, of course, the regular answer that a scientist will give you is that dinosaurs, Wittgensteinian assumptions aside, did once live; they have left real fossils, and what we experience in literature or film or the media or in imaginative flashes is merely exaggeration, simplification, momentary speculation, or outright distortion. In contrast to such mediated phantasies, however, there is this other, scientific, temporalized landscape of dinosaurs that is taken to be very real.

As mentioned already, geologists and palaeontologists generally refer to this time-space locale as the Mesozoic Era. This geological frame is said to have started roughly 248 million years ago, ended roughly 65 million years ago, and consisted of three periods, each with its own series of sedimentary geological sequences: the Triassic, Jurassic, and Cretaceous. The most oft-cited boundary-making devices or indexes in this second scenario-making case are the fossils in geological facies (i.e., strata).

In his 1970 presidential address to the American Society of Economic Palaeontologists and Mineralogists, geologist Digby McLaren emphasized how crucial boundary definition was to the palaeontological enterprise:

> Boundaries require definition, just as much as frogs [*sic*]. Correlation means time in the proper sense of the word, and is the central underlying primary task of geology. Life is the only, and will remain the only method by which boundaries may be defined and correlated on a worldwide scale to a degree of accuracy far beyond any other. To me this is one of the

> most important and challenging roles of palaeontology. To this all studies must ultimately be related, whether of morphology, ecology, or evolution. Without knowledge of time, there can be no lineages, and lineages are morphologically distinct units in an evolutionary continuum influenced by the environment. The proper study of life of the past must involve time, and we derive time from boundaries.[52]

Following these principles, morphological, ecological, and evolutionary reconstruction – a sort of "world-making" practice – is achieved within circumscribed boundaries in time and space, by way of temporally delineated strata within encompassing stratigraphic columns. Moreover, life and its lineages are understandable through time; time is understandable through boundaries; boundaries are understandable through the remains of life; the remains of life are understandable through their lineages – and so forth. This very contingent, circular, and reiterative process over the history of geological discourse tends to "harden" the most useful boundaries, making them more robust and consequently more enduring. Once established, a bounded, correlative temporality – the Mesozoic, for example – tends then to predict the fossils that will be found once one has a previous understanding of the age of the geological sequence out of which these fossils are drawn. Certain sorts of life predict certain bounded temporalities, and bounded temporalities, in turn, predict those sorts of life. To name and identify a fossil creature that indexes the Mesozoic, therefore, is to call the Mesozoic into being – the performative gesture – and so to reinforce the bounded "field" that contains the fossil, the creature, in space and time.

The specific designation "Mesozoic" was proposed by the British geologist John Phillips in 1841 (the same year that Richard Owen coined the term "Dinosauria"). It was intended for just such a cause of boundary definition; in this case, to emphasize faunal remains that indexed the geological sequences. The Mesozoic, or "middle animal" era, containing as it did many bizarre and gigantic saurians, soon obtained the normative vernacular designation as the "Age of the Reptiles" – as opposed to the Cenozoic (also Cainozoic) or "newer animal" era, which has taken on the vernacular designation "Age of the Mammals."[53] This simple hierarchy of successive life-worlds has likely contributed, for instance, to the cultured oppositionality of the reptilian against the mammalian in wider discourses (including such popular culture cases as the late 1990s North American, animated children's television program

"Beast Wars," which pits typically evil saurian-cyborgs against typically good mammalian-cyborgs). The name Mesozoic has stuck and expanded tremendously in public and scientific discourse ever since Phillips coined it, with an equally tremendous diversification of the stratigraphic, faunal, and floral diversity it contains. In effect, a framing has been imparted, much as in the semiotic process of category definition,[54] and with that a revisable "otherworld" of beings with their own histories and relations, defined in space and time, has been built up.

The phantasmatic slipperiness of this otherworld-making – as well as the incursion of what amount to Bakhtinian travel narratives – is quite apparent when considering, for instance, the notion of the "lost world" as used by American palaeontologist George Gaylord Simpson in *Discoverers of the Lost World*, his 1984 volume reviewing the history of palaeomammalogical study in South America. The book is subtitled "An account of some of those who brought back to life South American mammals long buried in the abyss of time," assigning, if only in jest, a rather godlike character to paleontologists' figurative capabilities to resurrect life. But the very style of the subtitle mimics and thereby preserves the phantasmatic twist of Doyle's subtitle of *The Lost World*: "Being an account of the recent amazing adventures of Professor E. Challenger, Lord John Roxton, Professor Summerlee and Mr. Ed Malone of the *Daily Gazette*." Inversely, Doyle's subtitle attempts to lend a sort of staid scholarly tonality, while Simpson readmits to his whiggish history a playful fictional tone. The two books may be seen as complex inverse plays on what counts as factuality and fictionality.

Nonetheless, Simpson attempts to restore the factuality of palaeontology after musing upon and *bracketing out* the fictionality of the lost Amazonian plateau written of by Doyle:

> So much for the fictional "lost world" … [however,] South America does indeed have a lost world. That world is not living on [isolated plateaus]. It is present in the vast extent, both in space and in time, of the geological strata that have been laid down over the hundreds of millions of years of geological time … [there are] … many lost worlds … changing constantly and thus becoming lost by extinction, by replacement, and by other changes as time went on.[55]

By a simple rhetorical move, Simpson – one of the most lauded evolutionary theorists of the twentieth century and co-developer of the "Modern Synthesis" in biology – has also *bracketed in* what we might better understand as the "true" lost worlds of the past. These will include, of course, such sedimentary sequences as the Cenozoic, the Mesozoic, and within the latter the Triassic, Jurassic, and Cretaceous periods.

Similarly, in an article on palaeobiogeography – the technical study aimed at reconstructing past life-worlds – Australian palaeontologist Ralph Molnar put it this way:

> [T]he past was literally a foreign world, and a "trip" into the Mesozoic would take us to a place unrecognizable except to specialists in the evolution and history of the earth.[56]

That is the important connection: the Mesozoic, as much a temporal domain, is also a place you travel to, and expertise makes it recognizable.[57] In this vein, Polish palaeontologist Zophia Kielan-Jaworowska stated what has come to count as a basic romantic motivation:

> No scientist familiar with the intellectual adventure of studying animals from times long past will have any hesitation in affirming that to travel millions of years back into the past, which is what palaeontological study amounts to, is much more fascinating than the most exotic geographical travel we are able to undertake today.[58]

These are rather colonial intonations, speaking as they do of travel, adventure, exotica, and privileged knowers. But in addition to the colonial language, all of this romantic, palaeontological time travel talk remains quite phantasmatic, indeed a journey to nowhere – and it provides direct examples of Bakhtin's chronotopes. It is a socially powerful, exclusive set of phantasies that is rooted in the intensity of practices held to back up the claims: here there is also a claim to evidence, to facts, to the matter that gives the constructed worlds that extra veracity to distinguish them as scientific worlds from those of public phantasy (which are somehow held, in opposition, to be less "grounded"). These tales of searching for material evidence expressly mix up references to time and space. In doing so, they effectively mobilize the spatial journey to the fossil locality as a journey of time – to study fossils is to travel across time. In turn, to travel to this bounded space/time is to find the fossil, map its location, impart privileged knowing in order to place it in a rational order of things, and return home with the valued object. The parallel with colonial subjugation is easy enough to recognize.

Centrally, however, there are important commonalities among all of the accounts I have presented so far, including the supposedly extra-curricular, out-bracketing accounts of palaeontologists mentioned earlier: they all produce and operate by means of a time-space geography where the saurian aliens live – in Burroughs's Pellucidar, in Godzilla's Monster Island, in Russell's speculative extraterrestrial or terrestrial worlds, in the *National Enquirer*'s saurian earth core, or in the palaeontologist's Mesozoic biogeography, the abyss of time. Like *Jurassic Park*, and indeed, like any of the multitudes of kitsch roadside attractions, theme parks, and museum displays presenting dinosaurs and their ilk, all of these sites, one way or another, *contain* these tribes of "fearfully great lizards."[59]

To get a more nuanced sense of how all these considered fictions and facts, sciences and spectacles of time and space have come together – and to point out some of the strange yet ubiquitous social and political work they have been able to do, especially in regard to constituting and privileging particular kinds of "subjects" and publics in line with certain kinds of dinosaurs – I now want to "travel," in an empirically detailed way, to the early twentieth century.

3 Land of the Fear, Home of the Bravado

Having suggested how the Mesozoic and Lost World chronotopes launch certain kinds of travellers' stories, I will use this and the next two chapters to delve further into the kinds of objects and subjects, animals and humans inhabiting and animating these stories and materializations in the early twentieth century. The politics implicit in the relations performed will be brought to the fore, as will the means by which this complex of natural/cultural action is extended more widely into both public and expert terrain. The questions I begin to think about are fourfold: Where is it that dinosaurs live? Who knows that place? How did they find out? And what kind of public (political) nature does this engender?

Tyrannosaurus and other giant, carnivorous dinosaurs come into physical being and are presented as central actors in these accounts – playing, it turns out, as key figures in stories of the imperial conquest of nature. I focus on the racialized, gendered, and classed arrangements of humans and creatures animated by such stories; hierarchies which in turn animate these lost lands in time, and even the very biological mechanisms that were understood to have led to these extraordinary, now-extinct carnivores. As such, it is the political phantasies and stories *animating the human network* of peformative relations that I foreground in this chapter by seeing how they appear simultaneously in the practices of palaeontology, museum display, and literature.

This approach will stage discussions in the two subsequent chapters, centring on the American Museum of Natural History in the early twentieth century. Chapter 4 considers the scientific and public discovery/coming-into-being of that most widely performed and performative creature, *Tyrannosaurus rex*, and the way in which it was materialized.

Chapter 5 then addresses the collapsing together of public-scientific techniques of nature-in-the-making that permit such highly animated, articulate phantasies of the natural to move fluidly between past-world and contemporary-world materializations. Put simply, these chapters are about how Mesozoic performativity goes public to generate particular political sensibilities, and vice versa – or, put another way, about dinosaurian nature as politics by other means.

The Mesozoic Animates Doyle's Lost World

The intensity of trading between literary, filmic, and scientific actions of the sort I have introduced is nowhere more visible than in the early part of the twentieth century. New fictional/factual spaces of the lost world and Mesozoic were found in natural history museum displays, in the palaeontological elaboration of the Mesozoic, and in popular literature, all of which refracted and rebounded off each other incessantly, and the most mainstream of which operated by some form of Darwinian logics. Two canonical expressions in these domains were, respectively, Arthur Conan Doyle's romantic adventure novel *The Lost World* (1912) and the palaeontological and public display complex of the American Museum of Natural History (AMNH) in New York. During most of the first three decades of the twentieth century, the AMNH operated under the direction of Henry Fairfield Osborn, a notable advocate of a variety of Lamarckian principles – which, as it turns out, helped to brace up his white, Anglo-Saxon supremacist eugenics philosophy, a philosophy which resonates with the ideas present in Doyle's novel.

I will start with an examination of Doyle's work as a literary study of "science as culture," but will move to Osborn's networks to exemplify the performative trade I have introduced already. The narrative of Doyle's 1912 *Lost World* novel may be abstracted as follows: an irascible English zoologist by the name of George Challenger presents some limited evidence at a meeting of the London Zoological Society of a virtually unexplored Amazonian plateau where time has stood still and prehistoric creatures still survive – a host of great dinosaurs, marine reptiles, giant mammals, flying reptiles, and even a "mysterious" white spirit beast among them. The young journalist who narrates the tale, Edward Malone, had been very impressed by Challenger's evidence, including a fragment of a bat-like wing supposedly from a pterosaur.[1]

For the Zoological Society, however, the evidence was incomplete and potentially contrived. With the proofs disputed, a new expedition is commissioned by the Society. The group includes another eminent zoologist, Professor Summerlee; a "great white hunter" figure and dandy, Lord John Roxton; plus the journalist, Malone – and is eventually rejoined by Challenger. The colonialist tale then spins through the journey to the lost world plateau, where dinosaurs are indeed encountered living variously in conflict or harmony with a hierarchically ordered array of humans and proto-humans. Ultimately – in the manner of Gulliver proving his sanity by the revelation of his Lilliputian sheep – the team returns with the final, indisputable proof: a living Pterosaur.

Along one axis, *The Lost World* performs much like a modern museum with its tale about wonders of nature, evidence, and the production of facts. However, and more to the point, it is also a tale of the imperial and colonial adventure networks of male bonding – that is, of an ardent homosociality.[2] That homosocial performance network helped in the early twentieth century to produce both the veracity of dinosaurian worlds and a more common knowledge of progressive evolution. In turn, it also authorized the work of scientists and the institutions of science.

Doyle's lost world operates through the idea of a rupture between the civilized and the savage, creating the sense of both evolutionary (i.e., temporal) and geographic (i.e., spatial) distance from the book's reader.[3] The journalist Malone, in keeping with the gender-normative practices of the book in its time and place, has joined the journey to show his fiancée, Gladys, his capacity for manly heroism. The rupture is clearly expressed in the narrative after the team arrives on the plateau. Here, Malone has ascended an adjacent pinnacle of rock and made a temporary bridge of a tree, which has promptly fallen into the chasm, leaving them "lost," as it were, in the remote world. Malone ponders the situation:

> By no possible means could we get back to the pinnacle. We had been natives of the world; now we were natives of the plateau. The two things were separate and apart ... No human ingenuity could suggest a means of bridging the chasm which yawned between ourselves and our past lives. One instant had altered the whole conditions of our existence.[4]

In due course, the evolutionary rupture yawning between them is implicitly filled in by a hierarchy of racialized players, all located lower down on Doyle's developmental scale from Challenger and his cadre of white male compatriots. They are aided by "faithful" Amazonian Indians up the river, as well as by their African slave "Zambo," who, as

Doyle puts it, has "an honest black face" and a "Herculean figure" – but then in turn they are betrayed by their Mestizo or (in Doyle's terms) "half-breed" guides. For Doyle, like many early twentieth-century European intellectuals, racial purity is prized and racial mixing is suspect – if an ongoing fascination.[5] To Doyle, the pure of race are honest and faithful while the miscegenated are faulty and amoral, contaminated. In addition to the dinosaurs, inhabiting the plateau are a population of indigenous peoples he calls the Accala, who are described as "little, clean-limbed, red fellows whose skin glowed like polished bronze" – again evoking a classical perfection – and as well a population of ape-human hybrid people who in contrast are reckoned as "malevolent," "bestial and ferocious."[6] When the two come into conflict, the gun-toting Englishmen ally themselves with the human yet uncivilized Accala and defeat the half-human, half-ape creatures. The colonial character of the narrative is undeniable. Challenger expounds on the event in vivid survival-of-the-fittest, supremacist terms:

> All the feuds of countless generations, all the hatreds and cruelties of their narrow history, all the memories of ill-usage and persecution were to be purged that day. At last man was to be supreme and the man-beast to find for ever his allotted place … These fierce fights, when in the dawn of the ages the cave-dwellers held their own against the tiger folk, or the elephants first found that they had a master, those were the real conquests – the victories that count. By this strange turn of fate we have seen and helped to decide even such a contest. Now upon this plateau the future must ever be for man.[7]

Challenger means "man" here in the doubly gendered and humanized sense. And if the masculine is the seminal force of conquest in the Lost World, then the feminine is the germinal centre of the landscape itself, the most interior secret of this land of mystery and wonder. When the central lake of the plateau is located, in standard imperialist style, it must be named. The young Malone is given the nod for this and chooses to name it for his fiancée, Gladys.[8] It was Malone who first gazed upon and charted the central lake after climbing the tallest tree on the plateau, producing a curiously cellular, even gynecological map of his dawn world (figure 3.1) – the central Lake Gladys being the final mystery plumbed in the adventure. Malone expressed his thoughts: "For once I was the hero of the expedition. Alone I had thought of it, and alone I had done it; and here was the chart which would save us a month's blind groping among unknown dangers. Each of them shook me solemnly by the hand." In this male rite of passage, Malone is accepted by his learned and powerful colleagues as the day's "hero" for the daring work of having climbed a tree in order to enable the making of the map, and with that the framing and entailing of the world. In a Foucauldian sense, through the power and mastery of the gaze and by means of the bounded mapping of terrain, the object world itself is produced. In the same action Malone is also made into a fully modern, masculine subject, though this was only one of "a range of masculinities" Doyle delighted in exploring, as Doyle scholar Joseph Kestner has suggested:

> By deploying a range of masculinities inflected by race, class, national origin, professional orientation and marital situation, and by engaging the entire range of male social relations (from comrade to husband), Arthur Conan Doyle presented one of literature's most powerful examinations/investigations of masculinity itself.[9]

The homosociality of Holmes and Watson or that of Malone, Challenger, Summerlee, and Roxton (and perhaps even Doyle's "faithful Zambo") marks out a psychic union from which an ideal, composite figure of masculine life may be extracted – a composite echoed widely in popular culture, as for instance in the 1960s *Star Trek* triad of Kirk, McCoy, and Spock. For Doyle, the most complete, singular figure of masculine subjectivity – one which exceeds even the scientific/rationalist figure of Sherlock Holmes – is Challenger, who alone and somewhat more ambiguously was described as combining both the primitivist excesses of a powerful ape-like body and a heightened intelligence producible only by the most advanced evolutionary processes known to the science of Doyle's day. Doyle would eagerly dress up and pose as Challenger on book tours;[10] it was clear that his Professor was, as a superhuman figure, "the embodiment of many of [Doyle's] male fantasies," as Jacqueline Jaffe properly noted (figure 3.2).[11]

Whereas Doyle may have begun rehearsing masculinity through the male circle of the Sherlock Holmes novels, that rehearsal blossomed fully into blatant, mimetic self-performance in Doyle's ultimate identification with Professor Challenger. George Mosse, writing on late nineteenth-century conditions of European and North American urban masculinity, suggested how such excessive performances would have been situated against popularized fears in Doyle's times:

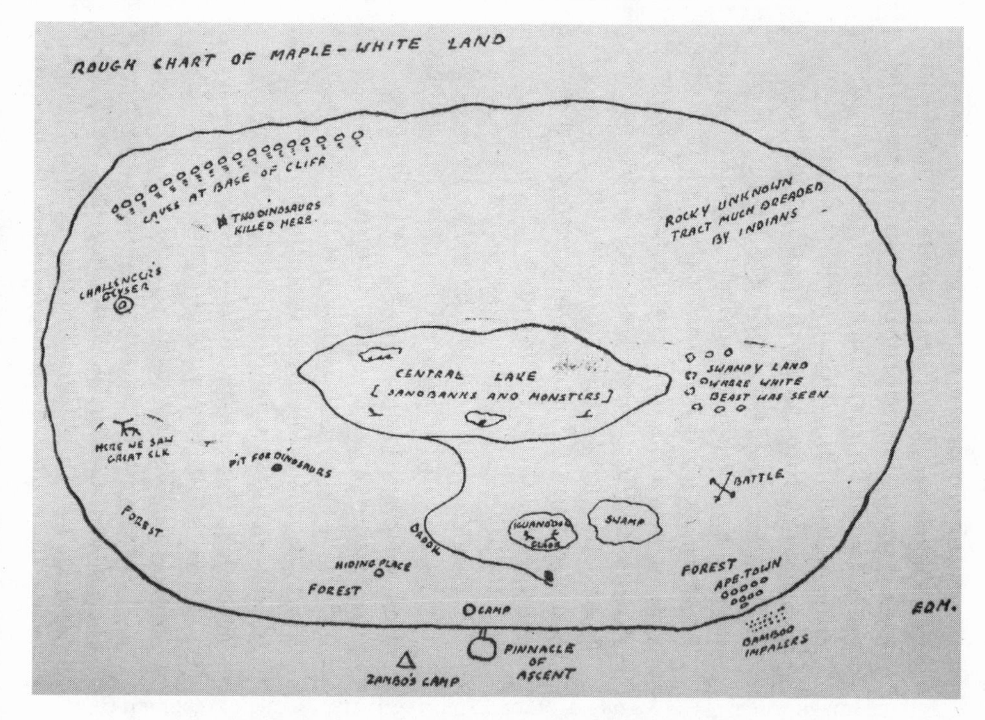

Figure 3.1. Malone's map of the plateau. Source: Doyle 1912b, facing p. 202.

> Modern masculinity was put to the test toward the end of the nineteenth century … The fin-de-siècle was [a turning point]: the years roughly from the 1870s to the Great War gave a new impetus to both masculinity and prolonged economic crises, and new technologies … added to the anxieties of the upper and middle classes by the end of the century … Just as important, such threats to individual health as syphilis, tuberculosis, and hysteria were becoming a general obsession … Under such circumstances the ideal of masculinity … had to be defended more strongly …[12]

And Doyle felt himself a victim to this sense of threat. One of the ongoing political causes in his life was his battle against British legislative reforms improving the marital rights of women, which Doyle took as an assault upon what he believed was the dominion of male privilege. Doyle's challenge of the Divorce Act was an act of anxiety over

Figure 3.2. Arthur Conan Doyle (centre) posing as Professor Challenger, with other mimicked adventurers from *The Lost World*. Left to right: Malone, Summerlee, Roxton. Source: Doyle 1912b, frontispiece, facing title page. Caption for original image reads: "The Members of the Exploring Party. From a photograph by William Ransford, Hampstead."

a perceived threat to Victorian, imperial masculinity. As witnessed by the central Lake Gladys – named for Gladys, who had challenged her Malone to find a manhood-proving adventure – the threat to the masculine transposed readily onto this feminized element of the unknown and mysterious, which if left unknown and unplumbed – and, literally here, *unmapped* – would leave an otherwise proper man incomplete. Drawing explicitly on sexual tropes, Malone notes that, with his colonial achievement,

> we should [now] return to London with first-hand knowledge of the central mystery of the plateau, to which I alone, of all men, would have penetrated. I thought of Gladys, with her, "There are heroisms all round us". I thought of McArdle, and that three-column article for the paper! What a foundation for a career![13]

Doyle's repeated returns to this event in the novel underscore how crucial a thematic and a throughline of the text it was. In more ways than one, it is the climactic moment for Malone, whose first-person narrative guides the text, and the moment appears as the resolution of the challenge posited at the beginning of the narrative. Relevantly, Malone is abandoned at the end of the novel by Gladys, who has married a solicitor in his absence, as if to undermine his youthful faith in women. Doyle's disillusionment is conveyed through this convention. Jaffe notes how, for Doyle, women rise to be reckoned as "at least equal participants in (and perhaps, given their position as moral arbiters of society, the prime exponents of) a society that has replaced adventure with superficial social restrictions, curiosity with fear, and aggression with meekness."[14]

For Doyle, the proper role of the feminine is to be "plumbed," "penetrated," and conquered as would be experienced in the mysterious plateau of primordial "nature" – not to act in the restriction of male agency, the threat he identified with Victorian socio-political change. In the final moments of the text, Malone chooses to seek out further male-bonding adventures with his new fraternity – at the same time, renouncing his associations with women. This, of course, leaves open the future to revisit and reproduce the thematic of male adventuring in the Doyle novels that followed this one. In the process, nature is thoroughly entailed by Doyle's very located, masculinist sensibility.

The purpose of Doyle's scientific adventure was mirrored by what he saw to be the purpose of the literary adventure, as well – to achieve romance, proof, and masculine completeness. In an 1896 speech, Doyle expressed this very anxiety about manly lack and the urge to wholeness attainable through the literary-naturalist imaginary:

> The man who does not care for the story is an incomplete man. The man who does not care for anything that has ever been or can be on God's earth is an incomplete man.[15]

Returning specifically to the dinosaurs of *The Lost World* again, they appear as a materially sourced, if phantastically animated challenge to be taken on by Doyle's team of adventurers in order to further assure the fullness of their masculinity. He effectively uses the saurians as an oppositional episteme of fear and encounter, a foil and contrast to his heroic English figures of bravery and manhood. The carnivorous dinosaur, in particular, is the utterly menacing opponent against which the resilience of an embattled, rugged, unrelenting imperial manliness could be played out.

But in addition, having been radically separated into an alter-geography, presented as an ultimate form, and positioned as the most brutal foe in nature to the accomplished, technologically capable European man, the carnivorous dinosaur could also become the exemplar against which an even more superior human attainment could be demonstrated. Malone, having been driven into a pit by one such big meat-eater, reflects nervously on its adaptive status:

> I recalled a conversation between Challenger and Summerlee upon the habits of the great saurians. Both were agreed that the monsters were practically brainless, that there was no room for reason in their tiny cranial cavities, and that if they have disappeared from the rest of the world it was assuredly on account of their own stupidity, which made it impossible for them to adapt themselves to changing conditions.[16]

So, we have this key adaptive point: intelligence is the key to success, to the ability to change with changing conditions. Those who have attained such an adaptive advantage are equipped to free themselves from the bondage of stasis which the bounded world of the plateau exemplifies – Challenger and his party will be able to depart for civilization, whereas the human and animalian natives of the plateau are implicitly doomed to remain, locked in their savage, perpetual war for survival.

This survivalist trope presents an apt entry point to the actions and homosocial cohort of Henry Fairfield Osborn, along with his most recurrent fixation – giant, powerful creatures, among them the great meat-eaters *Allosaurus*, *Albertosaurus*, and *Tyrannosaurus rex*.

The Lost World Animates Osborn's Mesozoic

A rather interesting thing about Doyle's text is that it can stand as a virtual template for the politically animated, scientific, educational adventure project which Osborn followed in his forty-two years at the American Museum of Natural History – from 1891 to 1908 as head of the Department of Vertebrate Palaeontology, and then from 1908 to 1933 as president of the institution.[17] What follows recounts some of the major points of correspondence linking the phantasmatics and performative actions of Doyle and Osborn, with attention to their drawing in and narrating of large, carnivorous dinosaurs engaged by a particularly configured cohort of scientific actors. For now, I want to keep attention on the performative lost world/Mesozoic and the intertwining of particular human and saurian players – this, then, serving as a working frame for a more detailed account in chapter 4 of the debut and full-scale articulation of that most fetishized of carnivores, *Tyrannosaurus rex*.

First, recall Doyle's troop of masculine adventurers. Osborn would produce similar homosocial networks that would journey to exotically attractive locales across the world. In contrast to Doyle, Osborn would achieve this in a much more forceful, authoritative, *institutional* setting – at first in the vertebrate palaeontology program, but eventually throughout the entire institution. Science studies scholar Donna Haraway demonstrated a similar effect in the zoological collecting and diorama programs of the museum under Osborn's direction; a matter, as it turns out, that has had significant relay into the history of dinosaur life-group reconstruction in museums, science, and film, and which I will take up in greater detail in the next two chapters.[18]

Through the AMNH and its institutional practices, and matching the bravado of *The Lost World* of Professor Challenger, Osborn helped to stage the very sorts of performances of the masculine that Doyle had narrated in the novel. Instead of the wealthy Great White Hunter, Lord John Roxton, Osborn had the gentleman dandy and "dinosaur hunter" Barnum Brown. Instead of the scholarly Professor Summerlee, Osborn had the highly regarded evolutionary scientists William Diller Matthew, Matthew King Gregory, Edwin Colbert, and George Gaylord Simpson. These lauded vertebrate palaeontologists were known to have undertaken the lion's share of descriptive work and basic technical drafting of texts which Osborn would then modify to conform to his own evolutionary visions. Rather than the journalist Malone, Osborn had large phalanxes of the American press corps on hand. Rather than needing the support of the London Zoological Institute, Osborn founded what would soon become the most influential palaeontological society in the world.[19]

Using his own personal wealth in many instances, or drawing on favours from wealthy patrons, Osborn was able to extend himself everywhere through his sometimes pliant, sometimes diffident group of male colleagues. In this and more, he was a disturbingly consummate *articulator* of what he felt counted in the natural order. He knew how to mobilize the actions of many, the fossils, the instruments of science, and institutional credibility in order to co-produce the matter-of-fact outcome of natural knowledge. The common ground for all was the Mesozoic; the stratigraphic, temporalized landscapes in which remains of dinosaurs could be found, resurrected, and reconstructed into visible, energetic, interactive beings.

Notably, Barnum Brown had come to be known – and is still reputed among dinosaur palaeontologists and popular media accounts today – as the "greatest" dinosaur collector ever.[20] Science historian Ronald Rainger notes that the Society of American Vertebrate Paleontologists of 1902 was the precursor of the Paleontological Society that succeeded it in 1909. Osborn is considered the founder of both of these, which in turn spawned the Society of Vertebrate Paleontology, the currently dominant professional association in the discipline. To the extent that Brown was also the best-funded collector of his day, owing to the support of Osborn at the helm of the AMNH budget, his reckoning as the greatest dinosaur hunter of his day would have to be true. Osborn, the son of the president of the Illinois Central Railroad, used his connections to railroad magnates such as Morris K. Jessup and J.P. Morgan to his advantage, minimizing many of the costs of palaeontology that would have affected Brown and his work.[21] Brown also found and collected the first specimens of Osborn's prized *Tyrannosaurus rex*, which, by the measure of many vertebrate palaeontologists following him, was the "greatest" dinosaur discovery of all time. In this as well, Brown had earned a distinction in the world of masculinist collecting achievement that would be difficult to surpass.

Recall Doyle's fixation on spectacular evidence as proof of his lost land. Rainger notes how Osborn took full advantage of the spectacle of giant vertebrate fossils:

> To the wealthy philanthropists who donated to the museum and dominated its board, fossil vertebrates were rare, large, and obvious facts whose display would increase the status of the museum and its benefactors … Fossil vertebrates also had sheer entertainment value and could contribute to public education. As the documentary evidence for evolution, fossil vertebrates could convey to the public the importance of nature and nature's laws. Osborn quickly and enthusiastically embraced those objectives.[22]

Now, recall the imperial conquest by "man" on Doyle's South American, dawn-of-time plateau. Osborn had assisted Teddy Roosevelt in his adventuring ambitions, supporting his organizing of the Roosevelt-Rondon zoological collecting expedition across the Paraguayan-Brazilian frontier in 1913, with scientific and logistical support from the museum's ornithology department.[23] Doyle, like Osborn, admired Roosevelt, whom he specifically compared to Challenger for having undertaken such a journey of discovery through South America.[24] Later, in the 1920s, Osborn sent out his champion expeditionary leader, Roy Chapman Andrews – reputed by some to have been the model for Spielberg's Indiana Jones[25] – to search for hominid origins, or what would be Osborn's "Dawn Man," in Mongolia. Instead, Andrews came back with fossilized dinosaur eggs and an array of other mammalian and dinosaurian fossils, which the institution nonetheless would parlay into a media frenzy. In imperial, colonial style, Andrews entitled his final report "The New Conquest of Central Asia."[26] As the intrepid hero of the Central Asian "conquest," Andrews was Osborn's ultimate man, who, as Tom Mitchell put it, "personified the Anglo-Saxon male potency" also embodied by Doyle's Professor Challenger.[27] Osborn could further extend himself through Andrews's actions. Andrews, as if he had just put down a copy of Doyle, would swagger out phantasmatics of romantic adventure at every opportunity:

> We stand on the threshold of a new era of scientific exploration ... In almost every country of the earth there lie vast regions which potentially are unknown. Some of them are charted poorly if at all, and many hold undreamed-of treasures in the realm of science. To study these little known areas ... to learn what they can give in education, culture, and for human welfare – that is the exploration of the future![28]

Andrews's 1920s ode to manly exploration reads like a direct reply to a lament from Doyle's *The Lost World* about the impending end of colonizable terrain:

> [T]he big blank spaces in the map are all being filled in and there's no room for romance anywhere.[29]

Recall Doyle's orders of racial purity and fear of contamination. Then note Osborn's presiding over the Second International Congress of Eugenics, held at the AMNH in 1921, while simultaneously developing a collection of murals for the "Hall of the Age of Man" depicting stages of racial advancement. Osborn, as if quoting Professor Challenger after the great battle on the plateau, would write that the displays demonstrate

> the struggle of man from the lower to the higher stages, physically, morally, intellectually, and spiritually. Reverently and carefully examined, they put man upwards towards a higher and better future and away from the purely animal stage of life.[30]

Recall Doyle's ability to depict and textually animate an entire world of powerful, otherwise extinct creatures. Osborn, with the financial aid of J.P. Morgan, sponsored the artwork of Charles Knight – without doubt, the most copied illustrator of prehistoric creatures ever. Knight's allegiances to the Osborn-Morgan network were firmly established through the artist's father, who had long worked in the financier's banking house, eventually as private secretary to Morgan himself.[31] Knight's work was the basis of several of the illustrations in the first editions of *The Lost World*, and then became the visual source for myriad Hollywood giant saurian films.[32] Under the close direction of Osborn and William Diller Matthew, Knight would produce some exceptionally energetic dinosaurs – notable embodiments of Osborn's conceptions of evolutionary advancement. Knight's illustrations would in turn become ubiquitous templates for entire skeletal mounts depicting the interaction of dinosaurs, most notably carnivores like *Tyrannosaurus* or *Allosaurus* pitted against each other, or against impressively large herbivores – or alternately, in Osborn's terms, "defensively" adapted herbivores like the horned dinosaur *Triceratops* (see chapter 4, figure 4.9).[33] The fossil evidence for these tremendous battles for survival amounted to bite marks on bones – no duelling dinosaurs in death postures had been found – and Osborn's imagining, supported by his AMNH cohort, was very particular here. All the same, these displays "meshed with Osborn's interests in glorifying the struggle for existence."[34] Again, Osborn's interests, along with the interests of those whom he held closely within his technical-scientific network, were extended into the very scene of the display that would then become the model for generations of highly imitative dinosaur exhibitionary practices (see figure 3.3).[35]

Though Osborn and Doyle moved through similar social circles, and while the extent of their shared sense of things was astonishingly parallel, it is uncertain whether the two ever actually met.[36] It is known, however, that the vertebrate palaeontologist Ray Lankester, director of the British Museum (Natural History) (1898–1907), maintained a correspondence with both of them, no doubt facilitating the circulation of their various conceptualizations. It is also known that Doyle derived much of his vision of the ancient world from the published popular lectures on the topic of "Extinct Animals" which Lankester had delivered prior to 1905.[37] Doyle's Challenger (as if Doyle himself was speaking) even refers to his "gifted friend, Ray Lankester," then recites precisely the text from one of the captions for an illustration of *Stegosaurus* in

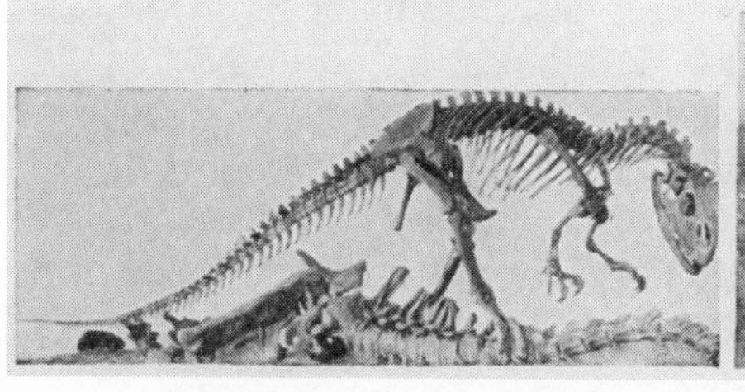

Figure 3.3. Osborn's behavioural translations, to mounted skeleton, to life restoration. Produced in collaboration with Charles Knight. Source: Osborn 1917, 213. Caption for original image reads: "Fig. 91. A Carnivorous Dinosaur Preying upon a Sauropod", and, "Mounted specimens and restoration by Osborn and Knight in the American Museum of Natural History."

Lankester's book, "Probable appearance in life of the Jurassic Dinosaur *Stegosaurus*. The hind leg alone is twice as tall as a full-grown man."[38]

Lankester's dinosaurian otherworld-making is visible enough in one of his very direct synopses of the Mesozoic time-space as understood through its faunal inhabitants:

> The great interest in regard to extinct reptiles centres in those which were so entirely different from the reptile of today that naturalists have to make separate orders for them. Many of them were of huge size. They flourished in the Mesozoic period and abruptly died out ... They are a prominent example of that kind of extinct animal which is not the forefather, so to speak, of living animals, but of which the whole race, the whole order, has passed away, leaving no descendants either changed or unchanged.[39]

Some basic commonalities of the phantasmatic Mesozoic that were traded by Doyle and Osborn (and many others) are readily discernible in Lankester's description. The Mesozoic is "entirely different from ... today," in effect a land of radical otherness. Indeed, the extreme alien character requires naturalists to "make separate orders for them," which is also to say that within the Mesozoic there are internal differences and lineages among dinosaurs. It was an organically whole, functioning time-space such that dinosaurs "flourished" there as an entire "order" or "race" of beings. And finally – notwithstanding the now conventional wisdom that birds are direct descendants of Mesozoic, dinosaurian ancestors[40] – for Lankester, the dinosaurs "abruptly died out" and were in no way to be considered as "forefather" to living animals. The totality of their extinction ensured that they were fully bounded from the world of later, contemporary natures. This vernacular account, of course, fits the standards of biostratigraphic boundedness which geologist Digby McLaren stressed as essential to technical evolutionary description.[41]

Here was the parallel rupture between the ancient and the contemporary, the pre-savage and the civilized, which Doyle also expressed through his figure of the plateau. Difference was expressed, first of all, within the time-space itself in a sort of ordering of saurian races that Lankester would then itemize in his lectures; and expressed, second, as a radical contrast to contemporary orders of life. The Mesozoic, like the lost world, was both chronotope and *heterogeography*, a parallel domain or land of difference populated by a panoply of animal kinds, the opposite of which was the contemporary domain of humanity in its diversity of human kinds (read, "races") in the discourse of the moment. This radical juxtaposition of the saurian "middle animal" world with the later, mammalian "new animal" world has come to undergird so much in constituting modern nature, superseding even the flow of transformation, of "unfolding," implicit in the terms of "organic evolution." Centrally, it poses the human against the animalian, allowing for the oppositional co-production of the "human" as a species.[42]

Lankester's description also demonstrates the way that scientific-phantasmatics of the Mesozoic, such as those of Osborn and his living cohort of narrators, are closely allied to literary-phantasmatics such as those of Doyle and his narrated cohort. Both manoeuvre and align with the organization of evolutionary-colonial space.

Extending from the common phantasmatics visible in the otherworld-making practices of Doyle, Osborn, and Lankester is the network of trading evident in the direct use and copying, by all three of these players, of Charles Knight's illustrations. Knight's *Stegosaurus*, first drawn in 1897, was then reproduced in association with articles by Osborn and William Ballou in the popular *Century Magazine* (figure 3.4a).[43] It was later modified by Lankester in his volume, and then roughly copied with the addition of an inserted human figure in the early editions of Doyle's novel (figure 3.4b).[44] Eventually, it was scaled up and presented as part of the London museum's displays, alongside skeletal mounts of the creature.[45] Visitors to the museum were exposed to recurrent visions

Figure 3.4a. Knight's *Stegosaurus* image. Source: Lankester 1905.

Figure 3.4b. "Maple White's" "The Monster," after Knight. Source: Doyle 1912a. Caption for original image reads: "Fig. 150. – Probable appearance in life of the Jurassic Dinosaur Stegosaurus. The hind leg alone is twice as tall as a well-grown man."

of this form of creature, associated as well with the composite of landscapes depicted in the life restorations and narratives cutting across the various media.

The Literary/Scientific as Colonial Nexus and Network

What the homosocial, Lost World/Mesozoic interactions of Doyle and Osborn point to are a highly entrenched set of practices and accompanying figurations. These now-familiar practices and figurations have been reiterated in science and literature and are grounded in a blending of colonial, racializing, capitalist, and masculinist phantasies of continual, unidirectional expansion and the continuous search for completeness. While they are historically revisable, such narrations are echoed in the wider practices of late twentieth-century science, as noted by anthropologist Sarah Franklin:

> Scientific pursuit is often described in terms of masculinity and adventure – as a domain of seminal breakthroughs, trail-blazing pioneers and uncharted territories. Such descriptions emphasize and valorize the enterprising activities of scientists as they busy about their colonizing practices.[46]

Such practised living of colonizing, exploration stories related to dinosaurs also provides the general contours of how embodied, material, and representational practices communicate across the socially erected, yet extremely permeable boundaries of literary and institutional-scientific domains. Through all of this, Osborn followed the imperial-colonial orders of practice which had similarly animated the action of Doyle's imperialist adventure: the class-structured masculine journeying, the mystifying, gendering, and knowledge-colonizing of alien space, and the return to the home-world with the proof – with the real thing.

The Mesozoic/lost world is an exemplary case of a natural/cultural *nexus* – or even twinned nexuses – within a heterogeneous *network* of social, technical, and material action. Wherever and whenever it is mobilized, it bounds in, fuses together, and organizes what otherwise counts as the imaginary and the material, the cultural and the natural, the phantasmatic and the real. It is literally a *nexus*, which English-language dictionaries commonly define in the doubled sense I am suggesting here:

> 1. a means of connection; tie; link. 2. a connected series or group.[47]

In other words, the Mesozoic/lost world is both a connecting point and also distributed across a network, appearing at and so connecting many points. In both ways, and because it is so very well distributed in this manner, it builds upon and stabilizes a well-articulated array of human and non-human resources.[48]

For both Doyle and Osborn, the materialized phantasmatic which was specifically articulated *with and through* the Mesozoic/lost world geography was the journeying of the romantic adventurer, scientist, collector, witness to nature, and big-game hunter into an exotic wilderness made available for colonization, manly attainment, and proof of the rationalist project. Such a geography also constitutes, positions, and disciplines the sorts of privileged subjects who might engage it: scientist, reader, museum visitor. It is, in a direct sense, a modernist political matrix of natural/human relations. Romanticism and enlightenment are also reconciled by this figure. They are part of the inbuilt narrative of the figure, the story that has made it potent. These attendant phantasies came to be performed and embodied in the masculine actions of living and literary figures alike. Those actions then articulated with, and increased, the materialization of the figure – most notably through the production of imagery, narrative accounts, museum displays, exhibited skeletal mounts, biogeographic maps, and so forth. While the full materialization of the Mesozoic would not be achievable, all these networked actions resulted in the effective concretizing of an integrated place in time. As it became more "real," the Mesozoic/lost world provided a potent visualizing, storytelling set-up, or a dispositif, and it converged around the visions and stories of its most influential, designing agents – most notably Osborn and Doyle, in my account. In this forceful, heterogeneous manner, the lost world and Mesozoic formed the *performative nexus* of a widely distributed *performative network*. Recalling Judith Butler's point once more, in becoming so widely distributed the very theatricality of the nexus is obscured, and "it conceals or dissimulates the conventions of which it is a repetition."[49]

Recognizable in the mutual, mimetic world-making actions of Doyle and Osborn is both the mobilizing of scientific practice in literary modes and the mobilizing of literature according to scientific modes of practice. Phantasmatic identifications move readily between and among these seemingly separate domains of human-centred knowledge production and practice, and indeed, between what is divided up and then counted as imaginary and real.

Feminist science studies scholar Susan Squier has examined the networked relation of literary and scientific practices of the late twentieth century.[50] In what may be seen as a contemporary counterpoint to the early twentieth-century Doyle/Osborn correspondences which I have been discussing, Squier takes up the "simultaneous publication in 1992 of two texts dealing with a global decline in sperm potency, P.D. James's *The Children of Men* and Elisabeth Carlsen's 'Evidence for Decreasing Quality of Semen during the Past 50 years'" – the former being a novel, the latter a technical article by a reputed endocrinologist in the *British Medical Journal*.[51]

In her analysis of these texts, Squier recognizes that these simultaneous literary and technical performances are much more than "remarkable coincidences," as others tend to suggest. She points out that speaking of "coincidence" suggests that "literature and science are stable and discrete categories." Instead, taking guidance from Bruno Latour's "seven rules of method," she argues that there is a network of human and non-human actions which connect such temporally, performatively coinciding literary and scientific productions.[52] Her purpose in doing this is to demonstrate that human reproduction is a "highly charged zone" of cultural traffic which conjoins literary and scientific practices, among many others. It too is a nexus. She suggests that the case offers an opportunity for practitioners of feminist science studies to "rethink … [a] primary genre-bound division between literature and science,"[53] and consequently, to develop models that "attend more fully" to the trafficking between the two. In one sense, it is a call for a methodical approach – if not a method *per se* – to address unruly flows of knowledge and matter in the constitution of what counts as the human, sex, gender, nature, and more.

Regarding such traffic, it was straightforward enough to show that Doyle had learned friends in common with Osborn – British Museum zoologist Ray Lankester, for instance. But more than a social network of transmission, what the bounded lost world and its counterpart the Mesozoic have presented is a visionary trading apparatus, around and through which everyday, habitual performances of the manly, the imperial, the capitalist, and the expansionist came to be enacted. Not only do such phantasmatics infuse and circulate in the lost world figure, but the figure itself also travels through public cultural and technical terrain, residing in the very materialized outcomes of these phantasmatics – from publications to display architecture, from technical diagrams to popular films. Notably – as will be discussed in the next chapter – Osborn would see to the development of a reptilian fossil gallery which portrayed his order of knowing, and an array of collaborative Knight-Osborn murals and paintings to further concretize his vision; a vision that modified upon and developed out of preceding historical flows.

Crucially, it can be said that the production of inter-grading figures like the Mesozoic and lost world – inhabitable only by a select group – installs a performative apparatus that readily launches and knits together narratives and practices which have long been associated with colonialism and imperialism. The effectiveness of that binding helps us to comprehend how the animating narratives may so effectively come to be accepted as authoritative when circulated as public culture. The non-human geography of the Mesozoic/lost world has become ever more concrete through performative reiteration. Even after the individual human figures have passed into historical memory, the Mesozoic/lost world remains as a socially embedded, complexly describable, circumscribed domain of nature. It bears in its very constitution remnants of its historically contingent interests – including those of Doyle and Osborn, who contributed so very much to its fashioning.

~

But beyond the outward performative actions of the expeditionary, literary, and exhibitionary networks I have been describing, how do such romantic, imperialist phantasmatics deeply permeate the conceptual substance of science, its practices of knowledge production and theorization, its material outputs? To consider this, I will turn attention to Osborn's most lauded ally, *Tyrannosaurus rex*. Like Osborn, Doyle's Challenger fully recognized and endowed an unsurpassable natural-cultural potency in the giant carnosaurs: "Among them are to be found all the most terrible types of animal life that have ever cursed the earth or blessed a museum."[54] Probably the most prominent museum to receive Challenger's blessing of such an accursed terror was New York's American Museum of Natural History in the period from 1902 to 1908, only four years before the appearance of Doyle's novel. Just as modern white man was reckoned as ultimate in the evolutionary order of the contemporary moment in Euro-American privileged social worlds, *Tyrannosaurus* became the ultimate in the evolutionary order of the Mesozoic for Henry Fairfield Osborn, who in 1905 described and dubbed this creature "King of the Tyrant Saurians."[55]

4 Animating *Tyrannosaurus rex*, Modelling the Perfect Race

Tyrannosaurus rex, the King of the Tyrant Saurians. The climax among carnivorous reptiles of a complex mechanism for the capture, storage, and release of energy. Contemporary with and destroyer of the large herbivorous dinosaurs.

Henry Fairfield Osborn[1]

With colossal bodies poised on massive hind legs and steadied by long tails, ponderous heads armed with sharp dagger-like teeth three to five inches long, front limbs exceedingly small but set for a powerful clutch, they are the very embodiment of dynamic animal force.

Barnum Brown[2]

Tyrannosaurus rex is certainly the best known, most iconic expression of dinosaurian embodiment. The Mesozoic/lost world is indebted to *T. rex* for its performative veracity, since *Tyrannosaurus* also stands as the public-scientific animal most indexing the Mesozoic, the Age of Reptiles, and particularly its terminal boundary; it is the creature understood to have become extinct at the very end of Mesozoic times – that is, at the end of the Cretaceous period, sixty-five million years ago. Indeed, Berkeley geologist Walter Alvarez conjured *T. rex*'s finality, supremacy, and phantasy purchase for Mesozoic natures in titling his best-selling 1997 book *T. rex and the Crater of Doom*[3] – a book tracking the interdisciplinary study informing the hypothesis that an asteroid striking the earth off the Yucatan coast brought about the demise of these one-time, gigantic "kings of creation," as Don Lessem and Jack Horner would refer to them in their popular non-fiction account of dinosaurs.[4]

While continuing to recall the linkages between Doyle and Osborn in their performativity of homosocial time-space travel, in this chapter I will discuss the appearance and material articulation of *Tyrannosaurus rex* in palaeontology and its now-canonical, normative restoration as a bloodthirsty killer, the most feared predatory animal ever to walk the earth. While there are other propositions that *T. rex* may have been a more passive scavenging creature,[5] the articulation of *Tyrannosaurus* as ultimate aggressive killer, hunter, and carnivore had its beginnings between 1902 and 1917, through the rigorous collecting work of Barnum Brown, the engagement with several partial skeletons, and more pointedly through the scientific imagining and fossil interpretations of Henry Fairfield Osborn during his tenure at New York's American Museum of Natural History (AMNH). For Osborn, *T. rex* – as type specimen, as model, as dramatic form – was the ideal figure, the right model for the job of bringing his race-motivated evolutionary theories to life.

To fully animate a biostratigraphic time-space, one has to populate it with animate beings, and *T. rex* offers a superb case in point. As with the chronotope of the Mesozoic, the finished form of *Tyrannosaurus rex* is delivered to us – with minor allowances for "errors" in scientific interpretation or missing "evidence" – as the pure and disinterested object of science, rather than as the very humanly animated, fossil-informed outcome of socially embedded technoscience that it always has been. Given that dinosaurs such as *T. rex* have been so extensively illustrated, animated, narrated, and with that, offered up unproblematically as extinct forms of animalian life, it is easy to overlook the fact that, in addition to the material fossils, it is such human players as Osborn and his cohort – their institutional-political location, their technical reconstruction practices, their imaginative work, and the circuits of social and political commitments informing them – that collided together, causing the initial and enduring performative iterations of *T. rex*. A consideration of these actions suggests the intricate ways in which the articulating and performative action of palaeontology – encompassing its theoretic formulations and its politics – works its way up to the dinosaur "reconstructions" that populate phantasmatic time-space worlds, and out to display, public, literary, and filmic circuits, and vice versa – all matters that I continue to press upon in this and the next two chapters.

Theorizing Energy, Advancement, and Aggressive Carnivory

The terrifying, carnivorous dinosaurs of Doyle's lost world emerged not long after the initial reconstructions of *T. rex* at the AMNH. Undergirding Doyle's lost world configurations, and the potency of these meat-eaters, were notions quite suggestive of Henry Fairfield Osborn's idiosyncratic yet historically consistent evolutionary theories. As noted earlier, Doyle's lost world imaginary could be a template for Osborn's. Conversely, if it were possible to have a text as the key source for Doyle's science, ideology, and theory in *The Lost World*, it might very well be Osborn's *The Origin and Evolution of Life*. However, Doyle's novel predates Osborn's 1917 monograph by five years, suggesting their common immersion – if not central roles – in the ethos of homosocial, racializing adventure-science at the time. Science fiction literature and scientific theories were clearly responsive to each other.[6] To understand the emergent, articulated, skeletal, and reconstructive materialization of *Tyrannosaurus* – and Osborn's ultimate dream of creating a battling "Tyrannosaur group" in the planned Cretaceous galleries at the AMNH – we first have to grasp the politically embedded, morally specific, theoretic formulations through which Osborn animated the past.[7]

Osborn's 1917 monograph bore a very telling subtitle which knitted well with the dynamic of energetic, manly interaction spun by Doyle: "On the Theory of Action, Reaction and Interaction of Energy." Though more noted as a palaeomammalogist, Osborn put a tremendous amount of scholarly and exhibitionary *energy* into the great meat-eating dinosaurs, assigned to and in large measure constituting the taxonomic sub-order Theropoda, presenting these creatures as exemplary figures of both evolutionary telos and climax within their particular geologic time – and most explicitly so in the terminal Cretaceous period of the Mesozoic era. In fact, Osborn put this group of dinosaurs to work for the modelling of his arcane propositions in relation to the "germ plasm," described by August Weissmann to be the unknown, essential substance of organic reproduction[8] – that wishful, mysterious, life-informing matter which Sarah Franklin and Nikolas Rose have recollected aptly in the term "life itself."[9] *Tyrannosaurus rex* was, in many ways, the ideal specimen and spectacle for Osborn, in that it articulated – or was made to articulate – "the total disparity between invisible energy and visible form," and the specific disparity between "the microscopic [heredity germ] ... as contrasted with the titanic beings which may rise out of it."[10]

Energy, vitality, action, preservation, advancement, purity, climax – these were keywords in Osborn's lexicon; elements of a theory, and of a mode of public display, which fended off all manner of decadence otherwise threatening the orders of power in the world as Osborn lived it. In Osborn's evolutionary plan, progress was crucial, and energy and velocity were indicators of evolution working well, as evident in this comment on the development of morphological characters in animalian anatomy:

> Although we may find that the course of evolution in one group of animals a character moves extremely slowly, it lags along, it is retarded, as if partly suffering from inertia, or perhaps, for a while it stops altogether; yet in another group we may find that the very same character is full of life and velocity, it is accelerated like the alert soldier in the regiment.[11]

The remedy to fend off retardation, stagnation, and inertia, then, was velocity, alertness, and individual action. At base here is a fear of senescence and obsolescence.

Both Ronald Rainger and Donna Haraway have pointed out that Osborn's social anxiety lay in fears about the demise of his clique of wealthy nineteenth-century New York families – the very clique discussed in the previous chapter, whose members helped finance Osborn's science, Osborn's AMNH. Rainger remarked that Osborn's agenda was formed by concern about the extinction of his ethnic group.[12] That group, known as the "Nordic" stock, was the principal subject of Madison Grant's eugenics-advocating book, *The Passing of the Great Race or the Racial Basis of European History*, for which Osborn penned the preface in 1916 – at the very time he would have also been preparing texts for his own *Origin* volume.[13] Osborn was explicit about his racial anxieties in his remarks about the value of Madison's work and of the relay of such anxieties into American political formation:

> If I were asked: What is the greatest danger, which threatens the American republic today? I would certainly reply: The gradual dying out among our people of those hereditary traits through which the principles of our religious, political and social foundations were laid down and their insidious replacement by traits of less noble character.[14]

Such a position was demonstrated by advancing parallel, and equally phantasmatic, evolutionary scenarios in which Osborn proffered models that could brace up the idea of the ascendancy of more "noble" characters, and the elimination of "less noble" ones.

Within the scope of biological study in Osborn's time, carnivorous dinosaurs could be put to work as exemplars of apparent evolution ending in the extinction of an otherwise great "race" of powerful creatures. Indeed, what better group of creatures to play one's anxieties against than those that seemed so dramatically ulterior yet somehow parallel to the "advancing" forms of humanity that concerned him? What could be said about their success and demise, in their particular chronotopic past domain, that would give guidance on projecting the success and preventing the demise of one's own "biological" group in the chronotope of the present?

Recalling the activation of chronotopes as a staple of palaeontological perfomativity, Osborn made ample use of palaeobiogeographic maps to situate fossil fauna. Biogeographic past worlds could be theorized and held in strict opposition to the contemporary human world, and in these past worlds he could offer stories or theories of ancient animalian action played out in a kind of time-space laboratory or theatre of the past (figure 4.1).

Osborn's notions of action and energetics were adapted from ideas of his teacher Edward Drinker Cope. Cope, who espoused a blatant scientific racism not uncommon in late Victorian times, had followed a rather exotic line of neo-Lamarckian thinking which he referred to as the theory of *kinetogenesis* – where continued use of certain parts of the body eventually led to the bodily acquisition and inheritance of functionally related characteristics, and hence, change over time – a positive use-it-and-gain-it corollary to the negative use-it-or-lose-it proposition.[15]

Given their interest in energetics, it was little wonder that both Osborn and Cope held artist Charles Knight in high esteem. Knight, whose interests were in creating vividly animated reconstructions, was someone who readily incorporated and elaborated the highly determined phantasy that vigour expressed power and adaptiveness. No better example of this is there than Knight's image of two leaping Dryptosaurs[16] – remains of which were first found in New Jersey in 1866 and described by Cope himself – depicting a level of energetics not seen again in scientifically endorsed illustrations until the last two decades of the twentieth century (figure 4.2). In Knight they had an artist who could visualize and virtually bring the creatures and the theory to life.

Osborn elaborated this theory by blending both Darwinian and neo-Lamarckian points. He theorized that the environment constrained or guided morphological changes – akin to Darwin's natural selection – in notably recurrent ways at different periods or eras, where the responsive actions of the creature, in line with Cope's thinking, communicated back into the morphology of the animal. This kind of "interactional selection" supposed evolutionary change to follow along lines of a potentially perfecting telos, with some creatures evolving directionally and more rapidly, others more slowly, and still others lagging in a course that would lead to eventual disappearance. He suggested possible communicating mechanisms ranging from hormones to "sub-psychic" behavioural cues.[17]

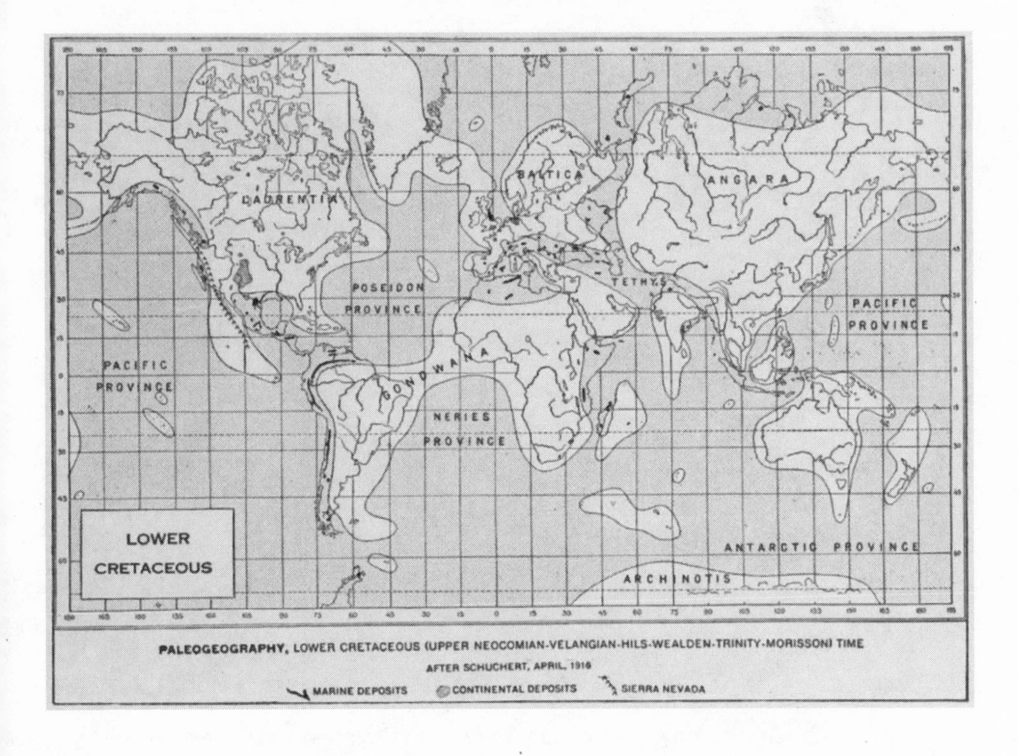

Figure 4.1. Osborn's Mesozoic (Lower Cretaceous) biogeography. Source: Osborn 1917, 217. Caption for original image reads: "Fig. 95. Theoretic World Environment in Lower Cretaceous Time." And, "The dominant period of the great sauropod dinosaurs. This shows the theoretic South Atlantic continent *Gondwana* connecting South American and Africa, and the Eurasiatic Mediterranean sea *Tethys*. Shortly afterward comes the rise of the modern flowering plants and the hardwood forests…."

Figure 4.2. Charles Knight's painting of leaping Dryptosaurs. Source: Image #35781, American Museum of Natural History Library (image first published in Osborn 1897, 14).

Unlike orthogenesis, which posited a guiding force directing evolution independent of selective forces exterior to the organism (e.g., natural selection), Osborn's proposal could retain a directional telos towards climactic forms through a model of interaction, with energetics as the pivotal agent of change. He projected this theorization – to which he would often refer as a biological "law"[18] – concisely:

> Throughout we shall point out some of the more notable examples of the apparent operation of our fundamental biological principle of the action, reaction and interaction between the inorganic environment, the organism, the germ, and the life environment.[19]

Osborn could deploy the great Mesozoic meat-eaters as the apex expression, in their times, of this "fundamental principle." Paramount among them was the exceptional creature *T. rex*, understood from specimens diligently collected and prepared for the AMNH by Barnum Brown, but which Osborn would take the lead in naming, in describing behaviour and functional morphology, and in guiding materialization into singularly illustrious public being as a fully articulated, mounted specimen in the fossil galleries of the AMNH.

Field Specimens to Public Spectacles

Although prior to 1900 there were a number of finds of isolated teeth and bones of what would later be recognized as belonging to the genus *Tyrannosaurus*, three major and dramatic finds of partial fossil skeletons were made by AMNH crews led by Barnum Brown in Wyoming and Montana in 1900 (AMNH 5866, 13 per cent of skeleton), in 1902 (AMNH 973, 11 per cent of skeleton), and in 1908, this being the most outstanding and compete specimen (AMNH 5027, 48 per cent of skeleton).[20] No significant *T. rex* specimens were collected again until the late 1960s (with a later florescence in finds and collecting of fine, though still partial skeletons from 1990 onward).[21] So, the first articulated reconstruction of *T. rex* into standing skeletal form was based on the three partial, mostly unarticulated fossil skeletons collected by the AMNH between 1900 and 1908.[22]

It was the 1902 specimen, AMNH 973 – consisting of only a lower jaw and some limb bones, plus a few others – that Osborn would describe in his first scientific paper on the creature (1905), making this the type specimen or holotype. It was at this time that Osborn also invented and assigned the now infamous name *Tyrannosaurus rex* (king tyrant saurian), "in reference to its size, which greatly exceeds that of any carnivorous land animal hitherto described."[23] A separate name, *Dynamosaurus imperiosus* (imperious dynamo saurian), was assigned to the 1900 specimen, AMNH 5866, which was briefly believed to be a separate creature.[24] This distinction was recanted after closer study,[25] and *Tyrannosaurus* and *Dynamosaurus* were then synonymized, with *Tyrannosaurus* becoming the name of palaeontological record, as it had been mentioned two lines earlier in the 1905 article – according to Linnaean rules of nomenclature, the first-published name receives priority. While, like *Tyrannosaurus rex*, the ring of *Dynamosaurus imperiosus* was fully consistent with Osborn's phantasmatic fixation on apex and kinetic formulations, the latter name would be set aside as a victim and artefact of nomenclatural proceduralism.[26] Indeed, had he simply mentioned *Dynamosaurus* first, then children everywhere today might very well be fixated on *imperious dynamos* rather than *tyrant kings* – Osborn's various phantasies embodied in either performative utterance.

The 1906 publication became the basis of the first major publicity on *Tyrannosaurus*, responding to the mounting of a composite of a pair of massive hind limbs that had been placed on display at the museum at that time as well. In December 1906, the *New York Times* offered a full-page exposée with a banner reading,

> The Prize Fighter of Antiquity Discovered and Restored: A Magnificent Tyrannosaurus Rex Set up in the Museum of Natural History Here. Last of the Great Reptiles and the King of Them All.[27]

The *Times* reproduced a drawing by an AMNH staff member, Mrs L.M. Sterling, of the hypothetical full skeleton in simple profile with the tail resting on what would be ground level and with a human skeleton to give a sense of scale. This depiction of *Tyrannosaurus* lacked the intended energetic life animation that Osborn desired and which he and his AMNH colleagues would eventually invest in their theoretic and public imaginings. All the same, this skeletal posture presumably served to signal monumental scale and general body configuration.

The first major two-dimensional life reconstruction of *Tyrannosaurus* illustrates the creature approaching a group of horned dinosaurs, and was painted in 1906 by Charles Knight under the direction of Osborn and Barnum Brown (see figure 4.3a). The image more or less fleshed out the relatively static skeletal reconstruction drawn by Stirling for Osborn's 1906 publication, and another one drawn earlier by assistant curator William Diller Matthew for Osborn's 1905 publication (figure 4.3b). While, as Don Glut notes, the poses in these renderings of the creature are probably informative of twentieth-century popular representations, my point is that it is in the later three-dimensional modelling work directed by Osborn (and discussed in his 1913 publication) that we find the fuller phantasmatic and theoretic commitments that would come to be materialized in the final *T. rex* mount of 1915.

So, it was the exceptional specimen AMNH 5027 – collected by Barnum Brown's field party between July 3 and September 11, 1908[28] – that would become the principal source for the eventual spectacular 1915 mount in the galleries. This specimen included an exquisitely preserved skull and the entire skeleton of the torso, and was missing only the forelimbs, back legs, and the end of the tail, most of which could be filled in by recomposition with elements from the two previously collected AMNH specimens, 973 and 5866.[29] The 1915 mount would combine both original and cast elements, as in the case of the otherwise enormously heavy skull, melding together one individual out of the remains of two.

Modelling Rex, Animating Apex

> Enlivened models animate imaginations, techniques, experimental strategies, research questions and pedagogical interactions. Embodied animations are thus more than aesthetic flourishes: such modes of body-work are a crucial step in luring scientists and their students into new kinds of understanding.
>
> Natasha Myers[30]

In the same sense suggested by Natasha Myers in regard to animated chemical models, the enlivened modelling of dinosaurs provides critical insights for understanding Mesozoic and dinosaurian performativity as a particular technique for animating life in a more general sense. Mesozoic performativity is a technique that enables the transfer of life animation knowledges (whether in theories or hypotheses, or in plausible analogies and homologies, even stories) of the scientist/technologist into constructive, worldly interactions.[31]

A frozen-in-motion, mounted dinosaur skeleton, whether on a full or modified scale, is in fact a model, as it *models* the imagining, the *informed* knowledge, of those who assemble it and put it in place. They must project the anatomical relation of the skeletal elements, pose the creature, position it in relation to likely viewers, and possibly set it in relation to other creatures or surroundings. They must decide on whether to fill in missing elements, or to apply conjectural constructions for what might be missing from the "specimen." Indeed, as is the case with the 1915 AMNH mount, in many circumstances more than one specimen – remains of multiple individuals – will be used to create a skeletal mount, giving the illusion that only one individual is drawn upon, only one individual represented. Finished palaeontological skeletal mounts therefore are complex materializations composed from interpretive, technical, and outright imaginative works of human and non-human collectives – scientists, technicians, available equipment, the fossils themselves, and much more.

Figure 4.3a. Charles Knight's 1906 fleshed-out reconstruction of *T. rex* developed under the direction of Osborn (detail from figure 4.9 below; see source there).

Indeed, modelling has long been seen as a highly productive technique for scientific analysis. Philosopher of science James Griesemer has suggested we might think about the place of models in scientific knowledge and practice in two ways: "the concept of a model is historical, and … we can appreciate the significance of 3-D models only in making and use."[32] Both of these considerations can be seen to be at play in the major full-scale materialization of *Tyrannosaurus* at the AMNH.

In this regard, as previously noted, the 1915 public exhibition of the articulated mount of *T. rex* [33] – effectively the world's debut of the full skeletal reconstruction – took place just two years before the publication of Osborn's *Origin and Evolution of Life*. A photograph of the monumental museum mount held a place of precedence in the frontispiece of this volume, harmonizing with the remodelled 1915 hall of vertebrate fossils (figure 4.4). The caption for the photo expressed in a single statement Osborn's deep confidence in the significance of *Tyrannosaurus* for his evolutionary theory. Full of triumphal verve, it reads:

> *Tyrannosaurus rex*, the King of the Tyrant Saurians.
>
> The climax among carnivorous reptiles of a complex mechanism for the capture, storage, and release of energy. Contemporary with and destroyer of the large herbivorous dinosaurs.[34]

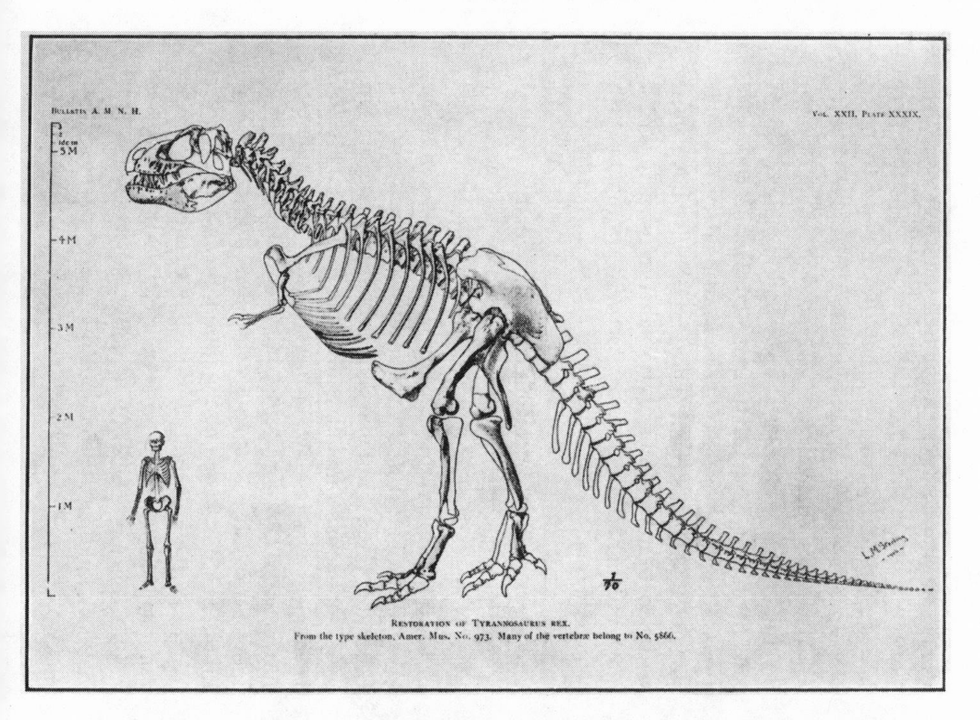

Figure 4.3b. William Diller Matthew's 1905 skeletal reconstruction developed under the direction of Osborn. Source: Matthew 1915, 51. Caption for original image reads: "Fig. 16. Skeleton of Tyrannosaurus in comparison with human skeleton."

While imposing enough, the modestly active stance of the skeleton – upright and mid-stride, mouth agape, with forelimbs outstretched and three-fingered claws (later understood to have one digit too many) flexed as if about to attack – refracts Osborn's much deeper machinations over the preceding ten-year period towards an even more theatrical posing of this enormous articulated skeleton.

Osborn had dreams of an even more highly dynamic skeletal mount of not one but two Tyrannosaurs, facing each other as if on the spur of some instinct-driven combat over the carcass of a downed, herbivorous *Trachodon*, a frozen moment of otherwise thrilling, animated interaction.[35] The dramatic scenarization was, in large measure, the consequence

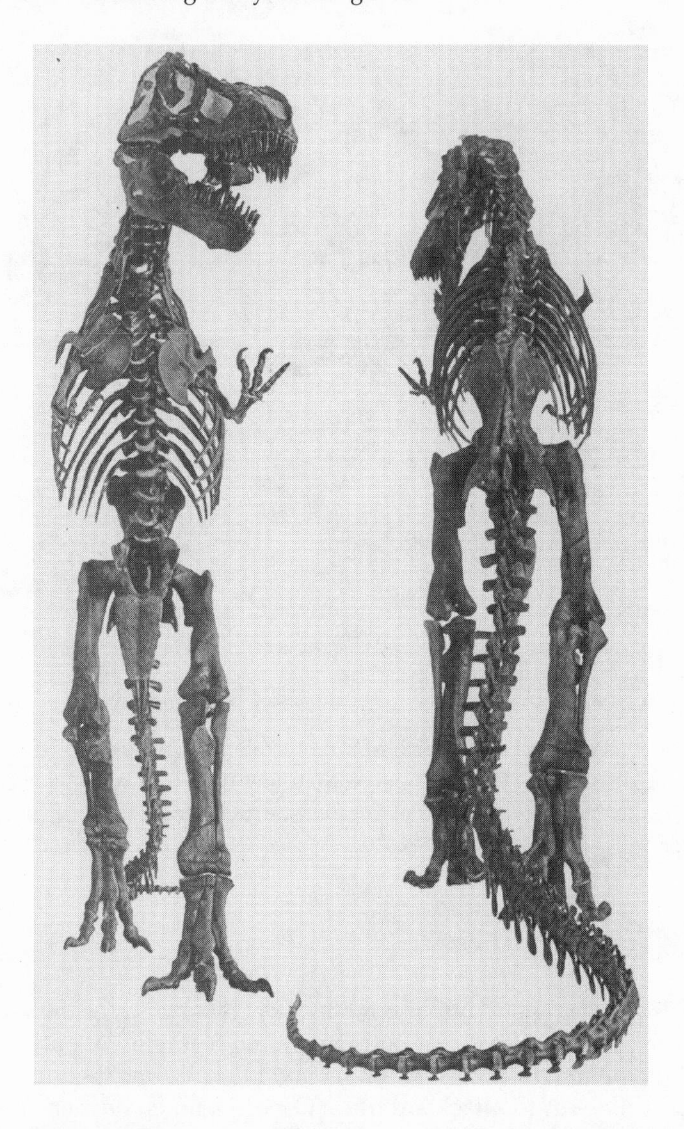

Figure 4.4. Frontispiece of Osborn's *The Origin and Evolution of Life: On the Theory of Action, Reaction and Interaction of Energy.* Source: Osborn 1917. Caption for original image reads: Tyrannosaurus rex, THE KING OF THE TYRANT SAURIANS. The climax among carnivorous reptiles of a complex mechanism for the capture, storage, and release of energy. Contemporary with and destroyer of the large herbivorous dinosaurs…"

of Osborn's intellectual domination over how this creature ought to be imagined in its past-world life moment.

A crucially important technical innovation helped Osborn, supported by junior colleague William Diller Matthew, to envision a wide variety of potential interactive postures. Erwin Christman, an artist and sculptor in the Vertebrate Paleontology department, created flexible, reposable, one-sixth-scale models.[36] Osborn noted,

> in preparing to mount Tyrannosaurus for exhibition a new method has been adopted, namely, to prepare a scale model of every bone in the skeleton and mount this small skeleton with flexible joints and parts so that all studies and experiments as to pose can be made with the models.[37]

As I discuss in chapter 6, this movable armature technique is very likely, then, the key precursor to later applications in stop-motion animated dinosaur films created by cinematography innovator Willis O'Brien, thereby opening the way for massive relay of scientific techniques of the animated Mesozoic creature-making and past-world-making into non-museological public spaces. Indeed the New York Zoological Society curator Raymond Ditmars advised Osborn on the several "studies and experiments" in the posing of the two models. Ditmars would later go on to direct one of the earliest stop-motion creature films in 1923 with the assistance of Willis O'Brien.[38]

Osborn favoured and highlighted the fourth of the experimental studies by including a photograph of the scene in his 1913 article (figure 4.5) and extolling the public-scientific virtues of the particular skeletal life group of these two gigantic, tyrant hunters:

> The fourth pose or study, for the proposed full sized mount, is that of two reptiles of the same size attracted to the same prey. One reptile is crouching over its prey (which is represented by a portion of a skeleton). The object of this depressed pose is to bring the perfectly preserved skull and pelvis very near the ground within easy reach of the visiting observer. The second reptile is advancing, and attains very nearly the full height of the animal. The general effect of this group is the best that can be had and is very realistic, particularly the crouching figure.[39]

Appealing to Ditmars's authority on the behaviour of predatory animals, and eliciting metaphorics of brute warfare, Osborn selected this particular scenario as a poignant, naturalistic moment of suspended

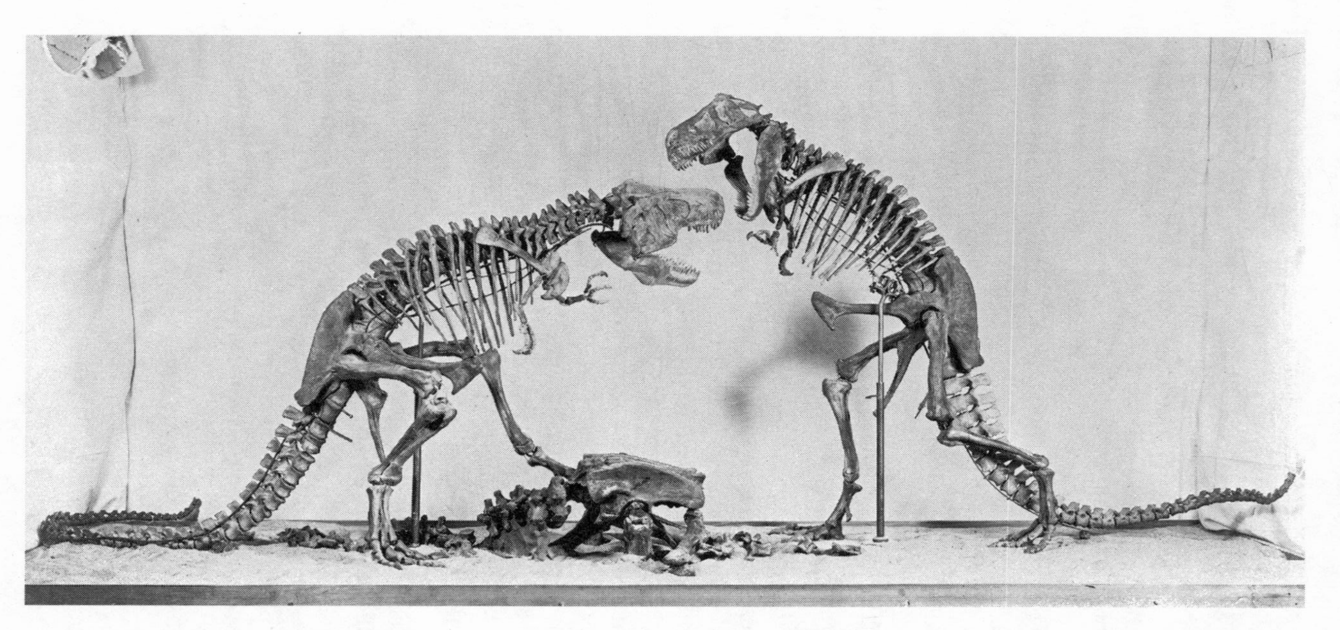

Figure 4.5. Osborn's modelled dynamical combat restoration group (1913). Source: Image #35576, American Museum of Natural History Library.

reptilian animation, with the animals depicted "just prior to the convulsive single spring and tooth grip which distinguishes the combat of reptiles from that of all mammals, according to Mr. Ditmars."[40]

In this same discussion, however, Osborn goes on to suggest severe technical and mechanical problems associated with the ponderous size and weight of the fossils – especially the skulls – when mounted full scale. It would be a daunting technical task. In 1915, Brown wrote of a future (but eventually unrealized) plan to mount the life-group pairing:

> When the new west wing is built the Tyrannosaur group will be the central exhibit of the Cretaceous dinosaur hall. Owing to limited space, only one skeleton can be mounted at present and this has to be placed temporarily in the quaternary hall of the fourth floor.[41]

Presumably, the space and technical matters presented the final impediment to realizing such a consummately theatrical scene. When it finally opened, the Cretaceous gallery would present the single *T. rex* in its assigned chronotopic context, but not in its conceived, juxtaposed, and animated life-group context.

Osborn did foreshadow the adjustments to the upright figure in the life-group model that, as it turns out, were applied to the gallery mount. Always careful to couch his concerns in terms of functional, anatomical constraints, Osborn remarked that

> [a] fifth study will embody some further changes. The upright figure is not well balanced and will be more effective with the feet closer together, the legs straighter and the body more erect. These reptiles have a series of strong abdominal ribs not shown in the models. The fourth position places the pelvis in an almost impossible position as will be noted from the ischium and pubis.

A juxtaposition of these two images (i.e., the model and the final mount), both presenting anterior (front) views, shows how the final mounted figure has the neck and head raised higher, now certainly attaining "the full height of the animal."[42] (See the pose in figure 4.6, and compare it with the pose in figure 4.7.)

The upright posed model became, with some small modifications, the visual basis for the AMNH's ultimate, solitary gallery mount in 1915, the image of which would also grace the frontispiece of Osborn's *Origin* book (see his figure 11) and then become the exemplar of his

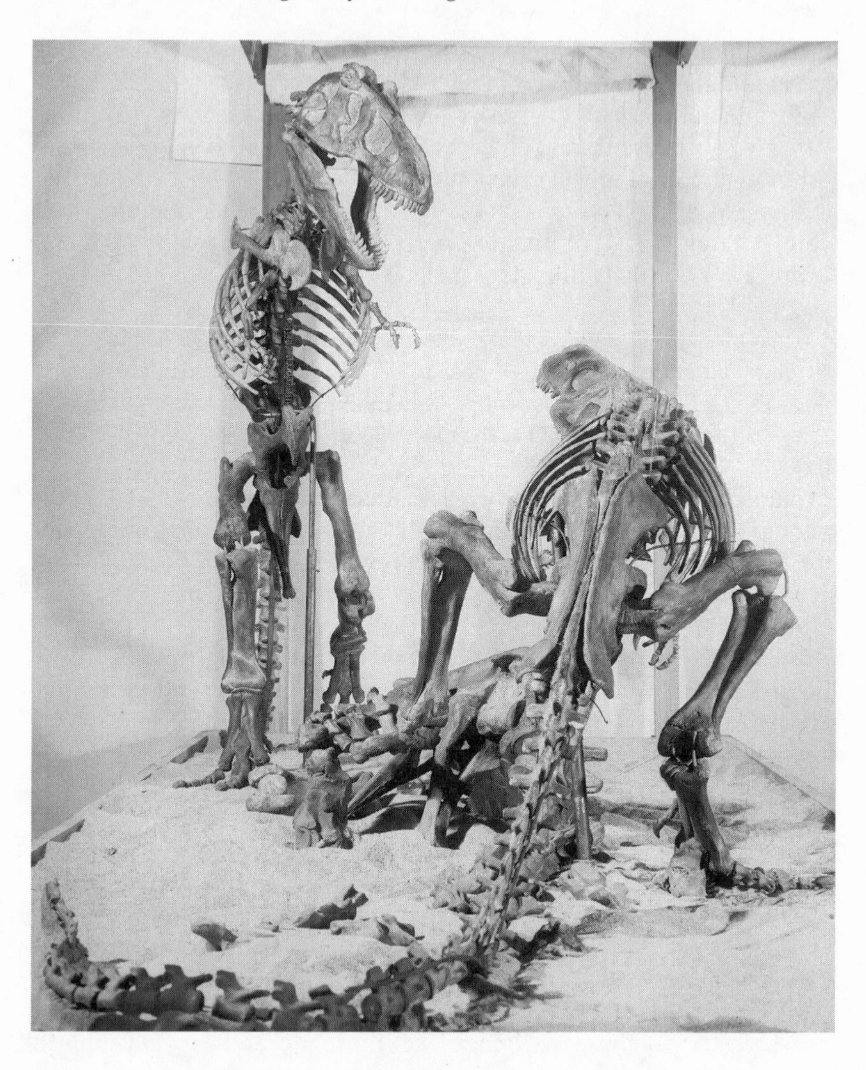

Figure 4.6. Endview of one-sixth-scale models of Osborn's combating *T. rexes* (1912–13). Source: Image #35582, American Museum of Natural History Library.

particular, imperializing, racializing form of evolutionary theory work. This, then, resulted in the iconic, scientifically endorsed museum materialization of *Tyrannosaurus rex* that would be set upon the world for decades to come. Such a stance would maximize the height of the animal in the architectural space as well, bringing the mount very close to touching the ceiling of the gallery, towering over other mounts in the same room, as is clear from a 1942 photograph of the mount on display (figure 4.7).

Exhibitionary, technical, theoretic, institutional, fossil, and sociopolitical impulses were consummately imploded to generate this phantasmatic materialization of Mesozoic "perfection." The creature was the largest, most fearsome of its kind and of its time-space. The mount, conjoined with its name, *Tyrannosaurus rex* – complemented even by its discarded precursor nomination, *Dynamosaurus imperiosus* – instates the "fact" of its supremacy. All of this operates in accord with J.L. Austin's notion of the act of linguistic, illocutionary performatives, which in their utterance bring the thing into being. The linguistic performativity is further galvanized by the reconstructive performativity in the skeletal mount's final comportment in the Cretaceous gallery, the creature's factualized time-space – all the more amplifying the creature's supremacy and ascendancy to the Mesozoic apex and ensuring that passing visitors would not doubt this point.

This materialization was described to the visiting public in successive editions of the AMNH *Guide Leaflet to the Hall of Dinosaurs* by Frederic Lucas, emphasizing the surpassing scale of the creature that so captured Osborn and his cohort:

> Towering above the others, his head eighteen feet from the ground, is Tyrannosaurus the well-named King of the Reptiles, whose terrible jaws and tremendous claws placed all contemporaries at his mercy …

While the imperious single *T. rex* would be the final materialization, the suspended-animation scene of duelling *T. rex* skeletons did receive some public exposure in narrative form, even though the model would come to be stored deep in the collections, away from public view. Brown offered this explicitly theatrical, cinematic characterization of the lost world/Mesozoic moment in his popular piece entitled "Tyrannosaurus, the Largest Flesh-Eating Animal That Ever Lived," written for the 1915 edition of the *American Museum Journal* at the time that the single mount was placed in the galleries:

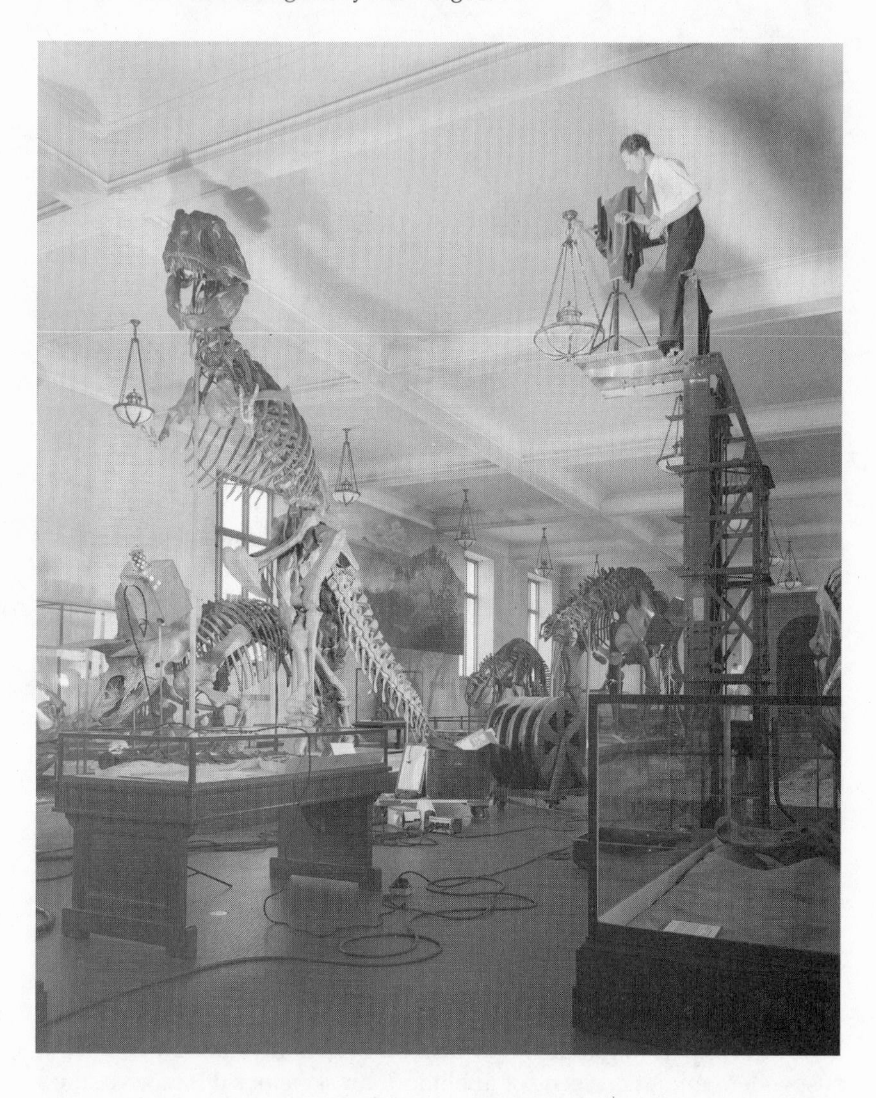

Figure 4.7. The Osborn model-derived, classic posed *T. rex* being photographed in the AMNH Cretaceous gallery (1942). Source: Image #315293, American Museum of Natural History Library.

> Dawn glows along the shore of a lagoon near the sea three millions of years ago [*sic*] in Montana. The landscape is of low relief; sycamores and ginkgo trees mingle with figs, palms and bananas. There are few twittering birds in the tree-tops and no herds of grazing animals to greet the early sun. A huge herbivorous dinosaur Trachodon, coming on shore for some favorite food[,] has been seized and partly eaten by a giant Tyrannosaurus. Whilst this monster is ravenously consuming the carcass another Tyrannosaurus draws near determined to dispute the prey. The stooping animal hesitates, partly rises and prepares to spring on its opponent. With colossal bodies poised on massive hind legs and steadied by long tails, ponderous heads armed with sharp dagger-like teeth three to five inches long, front limbs exceedingly small but set for a powerful clutch, they are *the very embodiment of dynamic animal force.* (Emphasis added)[43]

Here, in the moment of lost-world scenographic resurrection, Brown's cue to "the very embodiment of dynamic animal force" offers us a direct relay to Osborn's theories of energy and the hereditary germ. Put another way, Osborn's theory of dynamic animal force had become the model for *T. rex*, and *T. rex* the model for the theory.

This claim to reciprocity in conceptual and physical modelling may seem bold, but, in the history of science, both physical models and conceptual models have long been understood to inform each other – the physical model of the double helix, the metaphor of genetic "blueprints" or "life itself," and the abstract concept of genes are a well-known case in point.[44] Chris Kelty and Hannah Landecker, writing on the entwining of biological cellular theory, modelling, and animation, note how both "cell biologists and computer programmers have built machines and media to force a theory of cellularity to become visible – on screen and in time. In these different media the representation of the cell remains central."[45] Likewise – but working on a "macro" scale in comparison – by means of modelling and animation, Osborn and his collective worked to make his theory visible. And similarly, the representation of the complete, articulated, singular dinosaur in time and space remained, and remains, central.

Science historian Ludmilla Jordanova discusses, as well, the complex backward and forward referencing work that models can do, and we can certainly recognize Osborn's inter-modelling, his inter-performativity, in this kind of referentiality:

> "Model" is what can be called an incomplete concept in implying the existence of something else, by virtue of which the model makes sense. This

> "something else" might already be in existence or yet to come. It might be larger or smaller, more or less complete, sophisticated, or accessible. Models then, however verisimilitudinous, beautiful, or satisfying, always refer onwards. As a result there are interpretative gaps for viewers to fill in, the "beholder's share" in Gombrich's words …[46]

While any interactive model would be subject to further in-filling by any viewer, Osborn certainly strove to ensure that his concepts could live, so to speak, in these materialized skeletal presences, and so to close many of the potential gaps for in-filling.[47] His fifth paper on *T. rex*, published in 1916, was entitled "Skeletal Adaptations of *Ornitholestes*, *Struthiomimus*, *Tyrannosaurus*"[48] and provides an explicit example of how he worked phantasmatically, seeking ostensibly to identify the best comparative animalian model for understanding the possible life-ways of extinct creatures, and how he would direct attention to those models that best suited his faithful theoretic and, by implication, moral-political interests.

Osborn unquestionably had this paper in mind as he prepared texts for his 1917 *Origin* volume, as he cites and discusses the paper's "findings" to brace up his propositions.[49] This careful, comparative study focused principally on two creatures, specimens of which would become key, juxtaposed mounts in the planned Cretaceous dinosaur galleries. On one hand, it reassesses multiple theories of the behaviour of the diminutive *Struthiomimus* (literally, "ostrich mimic") as against that of the gargantuan *Tyrannosaurus*, the ultimate effect being to set up a binary that demonstrates the adaptive and evolutionary perfection of *T. rex*. Osborn begins by laying forth hypothetical analogues for modelling *Struthiomimus* behaviour:

> In view of the ostrich-like structure of the skull and of the pes [the foot], and the partly suspensory, partly grasping structure of the fore limb and manus [the hand] of Struthiomimus, it is very difficult to form a consistent hypothesis of the habits of this remarkable animal. It may be compared with certain of the lizards, the struthious birds, the tree sloths (Bradypus, Cholcepus), and the Aye-Aye (Cheiromys).

His argument goes through a range of animalian models or analogues, dismissing them one by one. Osborn settles on his own modelling, suggesting repeatedly how the "feeble" forearms and claws are more suited to pulling down branches (like a sloth) than to either the "raptorial" habits of seizing prey (like the earlier *Ornitholestes*, or living chameleons) or the digging habits of an anteater. He suggests the toothless beak is more like the "struthious birds" (e.g., ostriches, emus), suiting it for browsing on vegetation, and that the hind feet are suited to "cursoriality" or running, possibly with some defensive capacity, but certainly *not* for apprehending prey (figure 4.8). His tack, then, was to demonstrate that his theory of *Struthiomimus* behaviour is the proper one, most notably that the creature was non-raptorial – in other words, it was *not* a hunter of prey, but rather a herbivore, or at most a non-raptorial omnivore.

Whether Osborn's interpretation is taken as correct or not, his evolutionary-adaptive argument in comparing *T. rex* and *Struthiomimus* brings home his fuller motives, for it is in this juxtaposing of radically different creatures in form and scale that *T. rex* stands out. Here was the great hunter. In Osborn's own words,

> Tyrannosaurus is the most superb carnivorous mechanism among the terrestrial Vertebrata, in which raptorial destructive power and speed are combined; it represents the climax in the evolution of a series which began with the relatively small and slender Triassic carnivore Anchisaurus.[50]

Whereas *T. rex* represents "raptorial destructive power," *Struthiomimus* represents a "feeble" creature. While Osborn acknowledges that *T. rex* also has feeble forelimbs, he nowhere takes up the functional significance of this, whereas his argument about *Struthiomimus* rests upon it. In fact, the forearms of the *Struthiomimus* skeleton are virtually the same size as those of *T. rex*, on an animal one-fifth the length and a mere one-seventieth the weight, so arguably they are more physically powerful than those of *T. rex*. Yet Osborn presses on:

> The obvious ancestral resemblances cease or are masked by the widely divergent adaptations of *Tyrannosaurus* to exclusive carnivorous habits and aggression, and of *Struthiomimus* and *Ornithomimus* probably to herbivorous habits and defencelessness, compensated for, doubtless, by alert powers of vision and rapid locomotion.[51]

These conclusions, pointing to the binary, contrastive set of superior offensive, carnivorous hunters against more or less inferior herbivores, become all the more understandable when situated against Osborn's binarist theoretic formulations of Newtonian action and reaction, where this contrast would also be visible:

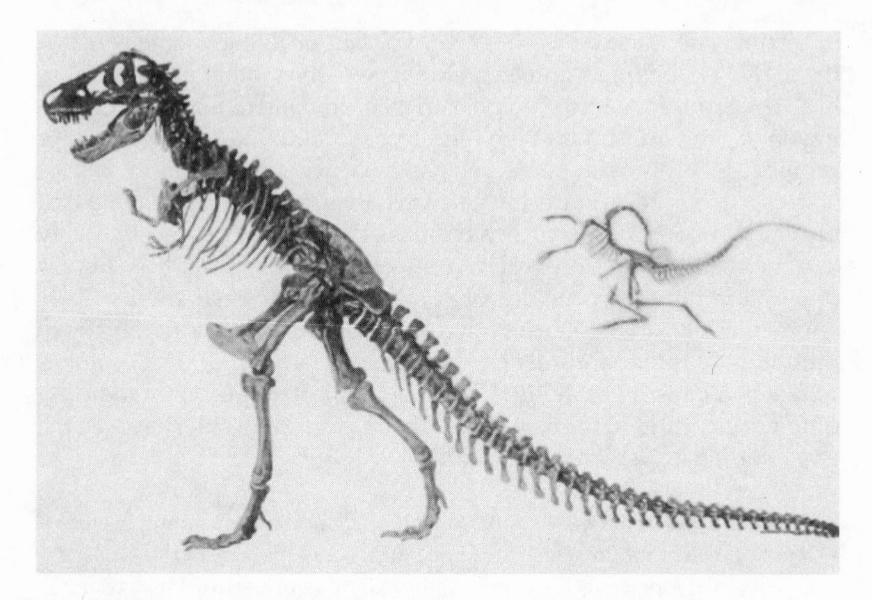

Figure 4.8. *T. rex* in skeletal comparison with *Struthiomimus* (1916). Source: Image # 218169, American Museum of Natural History Library. Caption for original image reads: "Fig. 92. Extremes of Adaptation in the "Tyrant" and the "Ostrich" Dinosaurs."

> … an example of the counteracting evolution of offensive and defensive adaptations, analogous to that which we observe to-day in the evolution of the lions, tigers, and leopards, which counteracts with that of the horned cattle and antelopes of Africa, and again in the evolution of the wolves simultaneously with the horned bison and deer in the northern hemisphere.[52]

This is the situated contrast that he marshalled materially in the Cretaceous galleries by juxtaposing *T. rex* (as a model of the supreme hunter) with *Struthiomimus* (as a model of the inferior, "feeble" herbivore or opportunistic feeder at best), and *T. rex* with other herbivores (as models of more or less defenceless prey). Through this contrast of the admirable hunter and the ready prey ("fair game"), Osborn reified the normative formulation of a nature constituted as a battle of the strong against the weak – a nature where the best of hunters reigns supreme. Indeed, this contrast bears out in his larger theorizations that would, in turn, unite with his thinking about the perfectibility of the human form, valorizing the dreamed ascendancy of the perfect masculine hunter along very particular racial lines.

Chronotopes and Ultimate Destroyers … to Better the Race

> The phrase, "author of a species," sounds faintly blasphemous. Saint Thomas Aquinas after all described God as the "author of nature," the drawer of species boundaries.
>
> Lorraine Daston[53]

> [R]elations of knowledge and power at the American Museum of Natural History should not be narrated as a tale of evil capitalists in the sky conspiring to obscure the truth … The concept of social relations must include the entire complex of interactions among people; objects, including books, buildings, and rocks; and animals.
>
> Donna Haraway[54]

One would be hard pressed to make the bald claim that Osborn "authored" *Tyrannosaurus rex* beyond its name, but it is clear that his designs and narratives of the operations of organic life were central, arguably decisive, in the elaborate co-fashioning of the creature we know so well today. Osborn's masculinist cohort of technical and scientific elite colleagues, the several specimens collected and prepared, the spaces of display, the allegiance to certain formulations of political practice and to certain practices of embodiment – all of this was in play in the performativity of the ultimate destroyer, *T. rex*. But, single or collective/complex authorship aside – and climax aside too – what of the vulnerabilities of Osborn and his *Tyrannosaurus*? What of this matter of anxiety over extinction?

I return once more to Arthur Conan Doyle's *Lost World*, in the comments about dinosaur intelligence – Doyle actually used the word "stupidity" – as a limit on adaptive and evolutionary potency. In a cognate way, Osborn wrote:

> This "king of tyrant saurians" is in respect to speed, size, power, and ferocity the most destructive life engine which has ever evolved. The excessively small size of the brain, probably weighing less than a pound, which is less than 1/4000 of the estimated body weight, indicates that an

> animal's mechanical evolution is quite independent of the evolution of their intelligence; in fact intelligence compensates for the absence of mechanical perfection.[55]

According to Osborn, the "mechanical perfection" of *Tyrannosaurus* was inversely proportional to its intelligence. Osborn would echo such a view – one familiar in Doyle's conjuring of the supposed bodily perfection and goodness, though mental inferiority, in his *Lost World* character "Zambo" – in a later racist-evolutionary assigning of demeaning, hereditary hierarchies based in an imagined physical, racial, intelligence calculus, applied to what was then not an uncommon classification of human groups. His tripartite racial (and racist) hierarchy of *Homo sapiens* – negroid, caucasoid, mongoloid – leaps so far as to suggest that "Negroids," for Osborn, ought to be reckoned as somehow less than human; indeed, in the following phrasing, even wholly different from *Homo sapiens*:

> The Negroid stock is even more ancient than the Caucasian and Mongolians, as may be proved by an examination not only of the brain, of the hair, of the bodily characteristics ... but of the instincts, the intelligence. The standard of intelligence of the average adult Negro is similar to that of the eleven-year-old-youth of the species Homo sapiens.[56]

Going in the other direction in relation to this classification – though he was never this direct on the point – Osborn's now-materialized, animated climax creature *Tyrannosaurus* could even stand as evolutionary counterpoint in support of imagined mental and racial superiority and perfectibility – indeed, it could be a support to the prized, white, Nordic genealogical stock from which he and many of his wealthy New England compatriots saw themselves as derived. Madison Grant's texts in *The Passing of the Great Race*, to which Osborn contributed both intellectual inspiration and an introductory chapter, clearly recapitulate Osborn's theories of energetics, vigour, sub-psychic cues, and interaction with the environment – resonating with his modelling of evolutionary climax – in reference to their common ideal of a superior Nordic stock within a Caucasian race. Grant – who joined with Osborn in lobbying successfully for US eugenics policy and legislation in the 1920s – believed that climatic surroundings were of relevance to "physical strength and mental vigor," with the Nordic climate having positive effects on its people. He even went so far as to delimit how the right climatic conditions ought to optimize, and select "naturally" for, the Osborn-like evolutionary energetics of a Nordic hereditary stock – which would also enhance, as he and Osborn would have it, their superior, even supreme capacity in conflict or battle.

A parallel world of climactic and climatic forces, of adaptive aggressive supremacy, would equally favour *T. rex* in its time-space evolutionary trajectory.

Like his and Grant's supreme, battle-ready, Nordic racial stock, Osborn had already performatively materialized his theories of natural struggle for existence via the giant, duelling carnivore-herbivore relation, again taking advantage of Charles Knight's talents to produce an array of "offensive," conquering carnivores pitted against "defensive" herbivores (figure 4.9). The Mesozoic struggle for survival, the struggle to progress, expressed itself as kingly battle. Osborn played this as the repeatable pattern of nature – the Mesozoic serving as a mimetic time-space milieu for performing a comparison of his own sense of ancient "reptilian" nature against that of contemporary "mammalian" nature:

> Thus in the balance between the reptilian carnivora and herbivora we find a complete protophase of the more recent balance between the mammalian carnivora and herbivora.[57]

Put another way, Osborn's dinosaurian hunter/hunted encounter could be taken as a *proto-performative* expression of the living encounters of mammalian hunter/hunted exchanges, and, one could readily argue, of the encounters between apical humans and their hunted counterparts.

As such, Osborn's biological explanation for his imperious tyrant beings of the Mesozoic could, by implication and analogue, echo and so brace up this white racial supremacism as a political-moral position – one worthy of his institutional program, whether in the public galleries of the AMNH or in his promotion of racial eugenics. Indeed, the entirety of evolutionary history for Osborn became a sequenced history of the natural attainment of superior over inferior animalian or organic forms. He continually celebrated the most spectacular of taxa; from his attention to giant saurians, to his research on the enormous and bizarre Tertiary Brontotheres, to his extended work on the evolution of the Proboscideans leading up to the modern African elephant, which came to be the central display feature in his Hall of African Mammals at the AMNH.[58]

Figure 4.9. Knight's reconstruction, under Osborn's direction, of the "Offensive and Defensive Energy Complexes" (1906). Source: Osborn 1917, figure 102. Caption for original image reads: "The carnivorous 'tyrant' *Tyrannosaurus* approaching a group of the horned herbivorous dinosaurs known as Ceratopsia. Compare frontispiece. The Ceratopsia are related to the armored Stegosaurs and to the armorless, swift-moving Iguanodontia. Restoration by Osborn in the American Museum of Natural History, painted by Charles R. Knight."

To cap his tale of ascendancy to superiority, Osborn even assigned the same language to humans as he had done for *Tyrannosaurus*. Where in the frontispiece he noted his tyrant king had been "destroyer of the large herbivorous dinosaurs," deep into the text he further noted that the Pleistocene extinctions were caused by hunters, that is, "by man, who through the invention of tools ... became the destroyer of creation."[59] In such a rationale, the products of intelligence, weapons and tools – like tooth and claw for creatures without intelligence – were the source of species superiority. For Osborn, the end of one ultimate power gave rise in due course, and inevitably, to its better.

All such evolutionary cases – that is, each chronotopic world of beings caught in the struggle to win, where some were thought to be naturally fitted to winning over others – could add cumulative, fact-like demonstrations to develop a phantasmatically elaborated principle of natural causes for his racial-political position. Recalling Doyle's lobbying for legislation against women's freedoms, Osborn sought to articulate these positions even further by lobbying successfully for the legislating of negative eugenics through established racial controls on immigration, and also, though unsuccessfully, for positive eugenics through "birth selection" practices.[60] Such were the practices that would preserve what his powerful male cohort considered to be the finest expressions of humanity – those of the selectively traceable germlines of Osborn's own cliques.[61] Rainger discusses how Osborn

> maintained that contemporary life, which allowed for the mixing of races and provided the opportunity for women to move beyond their separate sphere, was in conflict with nature's laws and could result in degeneration ... By recognizing the importance of race and the power of heredity, restricting different races [and genders] to their own domains of social and sexual activity, and allowing the New England stock to maintain its biological and social ascendancy, humanity and society could be preserved.[62]

As with Doyle, racial and species purity – the great Mesozoic carnivores being a case in point – was to be glorified along with masculinist ascendancy and the relegation of women to a lower political status. In this way, with racial and gender difference instantiated and rationalized by such imposed, "progressive" evolutionary thought – one could say a scientifically protected form of hate speech – racial mixing and women's agency could be considered repugnant and retrograde, something to eliminate with steadfast conviction.

But to query further: if *Tyrannosaurus* was such a perfect "life engine" for Osborn, what possibly would bring on its extinction? For Osborn, the small-brain hypothesis was not fully adequate. In addition, he argued:

> [T]he arrest of evolution among the Reptilia appears to lie in the internal heredity-chromatin, i.e. to be due to a slowing down of physico-chemical interactions, to a reduced activity of the chemical messengers [i.e., hormones] which theoretically are among the causes of rapid evolution.[63]

So, in theory, at least for Osborn, surging hormones were also a source of surging evolution, while lagging hormones were the cause of stagnation, retardation, and termination. For both Doyle and Osborn, intelligence was explicitly that which allowed man to better the great carnosaurs, to adapt to environmental change, to rise above the limits of biophysical constraints. For Osborn as well, the hidden, universal physico-chemical workings of the heredity germ and its chemical messengers fully articulated the process of potential human improvement – indeed, of the very delineation and perfectibility of races.

Osborn had marked the opening of his volume with the image of the "titanic being" *Tyrannosaurus*. Reciprocally, he saw fit to close his book with an invitation to future study of how to manipulate "the microscopic" heredity germ, borrowing on the historical authority of early modern philosopher Francis Bacon; he then quoted from Bacon statements that could be taken readily as an anthem for eugenics in the early twentieth century or, indeed, for biotechnological experimentation in the twenty-first century:

> "… it would be very difficult to generate new species, but less so to vary known species, and thus produce many rare and unusual results. The passage from the miracles of nature to those of art is easy for if nature be once seized in her variations, and the case be manifest, it will be easy to lead her by art to such deviation as she at first led to by chance; and not only to that but others, since deviation on the one side lead and open the way to others in every direction."[64]

Here, at the end of his work on the evolution of vertebrates, was Osborn's rationale for redirecting evolution, supported by a founding figure of modern rationality itself. This could stand as an ideal bridging point for those who agreed with his anxious, racist ambitions to preserve, insulate, and improve his own kind of rugged, American, self-made, yet highly evolved man. In Osborn's phantasmatics, there was indeed "a racial soul as well as a racial mind, a racial system of morals, a racial anatomy,"[65] and every living kind, from "monad to man," in every otherworld, lost or living, proved it so. Neither Arthur Conan Doyle nor Professor Challenger himself could have articulated it better.

5 Politics/Natures, All the Way Down

We inhabit these narratives, and they inhabit us. The figures and the stories of these places haunt us, literally.

Donna Haraway[1]

Politics is not just an arena, a profession, or a system, but a concern for things brought to the attention of the fluid and expansive constituency of the public.

Bruno Latour and Peter Weibel[2]

The phantasmatic contours of Osborn's theory-building and dinosaur reanimations tell us, quite directly, how that which we take as nature is always also a kind of politics. Many in science studies have asserted that modern science is, and ever has been, politics by other means, since it is always meant for particular publics and particular purposes. It is offered up by those who are positioned politically, and it assembles certain kinds of agents and things, non-humans and humans, into collective relations.[3] Indeed, in the reading I have been offering, there is also the recognizable embedding of politics in theory work, specimen collecting, skeletal reconstruction, technical modelling, powerful funding networks, display, and even (as we will see in the next chapter) cinematic and pulp fiction natures. Much of this embedding of nature-as-politics is animated, repeatedly, by means of the chronotopic impulse.

Even Barnum Brown succinctly expressed the chronotopic impulse in writing of the duelling *T. rex* experimental models where the aim, so seemingly apolitical, so matter-of-fact, was simply "to picture an incident in the life history of these giant reptiles."[4] We can now recognize the duelling *T. rexes* as a chronotopic modelling of Osborn's politics in skeletal form, if not in the flesh. The chronotopic impulse, as Brown's statement implies, is one that works at the most trivial or incidental scales (e.g., the battling *T. rex* moment), yet can radiate to entire imagined histories and geographies of life (e.g., the Mesozoic, if not an entire imagined panoply of phased-earth histories), and back down again, iterating, reiterating, adjusting as it unfolds.

In the discussion of Osborn's *T. rex* reanimations, I was concerned with how politics came to be *performed into* this singularly famous dinosaurian, especially when set against a range of supposedly inferior creatures. Here, I introduce the stronger proposition of *politics/natures* as simultaneous, conjoined, inalienable, yet malleable forces of the reality of science. I address the more encompassing issue of how and when *nature goes public*, to borrow a phrase from Corinne Hayden; that is, when politics are not just embedded in one natural place or being, but are distributed across many diverse places and beings that are often indexed with the term "nature," and how nature and the natural are always and already fully political.[5] To do this, I consider another forceful vector of nature-as-politics that attends to the fluidity of tropes, chronotopes, and performative effects across diverse discourses, sites, and forms of the natural. Just as it is possible to recognize some remarkable correspondences in the alternately literary-fictional and palaeontological-scientific practices of Doyle and Osborn, so we can find strong correspondences in other performative sites and stories of the human/nature encounter.

The Inter-Performativity of Expeditionary Hunting, Gorilla Dioramas, and the Age of the Dinosaurs

The metaphorics of big game hunting will be one of the connective tissues of actions I follow in advancing this discussion of the layering up of phantasmatic/material correspondences, where hunting combines with the purity of disembodied vision, progressive evolution, expeditionary travel, passages from boyhood to manhood, and masculinist conquest as key tropes in this very political action. On these points I am reinvigorating a conversation with the work of Donna Haraway, who considered the political imbrications of manly hunting in the AMNH's zoological collecting, conservation, and display work.

I will continue to frame the discussion around propositions of the performative nexus, of which we already have one in the chronotope and political nexus of the Mesozoic/lost world. To this I offer the second political nexus of manly passages through expeditionary "hunting," and a third: the performative and equally political nexus that is the elaborately crafted museum habitat diorama. Recognizing the overlaying of these nexuses then enables a discussion of how embedded politics in different forms of the natural, such as dinosaurs and apes – saurians and simians – come into coordination, thereby sedimenting and intensifying their political purchase as public science, their interperformativity of nature. When these are considered together, we begin to recognize the narrowly prescriptive political possibilities in largely routinized human/natural relations.

Looking ahead, I will take up the ongoing public science associations of saurians, simians, and humans in the next chapter. There I will trace how these kinds of politically animated figures move so readily through wider scientific-public and science fictional circuits, including vernacular palaeontology, Hollywood film, and pulp fiction.

Politics/Nature Nexus 2: Manly Passages, the Hunter, the Hunted

As discussed in previous chapters, the tropes of big game hunting and hunters reside squarely in the narratives of the lost world and Mesozoic (the first nexus I have indicated), especially but not exclusively so in early twentieth-century natural history discourses. Recall that, not unlike what they believed to be essential in the characters of the greatest of great white European and American men, Osborn's and Doyle's performative reconstructions of the great carnivorous dinosaurs *Tyrannosaurus*, *Megalosaurus*, and *Allosaurus* displayed vigour, the ability to conquer, to achieve a kind of imperial presence in their times and spaces. They could be and were rendered as the ultimate foes to humanity. In this, the dinosaurs themselves were antagonistic counterparts to the characters of Doyle's Challenger and Roxton, Osborn's expedition leader Roy Chapman Andrews and diorama builder and taxidermist Carl Akeley.

In such politically animated accounts, what these great, mechanically impressive saurians lacked was what these ideal men of science and adventure possessed: "brain-power," intelligence, and via the germ plasm, the inherited physico-chemical capacity to permit them to adapt and evolve further than their apex expression, the great carnosaurs.[6] The moral lesson in Osborn's supremacist natural order of things was to win the struggle once and for all, to go beyond *Tyrannosaurus*, the former destroyer of creation above all other "races" of dinosaurs, and so to become the new destroyer of creation, the greatest of great hunters. For Osborn, Andrews, and Akeley – for Doyle, Challenger, and Roxton – the ideal model of "hunter" was that of the intrepid masculine sportsman. The strictures adhering to this kind of sportsman's life were expressed clearly in the qualification rule for membership in Osborn's cherished Boone and Crockett club:

> A. No one shall be eligible for Regular membership who shall not have killed with the rifle in fair chase, at least one adult male individual of each of three of the various species of American large game ... [7]

Beyond living species of American big game (e.g., elk, bear, bison, bighorn sheep), the largest of large game imaginable in Osborn's vision had to be *Tyrannosaurus rex* – nominally, by Linnaean taxonomic gesture, also an "adult male." The hunter saurian, now, could easily become the hunted. In turn, such a move helps rationalize expeditionary science. The specimen which one *quarries* is recomposed quite literally as the hunter's *quarry*. Indeed, as legacy to this conflation of fossil specimen as hunter's prey, myriad popular and semi-technical books and articles on dinosaurs published over the last one hundred years have used the linguistic and performative resources of "big game" hunting.[8]

The metaphorics of the sporting life of the hunt as an animating, performative frame for palaeontology were put concisely by AMNH evolutionary theorist and lost world enthusiast George Gaylord Simpson: "The fossil hunter does not kill; he resurrects. And the result of his sport is to add to the sum of human pleasure and to the treasures of human knowledge."[9] The "health-giving" sport of fossil hunting was also referred to in a lecture delivered by Ray Lankester at the British Museum of Natural History. His training recommendations for young people echoed the scouting principles of Baden-Powell, and paralleled the hunting expeditionism of Doyle and Osborn, while interpolating those ideas within the journey narrative of visiting the museum itself as an effective portal to exploring the natural world:

> [T]he reader should visit many times the Natural History Museum, see the actual specimens, and by the aid of the illustrated guide-books get to know more details about them. And, if he or she have the chance and can go and hunt in some of the quarries or cliffs which are so often full

> of fossils, an endless delight and a health-giving pursuit is the prospect before him or her. Fossil-hunting with the hammer and chisel and a bag to be laden with specimens, is splendid exercise, and, if skillfully conducted, an exciting form of sport.[10]

Through the disciplining, health-promoting journey from doorstep to school to museum to field – and ultimately into the phantasmatic time-spaces of the fossil past – one could, in this much more passive encounter, even obtain specimens of the biggest, rarest, and most monstrous forms of life. Moreover, this action would yet preserve the imperialist, colonialist sporting spirit of conquest of the unknown, of the frontier. Lankester added, "there is always the chance – a good sportsman's chance – of finding 'something new.'"

Donna Haraway has demonstrated the sort of great white hunter motivations and tropes which animated the making of mammal dioramas at the AMNH – most notably the gorilla habitat group, featuring the taxidermically prepared male silverback known as the "Giant of Karisimbi."[11] These tropes were similarly extended into the languages and practices of palaeontological work by Doyle and especially Osborn, by their supporting cohort and by so many to follow. Whereas Haraway characterized the hunted-down and taxidermically prepared "primal ape in the jungle as the dopplegänger and mirror to civilized white manhood in the city,"[12] the mounted skeletons of giant, carnivorous dinosaurs could be regarded as ultimate enemy alter-beings. Moreover, their existence in the most remote places in time – hidden by geological sediment – invited, both performatively and allegorically, an even more masculine journey of discovery and conquest. This was a journey, however, where the dangers of the hunt were more decidedly phantasmatic, even for the palaeontological collector, as the terrible creatures were already quite dead.

As exemplified by Doyle's Malone, boys would pass into full-blown manhood by taking up such a journey to encounter nature's most frightening animalian figures. At the same time, men like Malone could retain their boyhood in the practice of fraternal bonding in nature's wondrous otherworlds of difference. These were phantasies explicitly signalled in Doyle's epigraph for *The Lost World*:

> I have wrought my simple plan … If I give one hour of joy … To the boy who's half a man … Or the man who's half a boy.[13]

Noting this epigraph, Joseph Kestner remarked on its articulations with the masculinist, colonial, outdoor disciplining actions of the expanding scouting movement in the British Empire and America:

> Doyle's epigraph establishes, more significantly, the symbiosis of gender modeling between men and boys which was the fundamental premise of Baden-Powell's scouting movement, with its close association of young men and scoutmasters. The near affiliation between acts of "scouting" or "tracking" and "deduction" in detective narratives, underscored by Baden-Powell in *Scouting for Boys*, reveals that this mode of popular narrative served to script masculine gender.[14]

Donna Haraway recognized the parallel call to manly passage through nature in the Roosevelt Memorial atrium of the AMNH. Its bas-relief friezes contain quotations from Theodore Roosevelt – T.R., whose life story is shot through with narratives of manly hunting[15] – uttering all the interests to which Henry Fairfield Osborn had aspired as the institution's leader. Referring to these reliefs, Haraway samples and comments upon Roosevelt's words, noting a sort of all-American, one-way passage from boyhood to manhood through the struggle with powerful forces of nature:

> The visitor – necessarily a white boy in moral state, no matter what accidents of biology or social gender and race might have pertained prior to the Museum excursion – progresses through Youth: "I want to see you game boys [*sic*] … and gentle and tender … Courage, hard work, self mastery, and intelligent effort are essential to a successful life." … The next stage is Manhood: "Only those are fit to live who do not fear to die and none are fit to die who have shrunk from the joy of life and the duty of life."[16]

White, classed, male rites of passage, sportsmanship, and struggle were the credo for both Doyle and Osborn, and so many influential men of their time. Their literary-scientific connective tissue was the same – the performative spaces of nature that were built, visited, conquered, and discovered by this extended and highly political community of living, once-living, and fictional men. Expeditionary hunting of large game helped to effect this nexus. Indeed, it worked as a performative, political nexus of nature in its own right.

In this way, and to resolve the puzzle posed in chapter 3 concerning whether Doyle and Osborn had ever met or had direct interchange, clearly there needed to be no such communication between the two.[17] Rather, they performed their politically charged knowing in an everyday way, and this was then distributed extensively: through the mass publication of their literary and technical works; through the mobilizing of their cohort; through common networks and shared stories of hunting and sporting life enthusiasts; through enormously popular museum displays; through public lectures and civic education; and in popular media stories. This incessant stream of performative enunciations and actions thereby delineated and held intact presumed lines of hierarchic evolutionary difference (including species, gender, and racial difference), difference that could so readily be reproduced and mimicked over the remaining decades of the twentieth century.

The romantic, evolutionary science-adventure, hunting-expedition tropes have continued at the AMNH, and were explicitly stated in its 1995 official anniversary publication.[18] AMNH president Ellen Futter introduced the book, writing: "Expeditions are the embodiment of this Museum's mission of discovery and understanding, throughout its illustrious past and in its continuing role at the forefront of scientific research." The fundamental rationale for this zealous mission, according to Futter, and adding the same sort of romantic mystification in which Doyle and Osborn revelled, was that "expeditions ... reveal the wonders and mysteries of life."[19] Famed Harvard sociobiologist E.O. Wilson wrote the foreword to the AMNH volume, adding millennial fervour to Futter's frontier vision of a knowable, capturable world "out there":

> Has the expeditionary spirit so well exemplified in the AMNH vanished? No, and I assure you it never will ... the exploration of the world continues as never before, in the field and in the laboratory. The promise of a new golden age is implicit in the task this great museum has set itself.[20]

By 2010, the AMNH was continuing to sponsor global field expeditions of its many divisions, and even offered expedition tourism for wealthy paying customers to travel with institution-affiliated experts on natural and cultural history tours to some five continents.[21] Here, those able to pay could immerse themselves in the nexus of chronotopic, camera-toting adventure travel, so aptly captured in the program's slogan, one that resonates with the actions of Doyle, Osborn, Andrews, and Akeley alike from nearly a century earlier: "*Who you travel with makes a world of difference.*"[22]

Politics/Nature Nexus 3: The Performativity of Akeley's Gorilla Diorama

To grasp the third performative nexus (to which we could add many others) and how it articulates with the Mesozoic/lost world nexus, it will be helpful to consider in a little more depth Donna Haraway's incisive analysis of the AMNH in her essay "Teddy Bear Patriarchy."[23] Whereas Haraway concerned herself with how nature, as populated with taxidermically prepared animals, came ready-made in the habitat diorama, I have followed a cognate action in the humanly worked populating of the Mesozoic via the battling tyrant kings and defensive herbivores of Osborn and company. Haraway offers a close reading of the diorama display and taxidermic work led by the museum's Carl Akeley. Akeley, like Barnum Brown, Charles Knight, and Roy Chapman Andrews, found a friendly environment in the museum under Osborn's aegis. Akeley's most enduring legacy has been the AMNH's African Hall, constructed between 1926 and its opening in 1936, consisting of an array of highly crafted, taxidermically exacting, dioramic habitat groups.

Haraway's descriptions concentrate on the gorilla group; its erect-standing, chest-beating silverback rising over the four other animals, together posed as "a natural family of close human relatives" and set against a painted, "Eden"-like, pristine landscape.[24] From here, Haraway demonstrates the complex manner in which the masculinist networks of the AMNH conformed this particular diorama into a "natural" scene. She elaborates how that scene is readable as an effective "looking-glass" for its presumed, idealized audience, the "white boy in moral state," pushing appropriately on the communicative intent.[25] A crucial point in Haraway's analysis of the diorama's representational realism was to emphasize how nature could be presented as given, as already made, total, and indeed as healthy, pure, untouched, and properly ordered. This sort of selective envisioning of wildness and purity was emphasized by museum officials and American policy makers alike as part of an attempt to promote the preservation of wilderness. Haraway tracked the philosophical points of Osborn and Roosevelt, and indeed of the Boone and Crockett and Sierra Clubs, as well as the New York Zoological Society in the early part of the century, and noted the prevailing commitment I have already pointed to: that "Conservation was a policy to preserve resources, not only for industry, but also for moral formation, for the achievement of manhood."[26]

In her extensive study of the habitat diorama, art historian Karen Wonders affirmed the policy-guiding role for the diorama at the AMNH as agent for American conservation ethics. She suggested that museums have an "urgent mission" to both protect and document endangered species, and noted that "[h]abitat dioramas served both these environmentalist purposes and in addition entertained the public by conjuring up illusionistic scenery that was an artistic tour de force."[27] Wonders also noted the role that the AMNH, and particularly Henry Fairfield Osborn, played in developing the habitat diorama as a major genre of nature presentation:

> The romantic idea that habitat dioramas preserved for posterity a view of the primordial wilderness as it had once existed before being ravaged by man was particularly strong at the AMNH ... Osborn was an active supporter of the habitat diorama and it was largely due to his influence that the AMNH became the world's leading proponent of such displays.[28]

For her discussion, Wonders described habitat dioramas in a manner that overlooks these constitutive politics, discussing them as "natural history scenarios which typically contain mounted zoological specimens arranged in a foreground that replicates their native surroundings in the wild."[29] Wonders's faith in the diorama as relatively innocent replicator of reality is reinforced by her remarks on how she sees the generic visitors engaging with such displays:

> In a diorama, the museum visitor can engage in a direct perception of the scene without the interference of technical devices that mediate and translate reality, predetermining both the pace and content of the information that is communicated.[30]

What Wonders does not take into account here is the very simple matter that the "direct perception" by the visitor is indeed through a "technical device." She elides the political, racial, and gender working and the eugenics association that Donna Haraway, Ronald Rainger, Tony Bennett, and my own discussion here have considered in addressing the AMNH. As with all dioramas, Akeley's is quite *predetermined* and exactingly technical, located within the very particular human and non-human network of the AMNH which brought it into being, which was also instrumental in bringing the finished, purified vision into the experiential space of the ambling visitor.

As with Osborn's *T. rex* mount, the diorama works simultaneously as object and as mediation made, by blending selected specimens in selective arrangements with scene painting, to look like a window and so, also, become a spectacle. It is yet another case of a performative *nexus* of politics/nature. By appearing as naturalistic and "true-to-life" as possible, the diorama does the work of effectively hiding the hierarchies of phantasmatic and performative arbitration that led to its luminous construction in the first place: the legacies of Roosevelt's sportsman's wonderland, Osborn's eugenics, Akeley's adventures of conquest, and so forth.

In her 1997 comments, returning to the diorama as technical form, Haraway described the complex tissue of practices and material circulations that came to produce outcomes such as Akeley's:

> Behind the dioramic re-creation of nature lies an elaborate world of practice. The social and technical apparatus of the colonial African scientific safari and the race-, class-, and gender-stratified labor systems of urban museum construction organized hundreds of people over three continents and two decades to make this natural scene possible. To emerge intact, reconstructed nature required all the resources of advanced guns, patented cameras, transoceanic travel, food preservation, railroads, colonial bureaucratic authority, large capital accumulations, philanthropic institutions, and much more. The technological production of a culturally specific nature could hardly be more literal. The intense realism of the diorama was an epistemological, technological, political and person-experiential achievement. Natural order was simply there, indisputable, luminous. Kinship was secure in the purity of the achieved vision.[31]

It is the question of pure, achieved vision that is crucial here, and of a highly particular yet monolithic conception of kinship. One of Haraway's most potent contributions has been to offer analyses that challenge disembodied, unsituated notions of vision that have been the practised foundation of supposedly objective scientific practice.[32] Vision-centred objectivity claims to set up a distance between the embodied knowing subject (who has vision, but is somehow apart from worldly action) and the object "world" (what is visible or made visible). What universal, canonically masculinist, vision offers to those who wield it as their authoritative practice is the Oz-like "power to see and not be seen, to represent while escaping representation."[33] The effect is to embed meaning while denying the technically delivered politics that are fully constituted in "achieved vision."

The explicitness of Haraway's description of the reconstructed political nature of the Akeley diorama indicates what is most important to me: the case for understanding the diorama as a political/natural apparatus or nexus can be (and is) much the same as that for understanding the Mesozoic/lost world, and other then-contemporary achieved visions or narratives of human/natural encounter. Dioramas, lost worlds, the manly hunt, expeditions of natural historical conquest, the Mesozoic – all are performative nexuses, locales made sensible by a normative projection of space/time, into which certain materials of the animalian may be assembled and animated, or for that matter deanimated, stopped dead. When one slows down to consider the histories, and indeed the technical and political matters that are mobilized within all this – not the least of which is the work of people obscured by the completeness and polish of this remarkable fetish of factualized nature[34] – then a more richly contoured tracking of its political materializations becomes possible. The idea that such a vision of the world – and one so articulate as that offered by Osborn and Akeley in the form of giant fighting saurians or mighty male simians defending their families – ought to be brought to those anonymous citizens who would not otherwise know it or see it takes us to the very nub of nature's politics, to the remarkable effectiveness of the politics of nature as it was practised at the AMNH in the early twentieth century.

The Overlay, Contest, and Intra-action of Politics/Natures

If we now consider, slowly, the layering up of these narratives, materializations, practices, sensibilities, etc., we can begin to recognize them as an accumulation of political/natural nexuses. We discern, at first, how things, worlds, and beings associated with them become ever more concrete; and second, how the constitutive politics come into coordination, sedimenting into a dominant formation – working together as an increasingly recurrent set-up, or dispositif, to use Foucault's term – a set-up of nature/politics.[35]

I will return to this proposition in due course. Before that, however, I wish to demonstrate the power in Haraway's approach, especially in regard to how astute it is in making plain the precipitation of political natures, and the stakes raised by our participation with and in them – that is, by our inhabiting of them. As I read Haraway, what she offers is an intricate study of the sedimentation of discourses, the "implosion" (her word) of narratives, phantasies, embodied practices, specimens, and technical workings, that bring about the dominant performative outcomes materialized in display – and which also then put certain limits on possibilities for public engagement with these displays. Her argument does not pretend to address the specificity and diversity of actions occurring elsewhere in the AMNH, or the complex of possible reactions from visitors encountering the displays and the institution *in toto*.

Haraway's discussion, as I read it, is concerned with the AMNH's capacity for generating a literal "technological production of a culturally specific nature";[36] a "nature" that was presented as singular and monolithic, even though we know that to be its historically contingent effect only. This "nature" was certainly dominant and had great efficacy at the institution. It worked to perpetuate certain hierarchic evolutionary logics and practices of modernist nature. Although Tony Bennett quibbles with Haraway's suggestion that the AMNH was wrought as a privately run institution funded largely by capitalist philanthropy, he does echo the analysis I offered in the last two chapters in regard to the deployment of the "germ plasm." He strongly refers to the AMNH's "selective memory" of unfolding natural history as one tethered to an economy of displayed difference, which, as we have seen for Osborn and his cohort, was highly racialized and elitist. As Bennett points out, the entirety of human history under Osborn's aegis must be "rewritten in terms of biology," and a particular biology at that.[37] Again, this is in agreement with what Haraway provides in her situated consideration of the precipitation of political/natural hierarchies in public expressions at the AMNH.

Bennett also calls our attention to Franz Boas's landmark work at the 1896 Chicago Exhibition and subsequently at the AMNH on ethnological "life groups," the theatre-like, dioramic views of human society groups equivalent to the mammal habitat dioramas, but which work against the supremacist visions offered by Osborn or Madison Grant. Exemplary among Boas's dioramic visions were the many Northwest Coast Kwawkwaka'wkw Hamatsa, "cannibal" life groups discussed by Aaron Glass.[38] Without digressing too far into the specifics of those displays, Bennett notes (following Nelia Dias) how Boas's ethnological life groups are readable as suggesting the freezing in time of the peoples represented in an apparently ahistorical moment.

But we know that this was not Boas's intent; instead, Boas was working to counteract deeply entrenched notions of universalist evolutionary history that could be turned to racist visions, presenting non-European peoples as somehow "lower" in a presumed scale of development –

along the lines of the very sort of logic we have seen in Doyle's and Osborn's narratives.[39] More pointedly, Boas's public engagement strategies to display the life-worlds of other societies by representing life groups as intricate, complex, and sophisticated was an astute move to counter the biological universalism that dominated the academic community not only at the AMNH (where he struggled directly with Osborn's vision) but in America generally up until 1905 – and beyond. Boas's nexuses of nature/politics were an answer to Osborn's. His life group dioramas were displays not of bagged big game but rather of the sophisticated practices of what he knew to be equally sophisticated peoples.[40]

Let's also recall the first moment of reconstructive precipitation of *Tyrannosaurus rex*. The year was 1905, and this year is crucial in thinking of the AMNH of Osborn, the AMNH I am writing about, and the AMNH which was the focus of Haraway's work. It was the same year that a deeply frustrated, yet unflagging Boas left the museum for a full-time post at Columbia University after disagreements with the museum's then-president, Morris K. Jesup, concerning displays that would favour a universalizing, staged, and hierarchic narrative of human development, rather than ones which would accord more distinction to the society being considered.[41] Of course, Boas is known to have continued advancing his vitally important and highly influential critiques in university venues for many years to come.[42] Osborn, on the other hand, remained at the institution and ascended to the presidency, replacing Jesup just three years later, from which post he held sway over the institution continuously for the next quarter century (i.e., until 1933). Haraway's time frame, 1908 to 1936, corresponds with Osborn's term with an allowance of an extra three years to 1936, when Akeley completed the gorilla habitat.

Although such struggles may have been taking place over what Michael Schudson refers to as "ideology," my contentions accord with those of Haraway in attending to the concreteness of materialization, more than the abstractness of ideology.[43] What becomes important is *which* technical workings and materializations in display *and* scientific work become the enduring, layered-up nexuses of natural/political relations, achieving dominant inter-performative effects. While the Osborn "vision of the world" may not have been total at the AMNH, it was certainly deeply entrenched and materialized, reproduced in multiple venues, and therefore was also the hegemonic vision offered – one that has continued to have widespread public-scientific effect. As Val Dusek points out, while agreeing on the importance of Boas's contribution and influence: "Contrary to the bright picture Schudson paints, Henry Fairfield Osborn's views – as well as many of his dioramas – live on."[44]

In sum, within the walls of the AMNH it was Osborn's and not Boas's politics/natures that won the day in 1905, with the blessing of AMNH president Jesup – the year that Boas left the institution to advance what would become a very political anthropology elsewhere. Crucially, this was also the year that Osborn published his first descriptions of *Tyrannosaurus rex* and *Dynamosaurus imperiosus*.

In consequence, as I have suggested throughout the preceding chapters, the emergent political natures – including the oppositional apparatus of saurians and simians – are much more than and different from what Schudson reckons as "social constructions."[45] Rather, they are the result of a complex of intersecting, mutually infecting practices – or, as Karen Barad put it, *intra-acting* practices – where certain fragments of living and once-living matter, combined with intricate technical works and materials in always-specific political conditions, are assembled in such a manner as to precipitate particular performative materializations of living and once-living worlds. Presaging Barad's point, what Haraway has done (and what I am striving to do now) is seek greater descriptive adequacy to capture the fluidity of the elements at play, as well as the outflow of that play into what become apprehensible as "life forms" and "forms of life" (to quote Stefan Helmreich).[46] This shifting from *inter*-action to *intra*-action, therefore, signals the co-constitutive relation between people and things, and the second-order "natures" wrought as a consequence. As Barad might note, had the collective of *intra-acting* agents in that moment been different, then "the world's becoming, in its ongoing 'intra-activity'" would also have played out differently.[47]

The Saurian-Simian-Human Syntax of Politics/Nature

> The glass front of the diorama forbids the body's entry, but the gaze invites his visual penetration. The animal is frozen in a moment of supreme life, and man is transfixed.
>
> Donna Haraway[48]

The AMNH's overlay and foregrounding of materialized spectacles of extinct reptilians and extant apes places these figures into a special oppositional arrangement and a progressive temporality that would, demonstrably, become commonplace in American public cultures of science over the rest of the twentieth century. In the oppositional arrangement, the gorilla as "simian oriental" stands as the human/animalian *near-other* in nature. It is set against the dinosaur "saurian alien," which stands as the reptilian/animalian *distant-other* in nature. When Haraway remarks on the experience of gazing into the gorilla diorama, stating that "culture meets nature through the looking glass at the interface of the Age of Mammals and the Age of Man,"[49] one could readily reply that *culture opposes nature in the face of terrifying monstrosity across the chasm between the Age of Reptiles and the Age of Man*.

What this amounts to, then, is also a shorthand public, museological syntax of politics/natures, blending certain anthropological, primatological, and palaeobiological chronotopic impulses. Johannes Fabian famously remarked upon the anthropological expression of time-space modelling,

> in the uses of time [that] anthropology makes when it strives to constitute its own object – the savage, the primitive, the Other. It is by diagnosing anthropology's temporal discourse that one rediscovers the obvious, namely that there is no knowledge of the Other which is not also a temporal, a historical, a political act.[50]

To be sure, there is no knowledge of the dinosaurian other that is not also a temporal, historical, and political act. The familiar mammalian habitat diorama seen in many museums around the world, and the scenographically bounded world of depicted dinosaur interactions equally familiar in museum settings, become syntactically related "stagings" of life. Rainger underscored this point in regard to the AMNH, where "for those that visited the museum it was a form of theater, a pleasure palace filled with dinosaurs, titanotheres, and other entertaining features."[51] What we arrive at, then, are the multilayered nexuses *within* the museum – from palaeontological worlds to wildlife habitat worlds – and a nexus *between* the museum and popular literature, and later popular cinema. The stories, creatures, chronotopes – and the tales of those who know and encounter them – move freely within and beyond the institutional frame, securing what is authoritative inside the institution, while permitting its phantasmatic effect to radiate among ever-wider publics, generating in its wake a dangerously narrow and prescriptive politics.

How, then, are "we" captured in this natural/political assemblage? The embodied extension of this syntax is the museum's publics; its visitors, patrons, its human interlocutors. For this sense of human constitution is produced by "publics" in the contemporary moment as they engage the expertly delivered simian oriental and its evolutionary precursor, the saurian alien. By this move, each of these figures – human, simian, saurian – is co-produced and co-present in the common, syntactically consistent chronotopoic world of the AMNH. Suddenly, "we" become material co-inhabitants of this materialized "conjectural history," of materialized phantasmatic history.

Thus, when Donna Haraway suggests, "We inhabit these narratives, and they inhabit us," she is being literal, explicit, especially when we consider how such narratives, when formed into natures and animal beings, become physical, public cultural co-presences with us in encounters with certain relations of time and space. We might then take literally and unproblematically these chronotopes as the time and space in which we conduct our lives. Unpacking the performative moves between big game hunting, dioramic visions, and Mesozoic and lost world chronotopes allows us to recognize the messy[52] intersections among, and trafficking between, these practices. It allows us to recognize that the nexuses, too, are *articulated*, though not in a particularly tidy or deliberate way (as compared, say, to the technical work of articulating a dinosaur skeleton). In effect, they are seamlessly conjoined as workaday practices. They are not merely susceptible to, but are also constituted by, and reconstitute and edit, colonial, racist, masculinist phantasizing.[53] Their syntax goes public – that is, it is made into the more or less durable realities we inhabit day to day. In this all of these practices and their outcomes become and remain deeply and extensively political.

Dispositifs, Implosion Zones, Nexuses: The Colonial Apparatus of Politics/Natures

> Purposeful action and intentionality … are properties of institutions, apparatuses, or what Foucault called dispositifs.
>
> Bruno Latour[54]

Gathering the threads of discussion together – and drawing forward my remarks on the Mesozoic nexus in chapter 3 – what all of this points to is the political assemblages drawn together when the multiplicity of performative nexuses merge and gain coherence through their relation to one another. The result is a *network of nexuses*, or conversely, a *nexus of networks*. Each nexus – of which I have only discussed three of many that might be discussed – is made up complexly of narratives, "objects," relations, visual techniques, human action. They are akin to Haraway's very elegant proposition of a natural/cultural *zone of implosion*; a provisional agentive-form and organizer of time-space with attendant lives – animated, reanimated, or otherwise. It is a zone where folks and non-human entities engage, and a place where

> the technical, mythic, organic, cultural, textual, oneiric (dream-like), political, economic and formal lines of force converge and tangle, bending and warping both our attention and the objects that enter the gravity well.[55]

Of course, Haraway also argues for the possibility of agency, of participating in such zones, such that alternative trajectories of "natural" possibility may ensue. In the case of the Mesozoic nexus, these actors and forces merged around a modernity-generated *topos* and *chronos* that produce and condition what can count as truthful visions of nature as a changing yet constrained dynamic – whether spoken of as systems, life, the gene, the biogeographic map, the landscape, the species, the Pleistocene, and so on. The nexus and the network form a kind of tissue, or perhaps, to borrow Deleuze and Guattari's term, a rhizome-like extension of these practices across heterogeneous and seemingly disjointed sites, in seemingly unrelated moments or places.

Taking such rhizomatically related nexuses or zones of implosion seriously, we are able to discern how it is that the chronotopic impulse is, in itself, always and already political. It is not merely that certain political notions are constituted *in* the materializations, but that the materializations come to colonize those who engage them, to draw them into encounters with them, to extend their narrative and phantasmatic purchase into their practices as well. They inhabit us and we inhabit them.

In addition – when we note how they are assembled together – the multiple, layered, messy peformative nexuses combine, working in much the same sense as Michel Foucault's *apparatus*, and more specifically as a *dispositif*. Giorgio Agamben has usefully summarized the idea of dispositif in his 2009 reading of a 1977 interview with Foucault on the proposition:

> a. It is a heterogeneous set that includes virtually anything, linguistic and nonlinguistic, under the same heading: discourses, institutions, buildings,

> laws, police measures, philosophical propositions, and so on. The apparatus itself is the network that is established between these elements.
>
> b. The apparatus always has a concrete strategic function and is always located in a power relation.
>
> c. As such, it appears at the intersection of power relations of knowledge.[56]

Yet Agamben, moving beyond Foucault, continues to propose how, out of European theological tradition through to modern praxes, we are delivered up a historically contingent, epistemic foundation for the emergent concrete and its moral content:

> I wish to propose to you nothing less than a general and massive partitioning of beings into two large groups or classes: on the one hand, living beings (or substance), and on the other, apparatuses in which living beings are incessantly captured. On one side ... lies the ontology of creatures, and on the other side, the *oikonomia* of apparatuses that seek to govern and guide them to the good.[57]

In terms of the simian-saurian-human syntax, the *politics in and of these creatures* are extended as constitutive in the dispositif of "natures" (e.g., the reconstructed Jurassic, Cretaceous, dioramic Karisimbe forest, etc.) such that *they both also become the politics of the people* – or, at least, those people who have access to the outputs of science, those who inhabit this set-up, this dispositif of the socio-natural. One effect is the way that these combine, through the creatures instantiated, to limit the possibility of what is thinkable as nature – and how folks might then participate in what is called nature – when these particular, networked natures become all the more concretized. As such, the politics of such creatures and natures are *our* politics too – so there is potential to repeat these politics, re-enact them, recompose them, or transform them, depending upon our interventions.

The moves to colonizing public life – from science to stage to literary fiction to cinema and back to science – will be made all the more explicit in the next chapter, which follows the easy uptake and trading of Mesozoic performativity in pulp literature, science fiction, popular cinema, and into later practices of palaeontology. I will suggest more about the collisions of these performative worlds of hunters, rapacious and racially partitioned masculinities, terrifying carnosaurs, powerful apes, and other gendered bodies and their entanglement in public science and fantasy entertainment over the twentieth century. It is here that the techniques of skeletal model-posing employed by Osborn and Brown arise again, becoming crucial for dramatizing past life. As will be seen, such techniques of palaeontologically supported cinematic action and world-making would help further link together the Saurian lost world, the Simian Eden-world, and the Human adventure-world. With that, these techniques and linkages perpetuate and extend the colonial, gendered, racialized phantasmatics animating these performative nexuses of politics/natures. They are, in brief, a thoroughgoing politics/natures, all the way down.

But here is actual hope: as we delve further, even this seemingly robust apparatus begins to totter, reaching the limits of its own coherence and stability, its own inhabitability.

6 Vestiges of the Lost World: Recirculating the Tyrant Nexus

Stop motion photography allows an artist to create life ... It is an intimate stretching of time which transforms the mass of metal and rubber to a literal extension of the animator, imbued with his own sense of style, grace, action, reaction, emotion, and idiosyncratic movement.

Jeff Rovin on stop-motion animators' creation of lives in linear space and time[1]

Now I will turn to life animation, in the familiar sense of film and its relation to dinosaur science – and, with that, consider some of the ways in which the Mesozoic/lost world apparatus has recirculated across the twentieth century. In the quote concerning mainstream cinematic animation that opens this chapter, we encounter a stream of key notions that could so readily have been drawn from the scientific and literary lexicons of Henry Fairfield Osborn and Arthur Conan Doyle: action, reaction, near-human cunning, godlike creation, movable armatures, creatures as extensions of the animator, stretching of time. Much that I have discussed in all of the preceding chapters revolves around this very particular sort of intricate, "intra-active" work of *bringing creatures and worlds to life*.[2]

Such creative reanimation of dinosaurs and their worlds has taken place concurrently in palaeobiological practices, in museums, in popular cinema, and in literary science fiction.[3] Whereas, to this point, I have traced the relocation of performative nexuses between literary, museological, and technoscience locales (i.e., as constituents of apparatuses, or dispositifs), I now turn to aspects of their movement and transformations over time.

In taking this up I momentarily lay aside some of the weightier analyses of the preceding chapter on politics/natures. I will use this chapter to follow some of the fraternal – even sophomoric – yet politically charged recirculations of the masculinist, racializing performative natures discussed so far. Here, I follow how they, on one hand, show up more or less *outside* the circuits of the technical-public realm of museums, and, on the other hand, are drawn more or less from *inside* but also *back into* the technical-scientific practices of palaeobiology over the twentieth century. This also foreshadows some of the juxtaposed practices that will be discussed in the latter half of this book. Leapfrogging from Donna Haraway's "Simian orientalism," I will expand my discussion in this chapter by way of a particularly curious near-object:[4] a somewhat quirky, but highly indexical dedication from the book *The Complete Dinosaur* that explicitly demonstrates the performative relation between vernacular cinematics and palaeontological practices. This helps to convey the persistence in performative circulations, and how phantasy is continually entailed, backgrounded, but then rematerialized in reconstructive, world-making, and genealogical practices of dinosaur palaeontology, through its ongoing conversation with popular spectacle production.

Cinematic Genealogies of the Saurian-Simian-Human Relation

Yet another important technical cue from Haraway's work on the Akeley gorilla diorama directs us to the easy move from stage set, to museum display, to cinematic rendering:

> Lit from within and surrounded by the panoramic views made possible by Hollywood set painting and the new cameras of the 1920s, the perfect natural group – the whole organic family in nature – emerged in a lush Eden ...[5]

As with simian-centred scenarization between museums and Hollywood, the history of animated dinosaur movie-making helps us to trace an extended genealogy of public-scientific trading of the Mesozoic/lost world nexus and its counterparts. This genealogy connects the Doyle/Osborn actions of the early decades of the twentieth century with the actions of dinosaurian palaeobiology and systematics in its final decade. While the intervening history is far richer than what I can convey in this limited forum, the shorthand genealogy I present here indexes quite clearly how the actions of dinosaur scientists have continued to interweave with those of public culture – especially populist film and animated dinosaur spectacle – by means of the performativity of the Mesozoic/lost world nexus.

The voluminous 1997 compendium of semi-technical dinosaur science and history, *The Complete Dinosaur*,[6] compiled by vertebrate palaeontologists James O. Farlow and Michael K. Brett-Surman, assembles together forty-seven essay contributions, most of them by leading researchers on dinosaurs and palaeobiology.[7] It includes a number of well-known dinosaur scientists and borrows from such widely ranging topics as history, biogeography, biology, archaeology, and even media and culture in its engagement with dinosaurs. As its title promises, the volume aspires to be "complete," sampling an astonishingly wide range of practices of Mesozoic world-making and performative action.

However, it is the dedication to this volume which provides a powerful sounding into the depths of dinosaurs and palaeontology within scientific and cinematic culture:

To

Ray Harryhausen

(whose work evoked a sense of wonder in many future palaeontologists)

and

Forry Ackerman,

and to the memories of

Barnum Brown,

Edgar Rice Burroughs,

Sir Arthur Conan Doyle,

Thurgood Elson,

Charles W. Gilmore,

and

Willis O'Brien:

You had an impact.[8]

These men did indeed have an impact, but some unpacking is required. All are "actors" in the nexus of Mesozoic/lost world performativity, participants in the resurrecting of science's monsters to vivid life in well-bounded geographies. While most are well known to established dinosaur palaeontologists in North America, only two were actually professional palaeontologists. Every one of the remaining figures is associated with the English-language film and literary industry of dinosaurian world-making. Heading the list, and particularly telling, is one of Hollywood's most lauded movie-monster animators from the 1950s onward, Ray Harryhausen – noted for his fabulous monsters in the Sinbad film and the original *Clash of the Titans* (1981), among many others.[9] Summarizing from here, the dedication names palaeontologists (Brown and Gilmore), lost world authors (Doyle and Burroughs), a B-film cult figure (Forry Ackerman), another highly acclaimed giant-monster film animator (O'Brien), and even a fictional palaeontologist from a monster movie (Thurgood Elson). This hybrid lineage leads ultimately to those on whom the figures had their impact: the editors themselves, two still-active twenty-first-century dinosaur authorities who are clearly prepared to recognize and honour the public cultures infusing the scientific cultures in which they participate.

It is fairly straightforward to elaborate this lineage by considering each named individual or fictional character and then articulating him more fully within the Mesozoic/lost world nexus.[10] I will use this partly serious, partly ironic dedication from Farlow and Brett-Surman to discuss the technical-vernacular trade (an extension of the political/natural apparatus discussed in the last chapter), linking the imaginary of museum-located palaeontology with the filmic and literary imaginary of the American leisure and entertainment industries.

Pre-Cinematic Animateurs

The two palaeontology professionals, Barnum Brown and Charles Gilmore, are famed fossil collectors who, starting from just prior to the turn of the twentieth century, provided a tremendous number of the material specimens and technical descriptions which helped to fill in the ever-increasing diversity of the ancient world of dinosaurs. We are already familiar with Brown, who worked for the AMNH under Osborn, while Gilmore collected for the Smithsonian Institution's Museum of Natural History. Both collected a wide variety of canonical specimens over their careers and were also central advisors in mounting many of the associated dinosaur skeletons at the AMNH and the Smithsonian, respectively.[11] As dean of American vertebrate palaeontology, Osborn, along with his animating phantasies, ever looms in the shadows throughout this history.

Of course, over much the same period as when Brown and Gilmore worked, it was the most famous artist-illustrator of prehistoric life-worlds, Charles Knight, who provided two-dimensional restorations to many major US museums. Eventually those visualizations – wrought in dialogue with palaeontologists Osborn, Cope, Brown, and Matthews, among others – informed interactions portrayed in museum skeletal mounts. We know already the importance of these same visualizations in forming the literary, narrative, and graphic imaginings in Arthur Conan Doyle's and Edgar Rice Burroughs's massively popular novels from the 1910s to 1930s, set on misty lands populated by apes, proto-humans, and dinosaurs – most notably in the Challenger and Tarzan novels.[12] Both Burroughs and Doyle, of course, are among the exclusive list of figures in the Brett-Surman and Farlow dedication. The simian-saurian human syntax was deeply fashioned by, and further embedded in, these world-making technologies of the early twentieth century.

Post-Cinematic Animateurs

Up to this point, and from previous discussions in this book, the network of figures, specimens, and productions to which the Farlow and Brett-Surman dedication connects ought to be fairly familiar. It is the remaining figures that I want to discuss in some greater detail. From here – and underscoring Haraway's remarks on the commonality in Hollywood set-design technologies and natural history museum dioramas – attention can turn to the movie-industry figures in the lineage. Those mentioned have been key players in film renditions of the Mesozoic/lost world which have come to inform the public imaginary of palaeontologists and non-palaeontologists alike. In 1914, Willis O'Brien – also known to film animation enthusiasts as "Obie" – inaugurated the technological tradition that would, probably more than any other practice, bring the saurian world widely into the public cultural imaginary.[13] That year, O'Brien took a page from Edweard Muybridge's frame-by-frame stop-motion work on animals and bodies in motion.[14] After first experimenting with clay figures of boxers, O'Brien moved to dinosaurs modelled around wooden, and then later metal armatures[15]

– this was six years after the duelling *T. rex* pair was being posed at the AMNH, for which Raymond Ditmars had first used such armatures with Osborn, William Diller Matthew, and Erwin Christmas. For his films, O'Brien posed and reposed the dinosaur figures, photographing them frame by frame with a newsreel camera. These images could then be projected at cinematic speed to produce the illusion of motion.[16] At long last, dinosaurs lived again.

Between 1914 and 1919, O'Brien had produced six comical animated shorts using a clay "brontosaur" along with other prehistoric creatures. These early works were simply voyeuristic, presented like dioramas or stage shows, without any visual portrayal of a rupture in time-space. In 1919 O'Brien produced the *Ghost of Slumber Mountain*, and Barnum Brown – his technical and institutional position at the AMNH granting him imaginistic privilege – was approached to advise O'Brien on the habits and movements of the diversity of dinosaurs that would appear in this film. Don Glut noted the dinosaur drama that ensued in the film's tale:

> A huge *Apatosaurus* lumbers through the water. Another O'Brien flightless bird, a *Diatryma*, primps its feathers, then devours a snake. Two *Triceratops* briefly fight one another, until their battle is interrupted by a large carnivore, the predatory *Allosaurus*, which vanquishes one of these horned dinosaurs and then stalks after Uncle Jack, before the entire adventure is revealed as a dream.[17]

Here, the modern viewing subject is led into the past world by means of layers of time-defying technology – the filmic animation, Uncle Jack's "dream," and an imagined instrument of vision, basically a "peep show" reminiscent of penny carnivals. The common phantasizing technologies of the imagination, the diorama, the camera, modelling, animation, scientific restoration, and street spectacles were all enfolded in this, the first film to directly consult a professional palaeontologist.

It is well known that O'Brien borrowed liberally from the imagery of Charles Knight – as many illustrators and animators of prehistoric creatures of the last one hundred years have done[18] – in these earliest examples of stop-motion animation of dinosaurs and prehistoric creatures, as well as in his later productions into the 1950s.[19] As vertebrate palaeontologists and museums increased the repertoire of Mesozoic inhabitants within their domains, so Knight created an ever-increasing diversity of forms in his art. Willis O'Brien transposed this diversifying of dinosaurian kinds in the animated lost lands of his many films from the 1910s through to the 1950s. His animation work was featured in the first filmic version of Doyle's *The Lost World* (1925) and in Merrian C. Cooper's film *King Kong* (1933).[20] *King Kong* played the otherworld/homeworld opposition explicitly, the homeworld being the same as that of the AMNH, New York City. In full Osbornian, AMNH style, the simian King Kong turns to face his *physically equal* but *mentally inferior* alien enemy in the form of *Tyrannosaurus*, along with other saurian kinds (figure 6.1).

Both *The Lost World* and *King Kong* were set in bounded world spaces; the former on a plateau, the latter on its ominous "Skull Island" – emulated later by Michael Crichton's "Isla Nublar," or "Clouded Island," in *Jurassic Park*. The 1925 and 1933 films exaggerated the temporal distancing created by the remoteness of the lost land, making use of a tree as bridge to span the metaphoric chasm of time between the adventurers' intended escape route back to their civilization and the deepest wilds of literally monstrous terror (figure 6.2).

Pressing upon public horrors as well, the 1925 *Lost World* film revised the novel's ending, bringing a huge sauropod back to London rather than a pterosaur that fitted into a suitcase. This began the Hollywood film thematic of setting a giant, rampaging monster upon peaceable urban spaces, which of course was repeated in *King Kong*, the giant ape now taking up the role of "destroyer of creation" – to borrow Osborn's somewhat awestruck turn of phrase – though not an instinctual destroyer like *T. rex*, but rather one defending "himself" after being taken hostage to New York to fulfil the profit-dreams of a Manhattan showman. While the audience is not meant to sympathize with filmically portrayed, base, cold-blooded carnivorous reptilians, in contrast the American-male-modelled, woman-desiring ape was played as a more ambivalent and, at moments, vulnerable character; one to whom the viewers' sympathies were meant, at least partially, to be directed.[21]

Underscoring the trading between museological and entertainment technologies, both films indeed used dioramic scene-making techniques to produce their framed world of evolutionary order for the viewer. O'Brien produced intricate miniature sets and matte-painted backgrounds, aided by ever more elaborate stage lighting. In both films – and borrowing the armature-based modelling techniques which zoologist Raymond Ditmars and Charles Knight had used earlier as a preliminary step in the process of developing a restored dinosaur scene[22] – O'Brien worked with sculptor Marcel Delgado to develop metal

Figure 6.1. King Kong battling *T. rex*, securing his prize, Ann Darrow, in the treetops. Source: *King Kong*, 1933, by RKO Films. Directed by Henry O. Hoyt, animation by W. O'Brien and M. Delgado. Everett Collection.

ball-and-socket "skeletons," covered by sponge-rubber "muscles" and latex "hides" – a sort of miniaturized taxidermic preparation out of synthetic materials.

But the reach into the contingencies of palaeontological knowledge goes farther. O'Brien and Delgado worked closely with Barnum Brown at the AMNH in New York, in fashioning the front limb of their animated *T. rex* with three digits, rather than the two-digit form that is accepted today and that had also at the time been proposed by Gilmore. Indicating this as a case of then "unsettled science," communication studies scholar David Kirby remarked: "Had O'Brien and Delgado chosen to speak with Charles Gilmore of the Smithsonian Institution, who argued that T. rex had two digits like Gorgosaurus, and implemented

Figure 6.2. Encountering dinosaurs in *The Lost World* Amazonian plateau. Source: *The Lost World,* 1925, by First National Films. Produced by Merrian C. Cooper and directed by Ernest B. Schoedsack, animation by W. O'Brien and M. Delgado. First National/The Kobal Collection.

his advice then King Kong's T. rex would be considered an 'accurate' representation today."[23]

Beyond Brown and Osborn's *T. rex* reconstruction, in a further resonance with the Osborn/Doyle moment, technique came to be wedded to narrative, as both O'Brien films played through the race-gender hierarchies familiar from Doyle's novel and Osborn's networks, replete with savage ape-people, technologically intermediate "Indians," and advanced, gun-toting, great white hunters. A new turn in the films, however, was the introduction of female characters into the adventure travel action, complicating the simpler attention to homosocial bonding with sexual desire – that is, specifically male-perspective heterosexual desire, which Laura Mulvey has noted as a common logic and affect of mainstream Hollywood to the 1980s, and which continues through to the present.[24]

In the 1925 *Lost World* film, the adventurers are accompanied on the journey to the lost plateau by a bold young woman whom Malone ultimately marries, replacing the demanding fiancée character, Gladys, from the original Doyle novel. *King Kong* focuses around the ape's fascination for the blonde character Ann Darrow, an actress (famously played by Fay Wray) brought to Skull Island to star in a film entitled "Beauty and the Beast" – the explicit conceit of the film. This fixation is a reification of masculinist, racializing eroticism surrounding the possibility of miscegenation and interspecies contact. That theme would come to be re-enacted both obliquely and subtly in late twentieth-century films on women primatologists, including National Geographic productions on Jane Goodall, or more directly in the Michael Apted film *Gorillas in the Mist*, adapted from Farley Mowat's quasi-biographical narrative of Dian Fossey's life.[25] And, in many mainstream films of the 1930s, such as *Blonde Venus*, these types of race-gender hierarchies would be examined in terms of their intersection with desire. In a cabaret scene from that film, the Germanic Marlene Dietrich performs in a gorilla suit against blackface female cabaret dancers, only to remove her simian garb and present herself in extreme blondeness, extreme whiteness to the predominantly white male audience, while African-American bartenders and service workers are explicitly posed, looking on, as if happy to be in these subordinate roles and witnessing this play of interspecies, interracial desire. While mirroring such science-crafted hierarchies as those of Francis Galton, Madison Grant, and Henry Fairfield Osborn with their biologically rationalized taboos against racial mixing, the playing of racial contact came to be used equally in the Dietrich role – as had been done with Ann Darrow in *King Kong* – as machinery that could effect desire.

Don Glut isolated and offered what he felt was the key O'Brien-animated scene in *King Kong*, which he also regards as "art," but which, more to the point, inaugurates the popular culture media genre emphasizing these over-produced, American male race-gender phantasies and desires.[26] In this scene, O'Brien's *Tyrannosaurus* is closing in on the vulnerable "tiny blonde woman," Darrow, who is cowering by a tree:

> Kong rushes to Ann's rescue, battling the scaly monster with near-human cunning and skill, until, his hairy hands tugging apart the tyrannosaur's steam-shovel jaws, he slays the dinosaurian adversary. Kong beats his massive breast triumphantly, but it was also a triumph for O'Brien, whose artistic peak was evidenced by this sequence, one of the most memorable in King Kong.[27]

O'Brien's scene (repeated to some extent in the film's 2005, $207 million Peter Jackson remake) and Glut's response to it consolidate many of the logics which were also available in Akeley's great "giant of Karisimbi" standing to protect its family and its Eden. The intended male-viewer identification with the "near-human cunning" of the gorilla also parallels the blended characterization of embodied intelligence and massive physical strength seen in Doyle's Challenger. The terrifying alien adversary found in Osborn and Brown's king tyrant lizard is repeated here, and here too are the gender-race-evolution hierarchies of the Doyle/Osborn interperformative nexus. As a dramatic moment of naturalized American, male-centred, heterosexual desire and identification, there can be little argument that it is indeed a "most memorable" moment from a film with a virtually inextinguishable place in the American cultural imaginary. The simian-saurian-human encounter would come to be replayed in film after film in the 1940s, exemplified in such serialized movies as Johnny Weissmuller's "Jungle Jim" films (figure 6.3).

Moving back again to the Farlow and Brett-Surman dedication genealogy, their leading figure was Ray Harryhausen, Willis O'Brien's protégé. Harryhausen may well have been the most viewed of saurian and non-saurian stop-motion monster animators in American film history (certainly until Spielberg and his technical associate Phil Tippett),[28] producing imposing creatures for the many, still highly circulated *Sinbad* movies; for the cult prehistoric monster "classic" *The Beast from 20,000 Fathoms* (1953); the palaeo-Western *Valley of the Gwangi* (1969); and

Figure 6.3. 1940s poster for the film *Jungle Manhunt*.

One Million Years B.C. (1967), which featured Hollywood "sex symbol" Raquel Welch.[29] On the latter film, Glut underscores what he feels made the movie successful: "The now-classic photograph of the shapely Welch in her brief animal skin lured most patrons into the theaters, but once they had taken their seats, Ray Harryhausen provided visual thrills of a different sort."[30]

These are, by now, very predictable elements of the visual-erotic-adventure makeup of these films and their associated libidinal and colonial programmatics, repeated *ad nauseum* in hosts of pulp and comic book fiction of the period and arguably still with us in the current moment (figure 6.4).

Many living dinosaur palaeontologists, now in their forties and fifties or older, have confirmed with me that they saw these films in their youth. These dynamic, animated dinosaurian presentations would have been available to them as geographic renderings and phantasmatic tales of the Mesozoic/lost world. They would complement the legitimated scenographies (i.e., expert, scene-making practices in dinosaur books and in natural history museum displays). Glut, as an intellectual colleague of many dinosaur researchers, expressed the sort of stance that palaeontologists might be expected to take in relation to such warmly regarded filmic pseudo-dinosaurs. He wrote that Ray Harryhausen's concocted Rhedosaurus in *The Beast from 20,000 Fathoms* "may not have been cited in any palaeontology book, but Harryhausen's creation is a visual joy, expertly crafted and animated and certainly the finest giant monster to appear on the screen since the original Kong."[31] That of course was prior to the *Jurassic Park* film's CGI animations of the 1990s. When considering the tremendous trading of filmic and palaeontological cultures, the cinematic Rhedosaurus has to be considered as phantasmatic kin to the palaeontological *Tyrannosaurus rex*, whose restoration had never been more accurately achieved in palaeontological terms than in films like *Jurassic Park*, or the 1998 animated 3D IMAX film *T. rex – Back to the Cretaceous*.

It is worth remarking now just how influential O'Brien's early work had been for Harryhausen, and in turn for animator Phil Tippett (to be discussed in the second part of this volume), who worked with Harryhausen and then later for George Lucas – so hooking together the twentieth-century trajectories of monster- and alien-world animators, through to *Star Wars* and other science fiction work from George Lucas and Steven Spielberg, and continuing to the present. In keeping with the King Kong motif, Lucas has likewise done his share of portraying

Jungle
COMICS
JAN.
No. 37
10¢
KAÄNGA
—JUNGLE LORD—
BRAVES THE UNSPEAKABLE
HORROR OUT OF AFRICA'S PAST
IN
"SCALY SENTINEL
OF TABOO SWAMP"
SIMBA—KING OF BEASTS
FANTOMAH—DAUGHTER
OF THE PHARAOHS
WAMBI—THE JUNGLE BOY
ALSO CAMILLA • TABU

DELL
15¢
KONA
FEB.–APR.
NO. 1256
MONARCH OF MONSTER ISLE
...Modern Man versus Ancient Beast!!

Figure 6.4. Composite of 1940s–1960s comic book covers: lost worlds of white, masculinist, heterosexual desire.

women as objects of desire and giant animalian consumption, exampled famously in the scenes of Oola, the green-skinned Twi'lek who performs as an exotic dancer for the bloated Jabba the Hutt, and who is then dropped into a chamber to be devoured by a captive alien monster. Mixing up and adding yet one more volley in this rapid-fire of ricochet cross-references through the history of alien/monster animation lineages, and as if to conjure the hidden "force" of Jedi and animation magics, one blogger remarked recently, "If Ray Harryhousen [*sic*] is the Obi-Wan Kenobi of stop motion effects, then Willis O'Brien was the Yoda (or at least the Qui-Gon Jinn)."[32]

Material-Phantasmatic Circulation

Finally, and continuing beyond their expressed listing of names, the editors themselves may be taken to complete the genealogy of Mesozoic world-makers, bringing it up to the late 1990s (and beyond, as both are still actively engaged in academic dinosaur science and museology). What Farlow and Brett-Surman do, in what was a simple, even playful dedication, is acknowledge the much grander and masculinist gestures of connection which link filmic, literary, scientific, and technical world-making practices. They suggest the mobility and malleability of the Mesozoic/lost world performativity to which I have been pointing. There is a heavy technical, personal, and phantasmatic trade occurring legitimately across these several domains, even if it may be a rather haphazard one. Moreover, that trading over the last century has continued to reiterate and transform the hybrid figure of the Mesozoic/lost world and its animalian constituents.

Farlow and Brett-Surman may have included the film-culture figures in their dedication in some limited ironic or parodic sense, but it is also clear that they are here out of an honest wish to pay homage to those *animateurs* who have influenced the editors' lives, desires, and thinking through dinosaurs, monsters, and other concocted inhabitants of lost lands. Indeed, their colleague and friend Don Glut writes consistently about the "homage" which animators and illustrators pay to one another by copying previous works of their colleagues – the animators borrowing heavily from Knight, from each other, and so generating an extended and significantly male genealogical cohort. Glut notes, with a precision achievable only by an expert insider to the discourse, how "Harryhausen paid homage to O'Brien by having the Gwangi, an enormous *Allosaurus*, scratch its ear, as had the *Tyrannosaurus* in [O'Brien's] premier scene in *King Kong*."[33] That took place in a 1969 production drawing on O'Brien's 1933 production, where O'Brien's *T. rex* was in turn indebted to the precursor work and materializations of Barnum Brown and Henry Fairfield Osborn from the three previous decades.

Connections across more than thirty years can be made by means of the simplest of gestures, including simple citational practice – even, indeed, by means of rumours. Remarkable networks of borrowing and exchanging of visions, tales, and associations can be made in the process. The most poignant connection, made by most living artists and dinosaur palaeontologists alike, is with Charles Knight and his influence upon their way of envisioning dinosaurs and worlds of the past. The cautionary note about cultural reproduction which I make here, of course, is that Knight's otherworlds were already phantasmatically configured; residing here, unsuspected, are the imaginings most notably of Osborn and Doyle, and, need I say yet again, a particular masculinist, imperialist imaginary.

Such gestures of homage help to weave the complex genealogies of the material-semiotic performances of commercial spectacle with those of quasi-commercial North American palaeontology into an elaborate, always shifting, relational fabric I have referred to as politics/natures. Living dinosaur palaeontologists do not simply and blindly reproduce the phantasies of Hollywood, or of Osborn's imperialist or racist natural histories. Rather, those phantasies are hauntingly present in the complexity of cultured knowing and materialized culture which inform palaeontologists in an everyday way, lurking like some phantasmatic *bête noir* in their world-making practices and figurative flows, permitted through the Mesozoic/lost world nexus.

Those phantasies of vernacular/scientific political natures also have typological consistencies. The worlds are populated by an enormous, interacting diversity of creatures. Primitive humans meet modern humans; simians meet saurians; scientists – almost always white men of privileged classes – resurrect and restore dinosaurs. Different technologies for travelling to, seeing into, or conquering the past are used. All of it plays off sensibilities of the real and the fabricated. Quite consistently, the Mesozoic/lost world as apparatus of transmission for such phantasies is conformed by travel, expansion, mapping, and evolutionary hierarchies; a conquering of the unknown geographic space/time. The impact of that, through the impact of all named in the

Farlow and Brett-Surman dedication *cum* genealogy, arises unselfconsciously, seemingly unavoidably, by means of genuine immersion in and recitation of these highly recirculated, yet thoroughly generated stories and worlds.

The genealogy traceable through the dedication, of course, is only provisionally reducible to such caricatures. In the dinosaur palaeontology and public science cultures of the most recent decades, there have been notable phantasmatic revisions, introductions of different dinosaurs, recalibrated gender models, alternative techniques of reconstruction, and revised evolutionary and classification propositions. I will be referring to some of this in the remaining chapters of this volume, triggering from the networks associated with the late twentieth-century dinosaur *Maiasaura peeblesorum*.

That said, it is important to bridge the period to the 1970s and beyond, and to highlight here how much has been written by palaeontologists and dinosaur populists in the last forty years regarding a shift in the palaeontological characterization of dinosaurs, often said to have started with the publication of a single paper by Yale dinosaur researcher John Ostrom in the late 1960s.[34] (Farlow, I should point out, studied with Ostrom for his PhD.) The dinosaur on which Ostrom focused his attention was, predictably enough, a carnivorous dinosaur, and is known technically as *Deinonychus* ("terrible claw") *antirrhopus*.[35] The beast has been described by palaeontologists as an energetic dinosaur, powerful, a group hunter, compact in size, and most notably, potentially endothermic – meaning its body temperature was regulated by some internal mechanism, more like mammals – more like humans.[36]

Moreover, the evolutionary history of *Deinonychus* has been linked to that of birds, invigorating a major turn in vertebrate palaeontology towards questions not merely of dinosaur-bird phylogeny but of whether we, as humans, are yet living with dinosaurs in the form of these mostly small, feathery, and yet highly social critters. The salient effect is that of dissolving the otherwise concretized spatio-temporal boundary of extinction and geological discontinuity, separating what was called the "Age of the Reptiles" from the "Age of the Mammals" and creating, therefore, a new continuity via dinosaurs-as-birds into the "Age of Man." I raise this point to index a general shift that began visibly in the 1960s, when dinosaurs became less alien, so to speak, both biologically and publicly; this was part of what Adrian Desmond called a "Revolution" in dinosaur studies. Multiple dinosaur lineages within the Mesozoic and extending beyond its bounds have been delineated, becoming more complex and detailed. Palaeoenvironmental analysis has become more thorough and extensive, in both global and local terms. Studies of dinosaur behaviours have become increasingly more elaborate, more technical.

With all these shifts have come new networks and cohorts, new associations of palaeontologists with new fossils, a revitalization of dinosaur palaeobiology, and an ever more vigorous "enterprise culture" linking the scientific with the public by way of market interests. Revised narrations have been brought to bear in the complex. Nonetheless, the Mesozoic/lost world chronotope persists as an overall ordering apparatus, and so do the giant meat-eaters that have fixated scientists and animators across the twentieth century.

Recollecting my earlier discussions of materialized phantasmatics, Bakhtin's notion of the chronotope has been quite applicable here, in that it incorporates the time-space features of the Mesozoic/lost world. However, it stops short of materialization. Perhaps as useful as Bakhtin's notion are terms that have come out of social and cultural studies of science over the last three decades. Susan Leigh Star writes of *boundary objects*, "which are both plastic enough to adapt to local needs and constraints of the several parties employing them, yet robust enough to maintain a common identity across sites."[37] Joan Fujimura discusses *standardized packages*, which are theory, method, and technical frames that scientists develop and use to organize and streamline their collective or distributed activities.[38] Bruno Latour suggests other sorts of boundary objects, which are more than mere figurative devices, or mediating metaphors. Referring specifically to what he calls "quasi-objects" – including such entities as machines, software, post-it notes (I would add dinosaur models and toy figures to the list) – Latour writes:

> [T]hey attach us to one another, because they circulate in our hands and define our social bond by their very circulation. They are discursive … they are narrated, historical, passionate … and never forget [they are] Being.[39]

While related conceptually, there is no particular term that quite captures what takes place with the Mesozoic/lost world, especially if we count such intangible modes of circulation as rumours, paying homage. One that comes close, but does still not quite include such intangible modes, is what Latour refers to as the "immutable mobile," for which Star offers this description:

> These are representations, such as maps, that have the properties of being, in Latour's words, "presentable, readable and combinable" with one another. Such representations also have "optical consistency," that is visual modularity and standardized interfaces. They are often flattened to make them tractable in combination. They have the important property of conveying information over a distance (displacement) without themselves changing (immutability). Thus, in contrast with a story told from one friend to another that changes with each repetition (like the old children's game "gossip"), immutable mobiles may be taken from one place to another, or sent, without substantial change. Maps, books, sets of specifications transmitted electronically, or readings from a meter submerged into the ground or stationed on Mars are all forms of immutable mobiles … But no mobiles are completely immutable …[40]

So, these are the kinds of "things" that circulate, along with certain kinds of reliably routinized stories of homage. The emphatic point I make is that the productive network of human and non-human agencies, even in circulating these immutable (or mutable) mobiles, has also experienced extensions and deformations over the twentieth century, though not wholesale revision. The space/time world as nexus of natural/cultural, scientific/public rupture and flow has remained intact. What this suggests is that human/non-human networks are simultaneously resilient and phantasmatically conformed, but also the location for distributed change. They remain sites where the flow of phantasy and materialization continues, conduits for the political interplay of serious and unserious dinosaurian resurrections, always susceptible to expanded colonization or recolonization. Strangely, it is the misplaced sense of completeness and concreteness of the nexuses of space-time worlds that often comes to obscure the articulating work and the phantasies that are so routinely at play. This is where the political and pragmatic challenge for science, media practices, and contemporary science studies emerges.

The resilience of certain formulations of dinosaurs is no better expressed than with *T. rex*, pointing to how effective the specimen-spectacle apparatus has been in conserving such formulations, even when palaeontological knowledge may suggest otherwise. While the kinds of actions I have pointed to here are commonplace in the trafficking of palaeontological and public culture, new circuits of dinosaur reconstitution have come into play. Over the last three decades, several palaeontologically supported reimaginings of *T. rex* have come about, most notably those that are more like the fast-moving "leaping lizards" vision of Cope and Charles Knight in their depiction of the leaping Dryptosaurs, originally named *Laelaps* by Cope (see figure 4.2). Now, it is almost a palaeontological convention that Tyrannosaurs and other large carnosaurs are presented in more horizontal postures, tails lifted straight behind them as they walk or run, counterbalancing the large skulls – a significant turn from Osborn's tail-dragging, imperious *T. rex*, though the creatures remain, for the most part, imperious. Geoscience educator Robert Ross and his colleagues have recently focused their concern here, as they are observing and confirming a "lag" between common college student and school-aged children's dinosaur imaginings and these revised imaginings, attributing much of this to the enormous recirculation of imagery and narratives in public culture and mass media.[41] Still, theirs is an abiding concern about *accuracy* of the dominant current palaeontological vision versus *inaccuracy* of past visions. It is an approach that yet seeks to bracket the phantasmatic from the verified, dismissing this potent ever-present force that is still operating, if with different resources, in the revised imaginings.

In closing, I come back to our genealogists, Farlow and Brett-Surman, for how they also remind us of the requisite modern impulse to hold fantasy and fact apart. The final section of *The Complete Dinosaur* focuses on "Dinosaurs and the Media," consisting of a single chapter written by none other than their colleague Don Glut. In their introduction to the section, Farlow and Brett-Surman strike the standard note of building boundaries between popular matter and technical matter by suggesting the former should be regarded solely as amusement, the latter as factual, testable, and quantifiable, "Read this chapter just for the fun of it. There won't be a test on this material. We promise."[42]

Perhaps, given how deeply interwoven the performative worlds have been, there actually should be a test, a fuller experiment in the connections and articulations? With this as a cue, my work is to continue taking up – and expanding – the test Farlow and Brett-Surman have saved their readers from: that is, rather, to actually put these purportedly non-scientific, yet fully science-engaged materializations further to the test. For that, in the second section of this book, I turn to an ethnographic account of *The Life and Times of Maiasaura*, the Good Mother Lizard – a performatively gentle, highly sociable, and maternally composed American-Canadian counter-story to the imperious, climactic, energetic ferocity witnessed in the life and times and in the lost worlds of the Tyrant Kings. As will be seen, the appearance of the Good Mothers in the performative worlds of the Tyrant Kings is both a continuation of the grander story-action and, even more, an opportunity to follow how it might be turned, and indeed, overturned.

But before moving to the Maiasaur Project, I wish to return in the last chapter of this section on Mesozoic performativity to a question still waiting to be addressed: whether and how scientists and their practices cope with, and continually adapt to, the ever-presence of phantasmatics.

7 Phantasmatics in the Systematics of Life

[T]heir physical lives run along the simple, linear track of time, but their minds move back and forth through the ages, jumping onto the tracks where time moves at a more complicated pace.

Peter Douglass Ward on palaeontologists' lives in space and time[1]

The question I return to in this chapter is that which I signalled in my opening chapter on Mesozoic performativity: how do scientists manage incursions of what is only supposed to be fictional into what is only supposed to be factual? The literary character of technical practice as performed by palaeontologists in their pursuit of new specimens continues. In many museum-located dinosaur presentations today, the visitor's attention is still typically directed to the outward journeying work, the spectacle of adventure, and the outcome in some wondrous specimen located in its framed world, which in turn gives some sense of special honour and authority to science and its practitioners. But in the opposite direction, away from or behind the scene produced – "backstage," as it were – the palaeontologist is acknowledged to work a domain that is far more difficult to scrutinize. It is the hidden domain of the anatomical, the systematic, the chemical, and so forth – all the stuff that goes on in the field, the laboratory, the comparative collections, and the conferences and symposia of professional palaeobiology. This is a bidirectional practice with non-expert museum visitors and expert scientists positioned on opposite sides of the "scene," which is also, therefore, a scene of translation.

Typological work is carried out within the containing time-space of a diorama, as much as it is in a palaeogeographic map reconstruction in a technical article, or again in a monster island scenario. The work of "scenarization" (i.e., translating ideas into scenes and scenarios) may be held in common, but the purported difference is that the scientist has a technical practice of morphological and phylogenetic analysis which authorizes the typologies which are developed, and that the typological work is supposed to be undertaken *a priori* – that is, before the scenario is developed. However, as soon as scientific practices are located within a much wider ranging public-scientific trading history such as those I have been discussing, then the question of phantasmatic recitation from all antecedent locales – expert, vernacular, and otherwise – comes into play.

As noted in the case of Henry Fairfield Osborn, the AMNH netted together public and scientific practices in an articulate manner, such that phantasmatic identification circulated everywhere – in the displays, the museum policies, the commitments of board members and patrons like Roosevelt, and of course into the very detailing of the workings of "life" via Osborn's hidden energetics animating the "heredity mechanism." Osborn's success and influence hinged on the harnessing of action in all these registers simultaneously and articulately. It was through these articulations that his eugenics agenda could also be advanced. And, of course, we are left today with a legacy of Osbornian relics in the form of the great dioramas of the AMNH, a matter of significant and ongoing public import. Scenarization via display and spatio-temporal performativity, as with the Mesozoic/lost world, was crucial to the articulation then, and continues to be so today, even though the Doyle/Osborn lost world narrative itself has receded somewhat in palaeontological practices as attention turns to the relations *in media res*, that is, in the midst of the ecological, phylogenetic time-space of the Mesozoic.

Having understood the importance of scenarization at the intersection of technical and public practices, it is possible to see the various accounts in the previous chapters of Phillip Currie's (and Burroughs's) Gryf, Dale Russell's Dinosauroid, Hans Dieter Sues's Toho monsters, and Farlow and Brett-Surman's science fiction commemorations in a different way. Those associations are far more expectable and commonplace, given the inter-performative history I have been tracking – in actuality, they are extensions of that history. At the same time, a notion of those associations as simple innocent externalities to the outcomes of science is put into question.

So how have such phantasmatics been mobilized, recirculated, resisted, and (just as important) *replaced* in the communicative and technical practices of palaeontologists at the end of the twentieth century and well into the twenty-first? What appears to continually be posited as the key distinction between popular and scientific world-making is that the latter is highly "systematic." In effect, scientists shift attention away from the scenario to the detail of its constituents and back again. To track this relationality, I offer another set of sketches.

Palaeontological Scenarization

One case of the managing of scenario-making can be seen in the way in which dinosaur palaeontologists position themselves in the actual working interface with artists. In an article on artistic restorations, Dale Russell implies that artists and scientists undergo an important set of highly productive epistemic exchanges in the making of a painting, sculpture, or diorama:

> In technical palaeontological writing, unconstrained speculation is not encouraged. Artistic restorations of dinosaurs and their environments nevertheless require a use of inference not unlike first-order approximation in astrophysics. Questions are posed during this process that provoke reflection. A major benefit of artist-scientist collaboration, these avenues of thought should not lightly be dismissed for they approximate the self-correcting exercise lying at the root of the advancement of knowledge.[2]

But these are not simply epistemological issues, as Russell seems to suggest in his point about the relation as a sort of "self-correcting exercise." They are equally articulation issues in terms of what they bring into coordination with specimens: technical writing, art-restoration techniques in modelling and painting, the artist and the scientist, and an anticipated use for the finished art piece as museum display, popular or technical book illustration, film animation, and so forth. All of these elements impinge in the making of the two- or three-dimensional scene of action, and the creatures animated within them. In the process, something else takes place: "first-order approximation."[3] Much as Chris Kelty and Hannah Landecker have told us, the emergence of such figures requires intra-action, hands-on work, and close working relations between scientist, artist, techniques of restoration, use of fossils, and much imaginative projection:

> Scientific and artistic objects, sea urchin embryos and science fiction mountains, are "realities that emerge from handwork," that of animators tinkering with machines and media.[4]

Nonetheless, Russell strikes a cautionary note on the need for palaeontologists to avoid "unconstrained speculation" in their technical writing. By having artists *produce the vision* in the literal sense, scientists do not have to take as much responsibility for that upon which they might really be imagining or speculating. As such, the artist is also cast into a role amounting to that of a scapegoat for scientists who may then play out their personally held phantasies of past-world scenarios. Russell takes the curious position in the following quote of a highly literate blind man, suggesting in effect that he has no vision without the artist:

> Artists are the eyes of palaeontologists, and paintings are the window through which nonspecialists can see the dinosaurian world. Palaeontologists usually do not paint and artists do not usually read palaeontological treatises; teamwork can be advantageous. A palaeontologist must discipline himself to assemble all of the available data needed and thoughtfully translate it from a technical vocabulary into the vernacular, and the artist must discipline himself to be palaeontologically (not compositionally) obedient. Words often fail, and a joint effort making scale models can replace words as a means of communication. When models are based on scale skeletal reconstruction it is amazing how the anatomical individuality of an extinct creature emerges.[5]

Of course it seems quite implausible that there would be such a radical difference between the palaeontologist and the artist. Russell deliberately overstates his position as though he has no conception of what a dinosaur looks like, and as though that is what allows him to discipline the artist, "the eyes of palaeontologists." Likewise, he understates the abilities of artists – most of whom have undertaken very detailed technical and personal study of skeletal anatomy and animal motion, which might indeed entail reading of palaeontological treatises.[6]

Tom Mitchell radically reverses the relation in Russell's comment, suggesting, "If you want to think of the art-science relation in terms of body parts, a better comparison would be to think of the scientist as the eye and the artist as the hand. Their collaborative relation would then be more like 'eye-hand coordination.'" Yet even Mitchell overlooks the more nuanced dimensions of what Russell is saying and doing here, and what commonly takes place in the relationship.[7] Russell points out that "words often fail," and the communication becomes a visual one. Reciprocally, one palaeoartist, Douglas Henderson, notes that envisioning relies on language in the strict sense: "Correspondence and discussions with willing palaeontologists prove the best opportunities for instruction and guidance."[8] The scientist and the illustrator exchange their matter-of-fact phantasies and their phantasmatic facts by all the communicative means available to them.

Russell gives the impression that "envisioning" does not take place in palaeontology, and so displaces the critique of unconstrained speculation and imagining. By taking this guarded stance in relation to art-making, and by then allying his socio-technical network with those of astrophysics and exobiology, Russell has articulately secured his credibility in producing the "thought experiment" mentioned previously – his speculative, phylogenetically convergent "Dinosauroid" (see figure 2.2). If his palaeontological networks may have failed to be supportive, Russell's astrophysical network would correct for this – and his collaborative phantasmatic being could then be fully wrought, its autonomy secured, along with Russell's professional position. Predictably enough, such an authenticated vision of an otherworld alien would garner the attention of mass media instruments from populist science publications to pulp fiction publications like the *National Enquirer*.

Russell's case may seem exceptional, but technical scenarization does indeed occur routinely in the most mainstream and prosaic of palaeontological practices, where relatively "constrained speculation" is quite permissible. An example is a very typical sort of descriptive article in disciplinary vertebrate paleontology: that of Jason J. Head's 1998 publication in the *Journal of Vertebrate Paleontology*, "A New Species of Basal Hadrosaurid (Dinosauria, Orthnithischia) from the Cenomanian of Texas."[9] The title alone is characteristic of the totalizing temporal-spatial world restoration: the species is *new*, it is *basal* (meaning ancestral in evolutionary terms), its biosystematic location is signalled, its geological time facies is noted (Cenomanian), and the location where the dinosaur specimen was found is given. The very creature described is prosaic enough, a duck-billed dinosaur, the kind of dinosaurs I have heard palaeontologists jokingly refer to as "cows of the past." While cows do have sociocultural and economic salience, they rarely get the dramatic play that lions, tigers, sharks, and king tyrant lizards do.

Head's article describes two specimens consisting of "disarticulated skull and isolated postcrania" (with illustration). Through an intricate set of citations and correlations, the article leads the reader through: (a) previous technical sources on the taxonomic group under scrutiny; (b) the stratigraphic situating of the specimen (with illustration); (c) its biosystematic positioning and relationships within the *Dinosauria* (with chart); (d) the reconstructed view of the skull (illustrated); (e) cladistic analyses of this hadrosaur (with cladograms); (f) proposals on the creature's functional morphology describing how it would have moved in life, etc.; and (g) suggestions of its biogeographic distribution in the world over time (with contextualizing maps showing positions of continents in Albian, Cenomanian, and Turonian-Santonian times). Each step takes the locally known fossil material and scales it ever up via a universal time-space pictorialization procedure. A montage of several illustrations aiding this cumulative world-production is presented in figure 7.1.

In short, Head has systematically "built up" the scenarization of this creature – from previous systematic phantasies to revised systematic phantasies – in bounded time-space, right down to the way the animal moved, approximately what its face – i.e., skull – looked like, where it lived in the ancient world, what its evolutionary placement and history was. And this was all done by mobilizing graphs, maps, an array of fossil elements, photographs, cladograms, and many citations of preceding technical papers. Each step is conjoined to produce completion, a sense of totality from the smallest feature of morphology of a single bone to the position of the continents around the globe over a period of twelve million years!

This is typical of palaeontological world-making as published in leading scientific journals. On the surface, it is very difficult to see any sign of Hollywood-style or any other sort of cultured phantasmatics in action. As one might expect, Head didn't cite a single Willis O'Brien film, or any Charles Knight paintings, and yet he produced for the reader through texts and suggestive graphics a vision of his newly named genus and species *Protohadros byrdi*. But finished, peer-reviewed articles where popular culture influence or "unconstrained speculation" is discouraged would be an unlikely place to find such an admission. For that, I have had to look elsewhere.

Purifying Life's Relations – Systematically

Contemporary dinosaur palaeontologists suggest that their work is relatively free of phantasies. This is achieved through the doubled action of focusing on technical detail, and then making claims that prescriptive storytelling and subjective interpretation have been cast out of the scientific enterprise. The case I present, which suggests how performative,

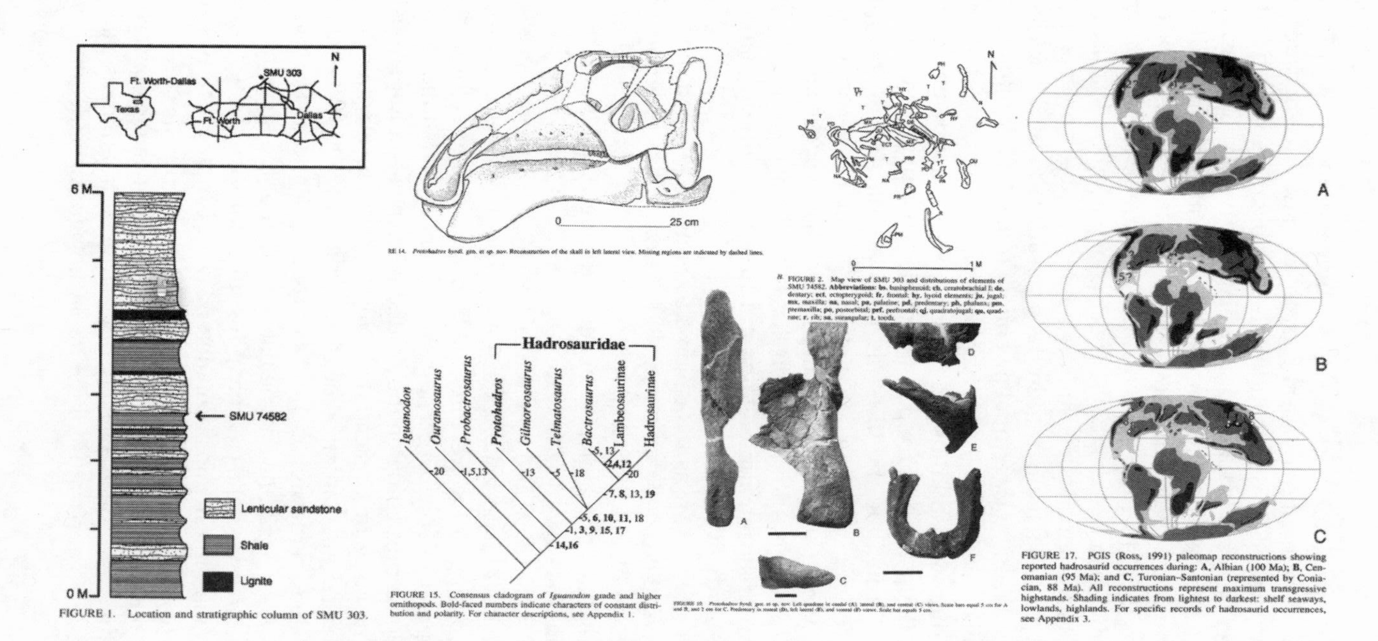

Figure 7.1. Montage of charts, diagrams, cladogram, maps, and specimen illustrations assembled from a scientific journal article locating a new dinosaur in time and space. Source: Head 1998. Reprinted by permission of the Society of Vertebrate Paleontology (http://www.vertpaleo.org).

phantasmatic dimensions of the Mesozoic guide the envisioning of scientists, is that of the methodological turns of the last three decades in the area of "systematics." Systematics is the organization of species diversity in order to demonstrate evolutionary history and relationships among biological organisms – biology's genealogical ordering of life.

The visual outcome of contemporary biosystematics is a revised sort of evolutionary tree diagram known as a "cladogram," which differs somewhat from earlier tree diagrams. This technique, called "cladistics" or "phylogenetic systematics,"[10] is now routine. Twenty-five years ago, it was a rarity to see a cladogram in the dinosaur talks at the Society of Vertebrate Palaeontology annual meetings. Today, slides with elaborate cladograms grace virtually every platform presentation, and new groupings of dinosaurian kinds are paraded out seemingly endlessly. (Incidentally, the cladistic group gaining the most platform time is *Theropoda*, which include the likes of *Tyrannosaurus rex* and *Jurassic Park*'s famed "clever" dinosaur, *Velociraptor*.)[11] Indeed, it is confidently claimed in a leading teaching text on the subject that over the last twenty-five years, "phylogenetic systematics has taken its place as the dominant paradigm of systematic biology and fundamentally influenced how scientists study evolution."[12]

The AMNH went so far as to fashion a $30-million palaeontological gallery, which opened to the public in 1995, based on this ordering.[13] It was also in this moment that the AMNH displays of *T. rex* caught up with the current palaeontological thinking on the creatures' life-posture. The fossils and casts used in the 1915 *T. rex* mount at the AMNH were finally dismantled in 1993 and then remounted for the museum's newer cladistics-based evolution galleries in 1997.[14] The new *T. rex* is now in the canonical posing of the creature that began to emerge in the 1980s, was made popular in Spielberg's *Jurassic Park*, and is so familiar today. The tail is no longer dragging on the ground with the head imperiously rising above the body, but instead the animal is tilted forward with the massive head aimed like an arrowhead, the spine aligned horizontally with the tail extending straight behind as a counterbalance to the head's mass, the legs now positioned at the balance point of the body, always ready, it seems, to propel the creature powerfully forward.[15]

This visual-anatomical transformation of *T. rex* is noted as a dimension of flexibility in Mesozoic scenarization, yet, for this discussion, it is rather the revisionary approach to displaying phylogenetic ordering of dinosaurs that is a more decisively radical transformation, as cladistics offers an altered apparatus for the translation and articulation work between specimen complexes and spectacle complexes. AMNH scientists have long lauded cladistics as "a new, more rigorous approach to systematics, organizing animals into groups based on the uniquely evolved characteristics they share."[16] The key here is that only unique, evolved anatomical characteristics – e.g., the presence of hair, feathers, wrist bones, a backbone, etc. – shared among a group of creatures can be used to diagnose their true, natural kinship, their common membership in a "clade," literally a "branching" group descended from one common ancestor, a genetic lineage. This contrasts methodologically with the older procedures, known as "evolutionary systematics," which took into account not only shared derived characters but also possible ancestral characters, a mix of possibly adaptive analogous characters, and ecological information. The so-called paraphyletic trees they produced were based on what Ernst Mayr called "grades" of variation, opposed to the new "monophyletic trees" based on narrowly delimited "clades."

The argument of "cladists" is that their technique is more "objective," based on real morphological "evidence"; it isolates truly "natural" groups, it is driven by strict rules of analytic consistency, and so it is less prone to arbitrary or subjective choices of spurious morphological characters on the basis of neo-Darwinian or other sorts of presumption. Figure 7.2 is an illustration of a phenogram (a tree diagram based on outward morphological differences) showing presumed relations of major lineages of Hadrosaurian dinosaurs, while figure 7.3 is the correspondent cladogram displaying relations of Hadrosaurians based on an analysis of shared/derived morphological traits.

In his influential – in certain ways foundational – 1990 polemic in a major collection on dinosaur systematics, University of Chicago dinosaur researcher Paul Sereno noted how cladists like himself "pride themselves in admitting as little evolutionary assumption as possible into their systematic methodology."[17] In this paper, which has also helped position him at the forefront of dinosaur palaeontology, Sereno uses the language of purification, admitting at the same time that "few practitioners, if any, would claim that cladistics has cleansed phylogenetic analysis of subjectivity," admitting the sole exception of how "distinguishing the descriptive subunits of morphology ... involves subjective decisions."[18] But the implication all the same is that some things are cleansed away, most notably adaptationist storytelling. While this politically interesting dismantling of progress narratives, one that displaces humans atop a now dissolving evolutionary ladder, may be aided in this reordering of life, Sereno is correct that the "cleansing"

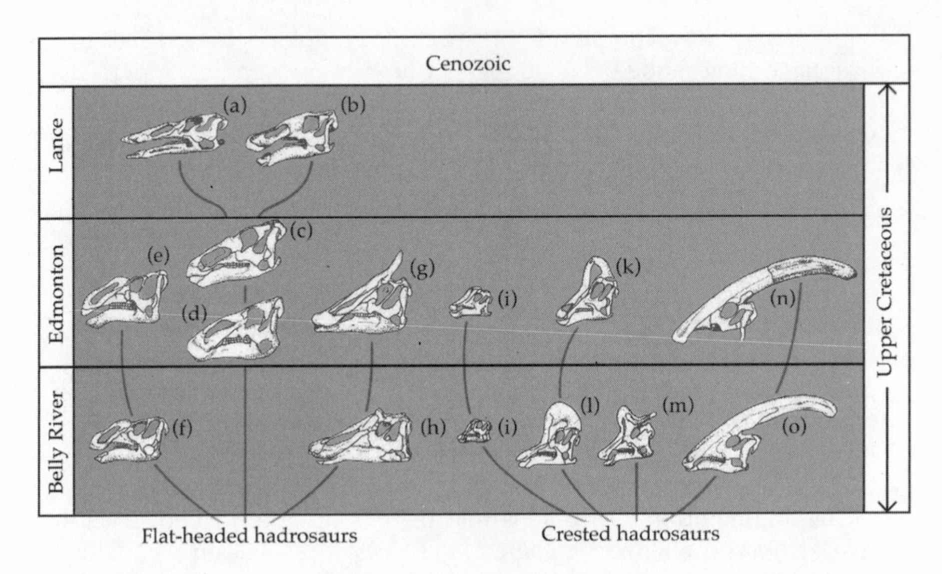

Figure 7.2

of human engagement is necessarily incomplete. Indeed, if we think of cladistics in itself as an elaborate human-generated apparatus, we can say that it is as much human as it is natural.

The question still remains, even if one does adopt cladistics analysis: where do the stories go? What happens to imaginings, fetishistic interests, scenarios? Do these simply evaporate? Andreas Henson, who was the curator of the Maiasaur Project exhibit considered in the remaining chapters of this volume, is an inveterate cladist himself. Henson provided what may be a partial answer to the puzzle in one discussion we had together:

> We no longer have people at scientific meetings standing up and telling imaginative if lovely stories about the evolutionary relationships they believe they have found … instead they present cladograms based on rigorously defined sets of synapomorphies [i.e., shared, derived features].[19] Delivering scenarios is the next step – once you have a cladogram, you can go, if you're so inclined, and look at features in a scenario, but there's an important theoretical distinction there, that you have a testable hypothesis of phylogeny, whereas a scenario, by its very nature, is less testable.[20]

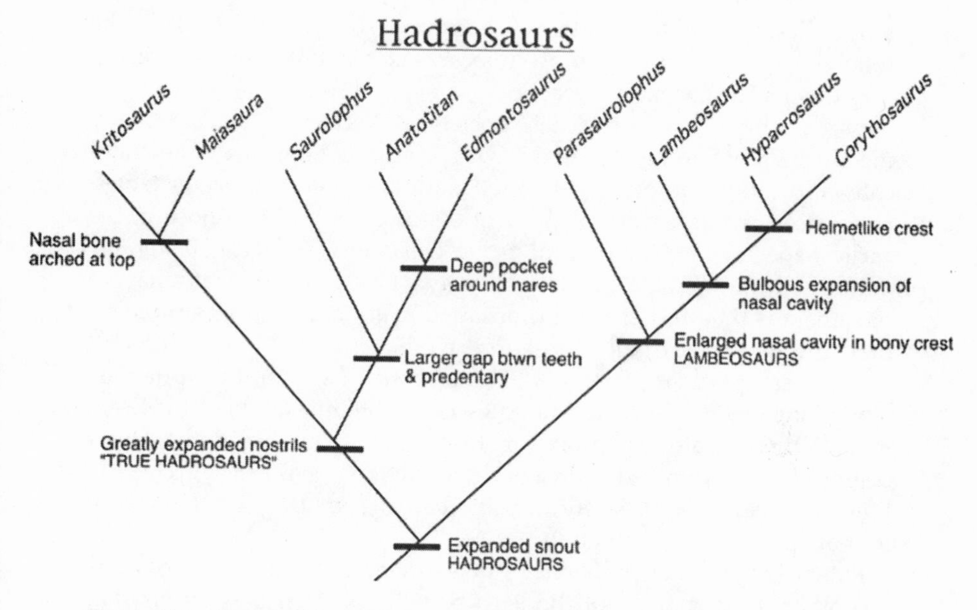

Figure 7.3. Figures 7.2 and 7.3 are two depictions of the relations of hadrosaurs (duck-billed dinosaurs): a phenogram and a cladogram, respectively. Sources: Simpson 1983, 38; Columbia University/AMNH course site for Earth and Environmental Science v1001, "Dinosaurs and the History of Life," http://rainbow.ldeo.columbia.edu/courses/v1001/ornithlab.anmh.html. Accessed September 15, 2014.

The effect here is that explicit storytelling, scenario-making, and such phantasmatics (apart from human decision over which traits to use for analysis) have simply been displaced out of the process of systematics to another locale. This is the same sort of displacement that Dale Russell made when suggesting that the "artists are the eyes of palaeontologists." One supposedly does technoscientific work in one register, speculative work in another.

To use a point on scaling from Marilyn Strathern,[21] all we need do is change positions, step back, to effectively "shift context," in order to see where and how the action of cladists is fully engaged with the phantasmatics of narrative, desire, world-making, and so on. I have already suggested how a fairly typical palaeontology article entails scenarizing tendencies. I would like to attempt a different sort of context shift in the following sketch, which simply relocates cladistic practices to recognize them as performative practice.

In a charged talk at the 1997 Society of Vertebrate Palaeontology meetings, Paul Sereno, yet again, with a projected barrage of aesthetically appealing, computer-generated cladograms and character tables, announced the discovery of the world's most primitive dinosaur, the earliest specimen of the natural group *Dinosauria*, named *Eoraptor.* He declared proudly to the excited assembly: "I fully believe that what we have here is the Mother of all dinosaurs." Suddenly, in this simple remark, "the Mother of all dinosaurs," we have moved into a different register, encountering Sereno's phantasmatic of origin, his "phantasmatic kinship,"[22] which he calibrates to his phantasmatic of biosystematics. The genealogy of dinosaurs for Sereno is a gendered one, and it begins from a feminized source of reproduction, the maternal. This is a literal parallel to Doyle's central Lake Gladys in his *Lost World*, the very font of his otherworld creation.

Extending this sketch, Sereno's investment in building cladograms has other dimensions, not the least of which is that it suggests that there are single common ancestors to be found in every lineage – from "Mitochondrial Eve," the counterpart "mother of all humans," to *Eoraptor*, the "mother of all dinosaurs." Seizing this potentiality has aided Sereno in his research funding successes, including support from McDonald's Corporation (on which the familial, Bambi-like search for the missing "mother" would not be lost), in addition to major national and international science funding agencies. This has enabled him to travel to distant locales – Mongolia, Patagonia, Niger – and his journeys of discovery have in turn become the topic of several television science adventure documentaries. Indeed, he has parlayed his "Indiana Jones" image, almost brand-like, into his fund-raising efforts, notably sampled on websites of the not-for-profit organizations of which he is a director. His personal website, even in late 2014, opens with an unabashed identification with field adventurism not far from the language of Carl Akeley, or Roy Chapman Andrews of AMNH fame:

> I see paleontology as *adventure with a purpose*. How else to describe a science that allows you to romp in remote corners of the globe, resurrecting gargantuan creatures that have never been seen? And the trick to big fossil finds? You've got to be able to go where no one has gone before.[23]

From a very different position and with altered commitments, Sereno's cladistics-rationalized work is as articulate as Osborn's had been at the beginning of the twentieth century – even though it does not mobilize Osborn's disturbing racial logics. Sereno's astuteness at marshalling, and indeed defining, many of the purification procedures of dinosaurian systematics has calibrated so well with his globe-combing fossil discoveries as to generate an articulate story of manly scientific achievement, if not indeed manly conquest. In turn, this has even transformed Sereno himself into something of a specimen for some rather different systems of classification. In 1997 Sereno was named by *Esquire* magazine as one of the "100 Best People in the World," not long after being chosen as one of *Newsweek* magazine's "100 people to watch for in the next millennium," and, the third laurel, as one of *People* magazine's "50 Most Beautiful People in the World."[24]

Though Stephen Jay Gould[25] has pointed out how the multilinear, bush-like envisioning of cladistic trees undoes older progressive representations of *Homo sapiens* as the most advanced form of life, phylogenetic systematics can be put to work to secure the position of dinosaur palaeontologists in the relation of specimens to spectacle. In delivering new dinosaurian kinds, this political-scientific network has provided ample rewards in return. Sereno has performed the Mesozoic differently, but with as much finesse as Osborn had done decades earlier. In this revised chain of articulations, systematics has traded complexly with phantasmatics, advancing and transforming Sereno's interests while also helping to secure cladistics as the proper procedure for establishing the order of natural dinosaur kinds.

The Rhizomatic Non-System: Recalibrating Systematics and Phantasmatics

Dinosaur resurrections are calibrated through an economy of imaginings conducted overwhelmingly by male players, and an expectable skewing to masculinist interests. Tempering this, male scientists such as Paul Sereno have begun to offer more opportunities for women. One example is Sereno's fundraising and community action arm "Project Exploration, a nonprofit science education organization encouraging science and science career pathways for girls and urban youth," opening potential shifts in the populating of scientific practice communities.

Yet other dimensions of phantasmatic identification include the familial practices of scientific nomenclature. Over the last two decades, there has been a tendency to assign feminine names to dinosaurs – e.g., the technical name *Maiasaura*, the "good mother lizard," discussed in the next chapters, or the vernacular "Sue" for the $8-million *T. rex* sold at Sotheby's auction house in 1997. There are continuing cases of the inscription of corporate and mass media in species names – e.g., *spielbergii* is now a dinosaur subspecies, paying homage to the filmmaker following his donation to Chinese researchers who collected the type specimen. Popular monsters make their way into scientific naming as in the case of *Gamera*, a fossil turtle curiously resembling one of Godzilla's giant enemies from which the name was derived.[26]

While explicit "lost world" narrations are arguably becoming more rare in contemporary palaeontological practice, their impact continues to appear in the margins of practice. References to Doyle's *Lost World* have been reinscribed, with ironic intent, in the naming of a new South American specimen of a Spinosaurid dinosaur that has been particularly tricky to position cladistically: *Irritator challengerii*.[27] The officially accepted name simultaneously indexes a string of ironies including the phylogenetic uncertainty, the fictional-factual locale of South American lost land discovery, and the memory of Doyle's counter-ego, the pompous professor who so bravely conquered the hidden plateau of hidden mysteries. While the persistent lost world phantasmatics may be more buried in Mesozoic world-making, the praxis of phantasmatic engagement itself continues, expressed through other kinds of concerns and figurations.

Cladistics in itself can be seen as an altered form of narration, and, rather than following conventional narrative patterns of plot development, works instead by the logics, as Robert O'Hara puts it, of "tree-telling."[28] We might also consider how systematics are phantasmatic projections in themselves. Andreas Henson explained to me how one leading old-school evolutionary systematist at a prominent North American university adopts cladistic method with the proviso (paraphrasing), "I really don't believe in the stuff ... I only use it to communicate with cladists." This palaeontologist is strategically deploying the systematics of his rivals phantasmatically. Many dinosaur palaeontologists work with cladistics in a "flavour of the month" fashion, still maintaining their more fully imaginative engagements simply because, as the American Museum's Mark Norell summed it up, "most people who study dinosaurs are interested in the more speculative aspects of dinosaur biology."[29]

Shouldn't phantasmatics, then, be thought of not as something "outside" the practice and contaminating it, but rather as something pervasive to it – a rag-tag but necessary fabric of systematic science? Phantasmatics – here seen in the form of worlds made, beings made, or systems made for that matter – could be seen as "rhizomatic," to borrow the botanical analogy from Gilles Deleuze and Félix Guattari, who write:

> The principal characteristics of a rhizome: unlike trees or their roots, the rhizome connects any point to any other point, and its traits are not necessarily linked to traits of the same nature; it brings into play very different regimes of signs, and even nonsign states. The rhizome is reducible neither to the One nor the multiple ... It is composed not of units but of dimensions, or rather directions in motion. It has neither beginning nor end, but always a middle (milieu) from which it grows and which it overspills.[30]

"Rhizome" suggests a continuously available tissue of connecting points, each of which may be a point of trading or resistance. The regularity of the forms traded and struggled over at each node signals, as much as anything, what may be a productive dimension of that which we call "culture," sustaining as I have suggested a flexible and controversial politics/natures.

It is in this spirit that Emily Martin also borrowed "rhizome" as a useful term, suggesting how, with this figure, one "can trace the convoluted discontinuous linkages between what grows inside the castle walls [of science] and what grows outside," which is precisely the sort of public/scientific, literary/technical movement I have been attending to here.[31] Martin used the analogy to aid in speaking of how certain action-eliciting, bodily training notions – of immune system "flexibility" and "adaptability" – circulate widely across seemingly disconnected social locations, such as the management regimes of large corporations, the rhetoric of practitioners in clinical immunology, and talk of infectious diseases among everyday folks in America.[32] She notes broadly the sorts of ramifications that might be occurring:

> To return to my examples there is no necessary spatially contiguous structural linkage between the corporate trainer I described and contemporary immunology. The links even if they could be discovered, might turn out to be ephemeral, accidental, transient. The CEO of the training company

might have learned about the current understanding of the immune system from any number of media – print or film – as easily as from his own allergy clinic or from the process of deciding whether to vaccinate his own children.[33]

Following from what Martin initiated, and which few have taken up since then,[34] what I try to bring into play is how a pervading and mutable sense of "phantasmatics" presents some characteristics of what is extended, circulated, and may provide the occasion for "ephemeral, accidental, transient" linkages and exchanges. In this sense, ways of ordering plus the resulting orders, and ways of world-making plus the worlds made, come into intricate interplay in the circulations taking place. Public and technical knowledges through dinosaurs and Mesozoic performativity are fully ramified in the rhizomatic sense of being convoluted, distributed, oddly connected. To reiterate what I stated in the preceding chapter: the performative nexus is widely distributed through the performative network.

What might a rhizomatic sense of how phantasmatics both underwrite and permeate systematic scientific work mean for ethnographic practice? Evolutionary or phylogenetic genealogies might be understood as phantasmatics in a systematic mode – a register of knowing amplified by scientific practice to authorize itself as different from public practice. Looking up close at systematizing practice, we see that phantasmatics animate the production of new groups and relations, or sustain fixations (as with meat-eating dinosaurs). Drawing back, we see that they infuse those productions located in institutional, market, and other public culture situations through which science and its entities are constituted and extended. Working in these two registers – phantasmatics as a requisite mode of systematics, and phantasmatics as multiplicities of rhizomatic connections – we might pay attention to those multiple exchange points to see how system overwrites or interdicts phantasm, or vice versa, and to see how the coordination of exchanges amplifies or cancels out effects. The task, then, is to identify the performative nexus (which incorporates action, phantasies, materializations, mediations), and to then track the variability of its use wherever it appears.

By risking this consideration of phantasmatics, I have been asking for a reconsideration of the decisive effects of that which is closeted as though it never happened – the disruptive stuff, the jokes, anxieties, angers, and prohibitions – acknowledging instead their rhizomatically distributed presence. Desires, scenarios, and phantasies of the biologist or the sponsoring corporate shareholder, for example – or flowing from the many possible incursions in the complex milieu in which the reanimated creatures come into being – may be brought into play.

But it is also important to hold to the generative promise in the phantasmatic-technical work of science. The field of possibilities for articulating specimens, scenarization, and phylogeny remains very much alive and open within scientific practice, especially as increasingly interdisciplinary approaches come into play. In the 2000s, the question of time-space and ecological thinking has become a concern for cladistics thinking and analysis, and here the primacy of specimens, aggregated into taxonomic groups as well as ecological complexes, has intensified – especially in the fast-growing area of Historical Biogeography, also referred to as Palaeobiogeography.[35] Writing on the effects of large-scale geological events such as continental plate movement and separation and their effects on ancient biotic communities, palaeobiogeographers have noted the relevance of their approaches to contemporary biogeography, and the "reticulate" transformation of biogeographic complexes over time:

> Since fossil taxa provide the only means for identifying simultaneously the temporal and spatial loci of evolutionary lineages, their importance for neobiogeography is greatly enhanced ... The value of temporal data has been widely recognised in historical biogeography. Cladistic biogeographic methods, however, have not formally incorporated "time." A theoretical perspective suggests that area and biotic relationships will change in a complex "reticulate" manner through time. This is because the disappearance of geographic barriers (and the concomitant mixing of biotas) is likely to occur as frequently as barrier formation and vicariance.[36]

Interestingly here, the idea of *reticulate* extension spoken of by Paul Upchurch, Craig Hunn, and David Norman, along with an ethos that dissolves boundary effects, echoes the idea of *rhizomatic* extension, and it is in the recognition of such possibilities that this anthropological work invites yet another interdisciplinary relationship. Indeed, such a rhizomatic extension in the disciplinary practice was signalled in a recent article by a group of Argentinian historical biogeographers, who also noted the importance of cladistics approaches, but additionally invited an opening up to further disciplines, that is, "to continue integrating historical biogeography along with other sources of information from other disciplines (e.g. ecology, paleontology, geology, isotope chemistry, remote sensing) into a richer context for explaining past, present, and future patterns of biodiversity on Earth."[37] Such are generative possibilities, especially in allowing for the addition of speculative work, looking back to the past and into the future, through the present, all of which requires an abiding capacity and practice of phantasmatic engagement.

Working against the otherwise disabling impulse to *bracket out*, my contention is that gaining a purchase on how that which is taken as imaginary *inhabits* that which is taken as empirical surely opens up new avenues for more reliable knowing, and *new approaches for articulating and transforming Mesozoics*, and other time-space propositions. I turn to such questions in the remaining chapters of this book, considering the articulations of a particular dinosaur, *Maiasaura peeblesorum*, at Toronto's Royal Ontario Museum.

PART TWO

Articulating the Good Mother Lizard

8 Articulating *Maiasaura peeblesorum*: The Life, Times, and Relations of ROM #44770

Approached last year by the team which recovered Cretaceous-era dinosaur fossils in northwestern Montana, ROM paleontologist Dr. Andreas Henson "saw it as an opportunity to show people how a specimen coming out of the ground is transformed into an object of scientific investigation and ultimately public exhibit."

1995 text from computer graphics industry magazine article on the Maiasaur Project[1]

The previous chapters on Mesozoic performativity set the stage – quite literally – for this second section of the book. Up to this point, I have been concerned mostly with the ways that Mesozoic natures and their inhabitants, notably great carnivores, come into public-scientific being, and how they are then recirculated as performative outcomes that are also natural/political nexuses. In the coming chapters I turn these concerns to how these natures are rearranged, recalibrated, transformed, or even cast aside, when new kinds of dinosaur specimens with different kinds of delimited capacities are brought into play in the specimen-spectacle apparatus.[2] The general move – from performativity to the more workaday action of articulation, disarticulation, and rearticulation – aids in following the natural/political composition of new and different kinds of dinosaurs, and with them, alternative sorts of human-dinosaur and dinosaur-dinosaur relations. It is also a move, therefore, that shifts some from the more past-tense study of trading between literary and scientific practices that informed much of the discussion in the previous section into the present tense. Rather than *Tyrannosaurus* and other large carnivorous dinosaurs, the key player in these next chapters is a herbivore that is also known to have lived in the Late Cretaceous period some seventy million years ago – but, being a herbivore rather than a carnivore, that lived in significantly different ways.

For this, I turn to a specific ethnographic study I conducted at Toronto's Royal Ontario Museum (ROM) intensively between 1997 and 1999, but which I further extended upon between 2000 and 2012. The ethnography addresses an exhibition that was built up around a particularly complete specimen (ROM #44770) of the Late Cretaceous hadrosaur (duck-billed dinosaur), *Maiasaura peeblesorum*. *Maiasaura* derives from the Latin "maia-", "good mother," plus "saura," the feminine for "lizard": Good mother lizard. The nomenclatural contrast with Osborn's king tyrant lizard is plain enough – and again only slightly obscured by Linnaean practices of using Latin and Greek roots to indicate intended significations.

Just as chapter 2, "Materializing Mesozoic Time-space," aimed to set forth the key concerns and propositions developed in the first part of the book, this current chapter sets forth the flow of discussion and propositions that follow in the second part. Readers will note that between the two sections there are a number of shifts both in focus and in technical approach. I will point to some of these briefly here before laying out more of the concerns that came to the fore in the ethnographic work that took place at the Royal Ontario Museum.

The first shift is that while attention to performative chronotopes necessarily continues in this section, the turn to articulation practices necessitates bringing those very practices to the fore. There is a genealogical point to be made here as well. While it has to be said that the lost world yet haunts the practices of Mesozoic performativity, and it crops up at many junctures in the chapters to come, the associated Doyle-like narratives have notably receded to the margins. To be sure, the *Mesozoic* remains the operative *chronotopos*, but its constituents are now seen to multiply and diversify, with altered stories and reanimations coming into play. With that, I attend to new concerns that accrete in the conjoined practices of those committed to scientific work and those committed to public work. As these new concerns emerge, many previous concerns discussed in the first part of the book recede – displaced by the choices and collaborations of researchers, technicians, museum specialists, new media producers, and the mobilization of updated and new technologies for exhibitionary work. The results are revised precipitations of Mesozoic dinosaurs and relations within a revised specimen-spectacle apparatus.

Although I gestured to practices of *articulation*, and notably so in chapter 4, "Animating *Tyrannosaurus rex*, Modelling the Perfect Race," I did so in a general sense, in part because my attention was on performativity, and in part because I did not have the advantage of close ethnographic engagement in dealing with past action. This intimacy in the case of addressing ROM's *Maiasaura* – a creature indebted to but clearly different from *T. rex* – affords the chance of tracing the action of articulation with redoubled nuance. Turning the concentration to articulation from performation (Michel Callon's word)[3] also has consequences for recognizing how we, as humans, relate to and with dinosaurs and others in the political/natures in which, or by which, we live. I will expand on some propositions concerning articulation in due course, but note here how the *very lived process of articulating* is foregrounded by ethnographic engagement in the very particular actions of bringing specimens, techniques, actors, and the tools of institutions into play together. Such actions, as I have noted already, generate outcomes that in their own right become not final *statements* of politics/natures, but rather *propositions* affecting the articulation of subsequent politics/natures.[4]

Indeed, articulation includes all the interstices of action informing and conforming "how," as the computer graphics industry article quoted at the outset of this chapter suggests, "a specimen coming out of the ground is transformed into an object of scientific investigation and ultimately public exhibit." Ultimately, then, this formal contrast (not an opposition) between what I offered in part 1 and what I offer in part 2 is also a consequence of moving from a more decidedly ethnohistorical account to one that is more ethnographic, one in which I as the writer was able to engage directly with the persons and the matter put into action in the mobilizing of a particular specimen, ROM #44770, the curatorial centrepiece for the museum's exhibit "Maiasaura: The Life and Times of a Dinosaur." This ethnographic engagement affords an amplified examination of the connections and relations at play in the composing up of dinosaurian natures. In turn, by way of the actions of scientists and others, it helps us recognize the curious result of alternative outcomes, now in the form of a differently composed creature, and a particularly feminized, biologically reproductive one (something occluded in Osborn's *T. rex* concerns). So, by shifting the technique and the moment of action being considered, we make gains in understanding the relations emerging as well as the relations at play, and how differently folks engage in, and act upon, the constitution of those relations.

Another move between the two sections is from natures and dinosaurs as components of literary/scientific engagements (Doyle, Osborn, pulp literature, popular film, museums, etc.) to an all the more nuanced, and notably more participatory, even ecological, set of actions among specimens, technicians, curators, exhibit-making practitioners, specific display techniques, field work opportunities, and much more.[5] This move is also a response to the pragmatics encountered: the drawing of publics into participatory engagement with exhibits is something that increased over the twentieth century, and something that continues to increase into the current moment in wealthy countries where governmental and private resources are put to use as an increasingly taken-for-granted way of generating museological heritage, and where personal digital media linking to social media derives its economic and political purchase from the thorough exercise of participatory engagement.

This added move is, therefore, an aspect of the political move that continues in this section. The political is readily recognizable in the gender turns of dominant dinosaur selection in public/scientific cultures, where feminine gender potentiality introduces new possibilities for telling dinosaur stories, for conjuring dinosaur worlds, for relating to and enrolling alternate publics, sometimes by other, often surprising means.[6] Masculinized creatures and masculinist circuits of production dominated in the first section; in this section notions and practices of feminized animals and outcomes come into play, in a historic moment at the end of the twentieth century and into the twenty-first when discourses of gender and agency were fast changing intellectual work, as well as public engagement. While gendering concerns are an easily noted point of animation flowing from these creatures, they also allow me to carry forward the techniques of tracing political/natures articulated as an action of the peformative nexuses *within and by* the dispositif associated with modern technical and public science. *Maiasaura*, then, is not just a gender turn. It is part of a compositional turn. Who it is that participates in this compositional work has changed. How they participate has changed. Which sorts of fossils are given priority in palaeontological prospecting has changed. And the natures composed, one way or another, have moved more or less in line with these changes.[7] In other words, the move from predominantly masculinized, marauding-hunter public creatures (following the Osborn/Doyle lost world performatives) to feminized, reproductive creatures in these two moments is a move in politics/natures.

Lastly, in the course of the move between two moments and milieus, I am also given the opportunity to *situate* two particularly relevant curators in their respective milieus: Osborn working with *T. rex* at the AMNH in the early twentieth century, and Henson working with *Maiasaura* at the ROM in the late twentieth, early twenty-first century. This then stages the possibility of discussing the vital participatory role of palaeobiologists themselves in the rearticulation of dinosaurian natures and of newly responsive worldly human/dinosaurian entanglements.

Coming across *Maiasaura* as a Study in Articulation

I have known of the creature named *Maiasaura* since the late 1970s when I met palaeontologist Jack Horner, around the time he first described and named the associated fossil specimens. However, I did not actually set out to do my research on *Maiasaura*. Rather, I came across the possibility of concentrating on this creature in 1996, via a particular specimen and exhibit, while I was seeking to undertake ethnographic work on the specimen/spectacle relation with the Royal Ontario Museum as my key field site.

Indeed, a good deal of the work informing the previous chapters, concerning "Mesozoic performativity" and politics/natures, came about while I was seeking to identify an ethnographic study at Toronto's Royal Ontario Museum to further consider the working of the specimen-spectacle complex upon dinosaurs in public. I recognized that in order to provide a focused study on how public and technical-scientific cultures traded, I would have to choose a well-delineated project or case where such interplay, such intra-actions, occurred, one in which a complex of people, specimens, and techniques at the museum would be involved. Apart from the operating of the Maiasaur Project exhibit, from 1994 and up until 2007 – when the ROM opened its newest permanent dinosaur galleries – no major dinosaur-related public programs were to be launched. A mounted cast of *Tyrannosaurus rex* was obtained and installed in the galleries in 1999, but that project was very limited in its scope and mobilized only a few players in the institution. As such, it constituted only a minor project and did not allow the wider ethnographic potential I was seeking for this project on human-dinosaurian, specimen-spectacle relations. Ultimately, I chose to focus my study on the Maiasaur Project.

The Maiasaur Project (see figure 8.1, exhibit entry) was the most recently installed dinosaur exhibition project at the ROM, having been opened in 1994 and running until 2006, and turned out to be a fairly radical departure in display technique for the institution. It specifically employed the most advanced techniques available to the museum in interactive and multimedia technology, while also putting on display an actual "working lab" in which museum technicians would prepare a fine specimen of the dinosaur *Maiasaura peeblesorum* (specimen ROM #44770), acquired largely for the purpose of this exhibition in 1994. Moreover, despite recent layoffs, the majority of staff who had worked in the development of the exhibition in 1994 and 1995 were still employed in the museum from 1997 through 1999 when I was in Toronto undertaking the main part of this ethnography. The staff I interviewed and conversed with represented several of the departments active in exhibition development, including the Palaeobiology section, Interpretive Planning, Project Management, Design and Production, Audio-Visual, Marketing and Communications, the ROM Foundation, Education, and, of course, Senior Management.

The exhibition was also intriguing in view of the museum's choice to focus on a non-Canadian specimen of a dinosaur kind that was repeatedly referred to as "friendly," relatively modest in size, and even noted to me by some staff as potentially "boring" to the public. This contrasted significantly to the inordinate historical emphasis in museums on large carnivorous dinosaurs, giant herbivores like *Brachiosaurus*, locally relevant finds, grand natural historical narratives, entire Mesozoic-era displays. This was an exhibit designed to be about "the life and times of a dinosaur," and a rather humble one at that, but which, nonetheless, had achieved significant notoriety in being the exemplar of a nesting duck-billed dinosaur that gathered in large herds, and to which accounts of parental care had come to be associated. The Maiasaur Project promised to be a focused, localized story, presenting a case study of an individual specimen that had popular display and scientific value, looming as a strong counterpoint to *Tyrannosaurus rex*.

No matter how dull duck-billed dinosaurs might be regarded in public circuits, the ROM had also historically amassed a collection of hadrosaur dinosaurs that was among the finest, if not the finest, in the world. This was largely a consequence of the ROM's history, reaching

Figure 8.1. Exhibit title panels and visitors viewing the pewter Maiasaur. Source: Photograph by Eric Siegrist; with permission of Eric Siegrist and the Royal Ontario Museum.

back to the 1920s, of collecting Alberta's Late Cretaceous fossil deposits, which were exceptionally rich in Hadrosaurine fossils.[8] Palaeontologists with an interest in Hadrosaurs regularly visited the ROM to study the fossils in this collection. Adding this outstanding specimen of a relatively newly described and in some ways palaeontologically "cutting edge" form of hadrosaur would serve only to enhance the ROM's collection and its reputation.

Still, not much was made in the exhibit of how this specimen added to an existing outstanding collection of duck-billed dinosaurs – this was to be an exhibit about this specimen, its life, its times, as it were. Museum staff concentrated upon four things to ensure that this project was unique and would be successful: (1) the choice and acquisition of an exceptional specimen noted as the "most complete" *Maiasaura* skeleton ever collected; (2) the use of interactive digital multimedia displays to "bring the Maiasaur to life"; (3) the in-gallery installation and live operation of a palaeontological laboratory for visitors to witness the work of technical preparation of the specimen; and (4) the presentation of the "story" of the life, times, and behaviour of a dinosaur understood to have cared for its young in large nesting grounds, and to have lived in large herds. That story would emphasize what the dinosaur's name denotes, "the good mother lizard." This was an exemplary project for considering the specimen-spectacle relation and the performative articulation of a particular dinosaur in its space/time.

Here, also, was this very loaded point about gendering, which, in anthropological terms, added more interest for me as a point for ethnographic consideration. Politics/natures were readily discernible in this project, as it mobilized an exceptionally complex array of human and non-human actors within the museum setting and the wider public environment. Several questions captivated me. What were the technical and personal actions that brought all of this into play together? What were the abiding concerns of those involved? What did the complex of interactions, and intra-actions, articulate into? In what way were they articulated to, or divergent from, actions and circulations that preceded them, or actions that exceeded the purpose-led action within the museum walls? What sorts of relations allowed the articulations to work, and what sorts of relations did they open up or close down? How, indeed, was such articulation done, and to what intended or unintended ends, if any? What sorts of saurian materializations would be composed, and to what effect?

Articulation: Ethnography, Science Studies and Tracing "Partial Connections"

> To see people passing artifacts between them, or to see the locations that hold these authors together, like casting on and off so many cats' cradles, holds promise for tracing networks.
>
> Marilyn Strathern[9]

At this juncture – much as I did in laying out such terms as "chronotope" and "performativity" in chapter 2 – I would like to introduce a number of technical points on the approach taken in this ethnographic study, an approach which is highly engaged with contemporary science studies. This continues to be that of following fossils, practices, instruments, museums, media, marketing, and their publics, which can be thought of provisionally as "networks" – made so, as Marilyn Strathern has put it, by an intricacy of "partial connections." By tracing these "things" ethnographically, I also trace the means and practices of articulation – not the obvious articulation of skeletal elements into an articulated skeletal whole, but rather the articulation of publics, fossils, and more so as to generate and regenerate, materialize and rematerialize worldly, dinosaurian creatures, and to generate at the same time new relations of dinosaurs and people.

As method, ethnography is a fluid set of techniques for following, in a more nuanced and intensive way, various nexus/network practices such as those I have set forth in the historical and genealogical accounts in the first part of the book. Ethnography becomes simultaneously a means for witnessing the messy complex of actions articulating specimens and spectacles, natures and publics,[10] *and* a further participatory means of articulating anew by means of its accounts and its interview texts to an engaged audience – that is, you, the readers of this text.

As Donna Haraway and Strathern have noted, this movement and trading can be likened in some ways to the casting, recasting, and passing of string figures back and forth among many actors. Ethnographic engagement allows us to witness and get close in with the intricate handwork, so to speak, that takes place in the fashioning, technical study, exhibiting, and funding of dinosaurs and their worlds, and as well with the complex public encountering with dinosaurs in these elaborately crafted worlds. So, where in chapter 6, "Vestiges of the Lost

Word: Recirculating the Tyrant Nexus," for example, I was able to trace recurrent phantasies and alignments of actors with such phantasies in the relay of science and pulp sci-fi film and literature, in the ethnographic texts following I am able to offer more intimate, though still partial accounts detailing the trading that takes place, and so results in particular sedimented, peformative outcomes.

I also wish to take up a number of propositions from contemporary science studies scholars, most notably from Bruno Latour, Joan Fujimura, Isabelle Stengers, and from Gilles Deleuze and Félix Guattari, that further address the matter of "articulation." Isabelle Stengers writes of a mode of intensive articulation when she discusses how persons, notably technicians or artisans, live and work in the "meso" (rather than the "micro" or "macro"), a kind of approach that is responsive to what makes things happen.[11] Her example is the metallurgist, who, in the experimental moment of working upon and shaping heated metals, is constantly responding – responsively, responsibly – to the way that the metals deform, bend, or display complex properties when acted upon in different ways, in different conditions. For Stengers, one is met with choices at every turn which, once deliberated upon and allowing a chosen practical action to be followed rather than another, will move the results in slightly different ways, creating new, sometimes surprising articulations, or sometimes more predictable ones.[12] Again, in exercising this choice in sympathy with the material, as with the metallurgist, one brings certain concerns to bear, drawing at all moments from the repertoire of engagements which have come to shape one's capacity to choose and that have informed the direction of the choice made, and then are applied to the shaping of an object or objects meant for the hands of extended publics. In the moment of choice, a thousand articulations may be presented, and a thousand articulations may be made or relayed.

Latour discusses a key aspect of Stengers's proposition of scientific participation and commitment to a *cosmopolitics*, noting "that a politics that will not be attached to a cosmos is moot, and that a cosmos detached from politics is irrelevant."[13] Cosmopolitics is the thorough, non-modern proposition resolving that which I have called politics/natures, the latter being a transitional proposition seeking to keep together what was separated by the modern divide of nature and culture. Participating in the "cosmic adventure" – that is, by *thinking with the Cosmos*, as Stengers has put it, or *practising in the meso* – is also to acknowledge and to become intensely responsive to the always unfolding processes of articulation.

Inclined in his thinking to the longer, mediated "chains" of articulation in technoscience, Latour describes "articulation" as the action "which occupies the position left empty by the dichotomy between the object and the subject or the external world and the mind"[14] – which, as we have already seen, can entail an enormous variety and intensity of human and non-human actions.[15] This is a much more distributed, non-compartmental notion of articulation than that which Marxian cultural theorists have suggested, where articulation cues more to alignment with the operation of hegemonic capitalism. The thematized work on articulation developed by Ernesto Laclau and Chantal Mouffe, Stuart Hall, Richard Middleton, Jennifer Slack, and others following on Antonio Gramsci has specific interests in questions of class struggle in modes of cultural production, that is, of how "situation" and "conjuncture" are articulated.[16] These are certainly analogous procedures. However, in keeping with my interests in political natures, rather than a delimited approach around the economic and human acts of production and cultural processes, I have found it more helpful to engage with science studies scholars where articulation, thought of in regard to the manifestation of worldly substance, removes the culturalist, economist brackets around the "natural" as it were, and attends to any potential articulation action that may be in play.

So, for instance, "articulation work" occurs, as Joan Fujimura has written, with the installing of sharable or "standardized packages" such as common software or communication and genetics processing techniques that will allow otherwise distinctive realms of practice to be interlinked and made meaningful to each other, co-generative with each other, allowing genetics and gene sequences at the same time to be isolated or generated differently.[17] Annemarie Mol, in her study of the differential engagement of physician researchers and patients with atherosclerosis (the condition commonly known as "hardening of the arteries") similarly challenges the received idea of a disjuncture of objects from actions, noting how:

> Passively rendering an object is not what science's systematic ways of working do. Instead, they actively constitute a traceable link between an object that is studied and the articulations that come to circulate about it.[18]

So objects, things themselves, are never immune to their articulations. What Mol, Stengers, Latour, and Fujimura offer to my own work are ever more intricate means for witnessing how something like the Mesozoic, or *T. rex*, or *Maiasaura* gets articulated into palpable being. This is, therefore, a deepening and an enriching of the action that was indicated throughout the first part of the book on Mesozoic performativity.

On the point of performativity however, and recalling from chapter 2 the geological definition of the Mesozoic as a sedimentological figure, philosophers Gilles Deleuze and Félix Guattari adopted the idea of "stratum" and "strata," in their discussion of "the geology of morals," to help describe what they call "double articulation": that is, the process by which forms and substances precipitate, or sediment, into being, a being that is always susceptible to later deformation and re-precipitation into alternate being. Using sedimentary rock formation to model their thinking, they write how:

> In a geological stratum, for example, the first articulation is the process of "sedimentation," which deposits units of cyclic sediment according to a statistical order: flysch, with its succession of sandstone and schist. The second articulation is the "folding that sets up a stable functional structure and effects the passage from sediment to sedimentary rock."[19]

Sedimentological process is their working metaphor, but they are referring here to *anything that can be thought of* as having form and substance, ephemeral boundedness, as they point out:

> It is clear that the distinction between the two articulations is not between substances and forms. Substances are nothing other than formed matters. Forms imply a code, modes of coding and decoding. Substances as formed matters refer to territorialities and degrees of territorialization and deterritorialization.[20]

So, putting this thinking to work, the Mesozoic as a public/scientific performative time-space in the manner discussed so far in this book is a kind of bounded, territorialized substance consisting of humanly engaged, formed matter, such as the biogeological matter of sandstones, fossil flora, and fauna, the latter of which includes *T. rex*, *Maisaura*, and other dinosaurs. *T. rex* and *Maiasaura*, thought of as formed matter, contain in their very manifestation *the mode of coding and decoding* which then articulates and so contributes in the constituting of the Mesozoic as performative time-space, performative substance. The procedure works reciprocally: the dinosaurs and the Mesozoic articulate each other, both acting upon each other in the emergence of natural coherence, and of course human actors are always near to hand, active in all of this unfolding. The crucial point is that seemingly unbounded forms or elements come to delineate a territoriality to some degree, a bounded substance like the Mesozoic for instance. Stretching beyond their metaphor, Deleuze and Guattari also note that "Strata are acts of capture, they are like 'blackholes' or occlusions striving to seize whatever comes within their reach," a proposition quite consistent with what I have discussed about the incorporative, colonizing workings of the lost world/Mesozoic.

Deleuze and Guattari are careful to emphasize the contingency of articulation processes, and of the metaphor itself: "Double articulation is so extremely variable that we cannot begin with a general model, only a relatively simple case."[21] In this sense, the action of articulating is complex, indefinite, emergent, constituting its milieu out of the constituents of its unfolding milieu as it takes place – much in the way that sediments accrete to create sedimentary strata, which in turn may erode into sediments all over again.

Thinking back to Marxian-Gramscian articulation, Jennifer Slack also points to its contingency, quoting Stuart Hall, who wrote of articulation in discourse as

> the form of the connection that can make a unity of two different elements, under certain conditions. It is a linkage which is not necessary, determined, absolute and essential for all time. You have to ask, under what circumstances can a connection be forged or made? The so-called "unity" of a discourse is really the articulation of different, distinct elements which can be rearticulated in different ways because they have no necessary "belongingness". The "unity" which matters is a linkage between the articulated discourse and the social forces with which it can, under certain historical conditions, but need not necessarily, be connected.[22]

With the point of Deleuze and Guattari at hand, along with Hall's on the "linkage between the articulated discourse and the social forces," I can restate the point I made in chapter 5 on politics/natures: "What becomes important is which technical workings and materializations in display and scientific work become the enduring, layered-up nexuses of natural/political relations, achieving dominant inter-performative effects." These technical workings are, as such, key loci of articulation work.

In the case studies that follow, I trace the transit of fossils into forms and substances, a double articulation. But the contingency point remains crucial, as it calls us to avoid making "articulation" into some kind of theory. Rather, in this ethnography, articulating and disarticulating actions are recovered from within the actions of the making of, and subsequent public engaging with, a palaeontologically authorized dinosaur exhibition and its central specimen, ROM #44770. In other words, I will allow *Maiasaura* and the Maiasaur Project to become my "method" for tracing the workings of articulation, rather than the reverse.

Yet again, a crucial point moving out from Latour's positing of human and non-human actors is to acknowledge that the Mesozoic and its dinosaurian constituents are contoured both by complex sorts of imagining (i.e., you cannot hold phantasies in your hand, until they are concretized into some "thing," but you can "think" them), as well as by complex material indices (sediments, fossils, etc.), and additionally by complex material outcomes (exhibits, publications, illustrations, maps). Articulation, then, has to take into account (a) the people involved; (b) artefacts, documents, or statements being passed among people (Latour's "non-humans"); and (c) a propositional *subset* of both the human and the non-human in phantasmatic formulations like the "Mesozoic" which circulate and are transformed in each passage, altered by the choices made in each passage as well. The shifts and amplifications in the Mesozoic nexus, in the physical form of the Maiasaur Project exhibition, will be considered in relation to these sorts of accumulating, passing, or enduring articulations.

The Laboratory of Ethnographic Engagement and Telling

So, the technique encompassing all the techniques I have noted is ethnography, which is always and already an experimental endeavour. The Royal Ontario Museum and the Maiasaur Project, as such, become the crucibles for this experiment in ethnographic engagement and ethnographic telling. This is in keeping with Latour's observation on how social science writing works effectively as a laboratory for tracing and detecting matters of concern:

> Textual accounts are the social scientist's laboratory and if laboratory practice is any guide, it's because of the artificial nature of the place that objectivity might be achieved on conditions that artifacts be detected by a continuous and obsessive attention … If the social is something that circulates in a certain way, and not a world beyond to be accessed by the disinterested gaze of some ultra-lucid scientist, then it may be passed along

> by many devices adapted to the task – including tests, reports, accounts, and tracers.[23]

Like Stengers's metallurgist, the ethnographer seeking to assemble a robust account is acting responsively to that which she or he encounters, engages, and in which they are circulating. These texts, in concert with what they obsessively attend to, are akin to the metallurgist's forge, or the technician's laboratory, engaging a series of experimental moments organized into chapters, which are in actuality experiences. Stengers pointed explicitly to this when she wrote that to experiment is "to signal a practice of active, open, demanding attention paid to the experience as we experience it."[24]

As the complexity of actions is always endlessly traceable, what I present here is necessarily a partial investigation of the partial connections – that is, the articulations – at play. In addition to the direct action surrounding the making of the exhibition, other dimensions of articulation that I begin to contour include detailing the flows of visitor knowledges inside and outside of the exhibitionary experience, and the flows through the scientific action of the curator in relation to palaeontological history and practice. Indicated as well are the flows through management and board of trustee action; training regimes for museum staff; the selection process for collections management; regimes of communication among museums; governmental policy flows; shifts in institutional history; the heterogeneous economic apparatuses of the state, and so forth. Although some of these interlinked trajectories are indicated generally in the discussion, each would constitute very consuming loci for research in their own right.

Readers will discern three recurrent techniques of experimental recovery:

(1) following heterogeneous actions and networks to recover articulations among people and things as well as the emergence of new mediating entities, or "artefacts," enabling such articulations;[25]
(2) watching for further nexuses as key "zones of implosion" where dinosaurs precipitate materially/imaginatively, and where intensive trading takes place; and
(3) recovering from these the dominant political relationalities[26] (e.g., kinship, gender, affinity, responsiveness, care, market effectiveness, etc.), and, as well, those that are marginal or emergent.

With these techniques, my intention is to achieve a reliable experiment[27] in assembling signal actions of articulation, and with that a partial account of the reorientations in Mesozoic, dinosaurian political-natures in the late twentieth and early twenty-first centuries, via the "Life and Times" of "The Maiasaur Project" and its genealogical forebears.

In preparing the ethnographic accounts of the Maiasaur Project, I have worked with a variety of source materials. Notes taken during several walks through the exhibition space during public hours constitute one principal through-line in the description. My intention in these notes was simply to consider the exhibit in its architectural space as a production in three dimensions – that is, the physical exhibitionary installations – including interactive components, display cases, a computer graphics theatre, a soundscape, a mounted skeleton, textual "didactics," lighting conditions, sponsor acknowledgment panels, etc. In some cases I recount what I witnessed of audience-exhibit engagements.

Those accounts are augmented by descriptive notes resulting from a review of an archival video record of the exhibition supplied to me by the audio-visual staff of the museum, which followed a path through the exhibit that the designers conceived visitors as following – even though, as I will point out, this path was only sometimes followed by visitors and only approximately in those cases. In addition, I conducted a number of unobtrusive individual "tag-alongs" of visitors through the exhibit to witness a variety of engagements with the exhibition.

More than forty individuals were interviewed and approximately seventy hours of recorded interview tape compiled in the course of the research. I sample from among these, but key in especially to those interviews that provide a strong sense of what (or who) generated relations, articulations, and disarticulations between people and fossils, fossils and displays, displays and people, etc. The curator, a host of visitors, the interpretive planner, the exhibit programming manager, the digital media producer, and the principal preparator of the specimen were key sources. I also conducted and draw upon approximately twenty interviews with a diversity of visitors to the exhibit.[28] Others I draw on are interview commentaries of one of the designers, the marketing coordinator for the exhibit, the principal sponsor of the exhibit, the production manager, and several other technical and curatorial staff of the vertebrate palaeontology section.

The intensive ethnographic research took place at the ROM between July of 1997 and June of 1999, and I returned several times to the ROM between 1999 and 2012, visiting the Maiasaur exhibit, or the remnants of it as it gradually came to be taken apart, its elements either distributed to other displays or removed to storage. These later shifts at the ROM and with the Maiasaur Project make their way into the discussion at various junctures.

While my ethnographic commentaries throughout continually trouble the very question of what counts as a "successful" exhibition, this in no way diminishes how impressed I have been personally by this exhibit and its effect upon those who engaged with it in so many ways, and by the sincerity and diligence displayed by each of the individuals in working to create it. I am deeply grateful to all those at the ROM who provided information, materials, insights, anecdotes, or commentaries and agreed to be interviewed on the Maiasaur Project and the museum in general. Pseudonyms are used for all the museum staff. The degree of candidness exercised by most people in the museum was notable, and it is that candidness which allows this ethnography to be rich enough as a reliable, if partial account of the way the heterogeneous network of actors and actions unfolded. There are undoubtedly several individuals interviewed who could offer a rich account of what took place in development and the outcome that is the Maiasaur Project. Their individual accounts would bring into play other highly valid concerns and actions which I either missed, left out as they were peripheral to the concerns being followed, or recount in a manner quite distinct from what each of them would offer. One of the points of ethnography is not simply to attempt to generate a multi-perspectival account, but also to risk wading through it all with the ethnographer's concerns always in play – this is the experimentalist's commitment.

Trajectory: A Composite of Retellings

Each of the chapters deals with the Maiasaur Project and the specimen it "centred" on – ROM #44770, the specimen animated by the spectacle, so to speak. The tale of the exhibit is told and retold a number of times along the way. This is not a history of the ROM, but rather a multilayered account about a single exhibit and its development within a wider range of social, cultural, material, media, and technical cross-currents. As the discussion proceeds, points raised in previous chapters are recalled and brought back into the mix, at times troubling and complicating previous propositions as a result of their resituating against different concerns and actions. As such, some guidance on the temporal framing will probably aid in reading this.

While I also continued to visit the exhibit and the ROM between 1999 and 2012, noting salient changes, the discussions of the Maiasaur Project in the coming chapters resolve on three time frames: the period from late 1993 to mid-1995, during which the exhibition was proposed, planned, and produced; the period from June 1995 to October 1997, during which the exhibition with all of its planned components was in place and operating; and the period from late 1997 to early 1999, during which one of the exhibition components, "the Working Lab," had been replaced by a standing mounted cast of the specimen which had been fully prepared in the first two years of the exhibition. My in-person ethnographic involvement took place within the last of these three time frames, though I make references to all three of them.

The in-gallery accounts, therefore, are of the exhibition in its "post-laboratory" phase. As such, the discussions of the laboratory are drawn from interviews with those who worked in the lab and with members of the exhibit development team, including the curator. I also made use of a (previously mentioned) video recorded when the exhibit opened, which provided me with a fairly clear sense of how the space and its elements operated at that time.

A second matter relates to the temporality of interviews. The comments made by museum staff in the interviews are mostly retrospective – reconstituting the activities of the planning and development phases or of the exhibition in its first two years of operation, with the lab in place. Those interviewed also spoke about the exhibition and other activities in the "working present" (i.e., the ethnographic present) of the interview. The museum visitor interviews all took place in the museum – most of them in the mezzanine landing just a few steps outside the gallery containing the Maiasaur exhibition.

A third matter is that the chapters proceed not so much in a direct linear fashion, but rather in a sequenced, yet cumulative fashion. Each recounts the exhibition through the understood orientation and concerns of different persons involved – first, curatorial concerns are highlighted (chapter 9), then marketing and promotional engagements (chapter 10), then exhibit planning and development (chapter 11), moving to a discussion of the Working Lab (chapter 12), followed finally by two more lengthy chapters sourced in my gallery observations and participatory engagements (chapters 13 and 14). My early comments accompanying the curator's overview are of a general sort that, to the reader, might appear somewhat uninterrogated and open-ended.

However, in subsequent chapters, which work as additional iterations of the Maiasaur Project, the evocation of the density of translations and transactions continues to build. With each move, different concerns, actions, and articulations are contoured, as well as their effects. My intention is to allow the reader to revisit the process and the exhibition and along the way come to a fuller, compounded sense of the complexities at play. Of course, the shifts and concentrations in relations will be emphasized, as will the manner of responsiveness of the actors. Two chapters, 15 and 16, close the book by returning to the challenge and promise of working as a curator committed to advancing knowledge of a particular dinosaur in increasingly market-driven museum activity.

Of any individual, the curator is referenced most. By returning repeatedly to his commentaries and concerns, it becomes clear how his intentions were gradually complicated by the multiplicity of players at work – how, in Isabelle Stengers's words, he was repeatedly confronted by "infernal alternatives"[29] to support the work of palaeontological study along with the specimen on offer, or to emphasize the marketing concern for making consumer-friendly spectacle. The exhibition, the sedimenting of all the translation practices, emerges as the final outcome. With the details of translations and contingencies presented, I then offer comments as to what has precipitated, what the resulting articulations and relations were, and how Mesozoic natures and orders of life are transformed in the process, how an altered political/nature ensues. Through all of this, the risks of the market pressures are ever present, impinging upon how Maiasaur is constituted as specimen-spectacle being.

The currency and generative promise of the sections that follow have been presaged by Stengers. Such articulating, compositional, work – in the "meso" – harnesses the potency of "creating the possibility for interstices that give a different texture to our world, that give us a chance to live out the challenge posed by the intrusion of 'Gaia' (earth)" upon our lives and actions, allowing for alternative, more liveable politics/natures to unfold.[30] At the end, I worked to honour young Amy's wondering and wonderful precept of dinosaurs, as the story of us. Since we are involved, always and already, and since alternative dinosaurs are always arising and unfolding through our involvement and concerns, then we begin to understand the importance of acting with care about our shared story. In the labyrinthine paths that lead from Tyrant Kings to Good Mothers, I am left at the end with a moral-political question from which to open to new speculation. The question at the end is not so much *"Do we care?"* with or about dinosaurs and their political natures. The complexity and robustness of our articulations of dinosaurs and their Mesozoic worlds uncontestably demonstrate this. The question, in response to the curator's lament, is *"How do we care well?"* as dinosaurs and their Mesozoic natures continue to come into being, as the civil/natural collectives in which we participate seek with renewed urgency to make shared, liveable worlds, whether of the past, of the present, or of possible futures.

9 "A Real Sense of a Dynamic Process"

A high tech exhibit illuminates the scientific process ...

Designed to expose the behind-the-scenes process of research, this 3500 sq. ft. display combines the cool modernity of scientific analysis with the theatrics of dinosaur reconstruction. A unique working lab allows visitors to watch paleontologists at work on an actual Maiasaur specimen.

1997 text from Graphics Design Annual, profiling the Maiasaur Project[1]

"From Bones in the Ground to a Neatly Mounted Skeleton ..."

To open my multiple tellings of the working articulations of the Maiasaur Project, I begin with the concise retrospect offered to me by the ROM's dinosaur specialist, Dr Andreas Henson, vertebrate palaeontologist and curator of the exhibit, as well as professor of zoology at the University of Toronto at that time. The curator, more than any one figure in the making of an exhibit, is meant to shoulder the largest responsibility for the knowledges imparted and produced through an exhibition, and it was Henson's proposal to create an exhibit that included a working lab, which, in his words, would offer "a real sense of a dynamic process." By attending to his accounts, I am also able to take up the larger question of the efforts in "trying to be a scientist," which will be addressed again more fully in chapter 16.

The quotes in the following pages are drawn from a series of interviews with Henson in the early period of my research at the ROM.[2] My comments and the selected quotes serve to signal some of the concerns and questions that will be discussed in subsequent chapters. They set out several of the benchmarks for tracing emergent articulations and relations in the unfolding of this project.

In these early interviews, Henson laid out much in his story of the exhibit. I approached the project as an opportunity to retrace how scientists, staff involved in the complex work of a museum, and visitors to an exhibition all draw upon fossil matter along with other displayed matter to conceive and actualize the "dynamic process" of palaeontological reconstruction – what it could mean to "flesh out" a dinosaur. As such, this project gave me the opportunity to explore the interesting correspondence between two modes of articulation: the practice of *articulation* as spoken of by vertebrate palaeontologists in reconstructing fossil skeletons or fleshing out extinct creatures; and the practice of *articulating* done by non-scientists and by things that becomes so vital to the unfolding of politics/natures. My discussions with the curator allowed for a teasing out of different senses of how articulation might work, how it was anticipated, and where it would present serious challenges in the eventual making and engaging with *Maiasaura* in this exhibit.

What is also interesting about this opening interchange is that it offers the curator's retrospective with the finished exhibition already in mind – stories of the complex unfolding of materialized phantasmatics – including key elements in the history of the exhibit's making. Dr Henson expresses significant pride, and has a considerable sense of authorship for the exhibition, from obtaining the specimen through to ordering the display content – its animations and means of communication.

Hindsight

BRIAN NOBLE: I would like to recount the history of the display … I understand that Bearpaw Palaeontological Inc., Vernon Runyon, collected the specimen and the ROM bought it from him … right?

ANDREAS HENSON: He served as an intermediary for the people who had found the specimen, and collected the specimen … Blackfoot Nation in Montana … members of the Flamand family … so when fossils are found they go to tribal council and ask for permission to collect it … And Bearpaw Palaeontological acted as intermediary for a buyer …

BN: And when was that?

AH: I believe 1993 … shortly after I started at the ROM … I started in 1992 …

This project, then, would be the first major opportunity for Henson, as the newly hired curator of fossil vertebrates specializing in dinosaurs, to make his mark through a public program at the museum. A long-time administrator in the museum, Beth Jameson, later suggested that this would have been a much more difficult task had Henson not been new to the institutional structure, its cumbersome organizational divisions, and the carrying forward of rather customary institutional relations. The administrator commented:

> [G]etting a dinosaur from the ground to the public, that all needs to be in one area … you can't have it divided … and yes, the Maiasaur Project was like that because it was a team … but they also had Andreas, who was new, he didn't have any baggage … he didn't know of previous problems … he wanted to make his mark … It would never have come off if it hadn't been for that …[3]

By dint of the curator's newness to the institution, according to the administrator, he was less prone to being captured by its conventions, and so might introduce something exceptional in the execution of this exhibit. Notably, the Maiasaur Project exhibition was the product of a team effort, an organized collective approach to exhibition development, as may be found in various forms in major Western museums of nature, science, and culture today. Andreas Henson continued with his account:

> They're the only family that goes out and collects fossils … two older brothers and her parents … Sherry [Flamand] discovered this one …[4]
>
> … So, to give you our perspective … we had this offer to buy this specimen … Now I'd been looking for some time to do an exhibit project that went beyond the usual dinosaur exhibit … What you generally get in a dinosaur exhibit … you put up a nice skeleton, which in itself is an object of aesthetic value, and to a few individuals an object of appreciation of the history of life on earth … But basically it's a very static experience … You're taken in, you're awed by it, kids go through and say, "Oh yeah, it reminds me of my green picture I have in my kiddie dinosaur book" … But basically it's a very static and ultimately not very satisfactory experience …

Static, "off-the-shelf," highly recirculated pictures and experiences, in Henson's understanding, lead folks astray from the potential for interesting turns offered in palaeobiology, and the dynamics of its modes of generating new knowledge. Henson continued,

> [S]o, I was interested in developing an exhibit that actually had a dynamic component to it ... that there was a lot of research behind ... in fact it took a lot of process to get to that final mounted skeleton ...

Henson was now beginning to refer to the actual set-ups of palaeontological practice that effect articulations that make an "interesting difference," as Bruno Latour and primatologist Linda Fedigan have pointed out.[5] His point, that "it took a lot of process to get to that final mounted skeleton," signals the complexities taken up in the next several chapters as they explore the set-up, the dispositif, of the nexus-in-the-making. These also became the complexities encountered in the making of the exhibition as a whole. Henson reduced the point to some simple questions to which it is the curator's responsibility, literally, to respond:

> [A]s a museum professional, you get a lot of questions associated with any exhibit ... people want to know "where's it from," "what does it tell us," "why do you have it," "what are the special problems in getting it" ... and then interpreting it ...
>
> ... And since the overall mandate of a natural history museum should be to explore, discover, and interpret nature ... All of these things you would want to have confluent in one exhibit ...

This, then, was a statement of the curator's basic rationale – his commitment to how science and museum professionals are to share their offered natures with visiting publics, to bring them into the dynamics of knowing as he might experience it. He was well aware of the mediations necessary for realizing the exhibit. Outlining his own approach to the work of dinosaur world-building and knowledge-building in exhibits, Henson explained:

> So, I basically designed the idea of having an actual lab on the floor where technicians worked preparing a dinosaur from its field jackets, interacting with the public in the process, but then building around it the whole infrastructure – the history of the specimen, where was it from, what ancient world did it represent, what did the world look like at the time, what *can* we know about dinosaurs?

Henson was beginning to intimate some of the challenges he would meet in attempting to "be a scientist," the "we" of "what can we know," about how to amplify scientific knowledges garnered through careful technical practices. He elaborated on his questions:

> ... But also, how does this dinosaur fit into context? Like, how does it relate to other similar dinosaurs, how does it relate to its ecosystem ... Basically, what you have to do, then, is have the facility of preparation, we have to have an exhibit to show the actual assembly and mounting of the skeleton ... but then show the scientist at work ... the background being specific to the one specimen, but also more general ... the ancient ecosystem, "What did North America look like at the time?"
>
> ... The way this was achieved was by a combination of, *one*, actual fossil specimens and casts; *two*, the preparation facility; *three*, video recordings from the side talking to the actual discovery – me walking around in the badlands, providing context: "Today it's badlands, but seventy-five to eighty million years ago, this was along the margin of a lowland coastal plain along a huge seaway dividing North America" ... but then, *four*, also interactives, like, Hadrosaurs are basically very stereotypical animals ... yet there's an enormous amount of variation in the skull ... "What does it do?" ... maybe the crests were used to produce sound in some of them ... in some cases you don't have crests but you have these incredibly enlarged narial openings ... "What is the possible significance of that?"

These last points, a progressive series of questions attending to central issues in palaeobiological reconstruction associated with the duck-billed dinosaurs, focused particularly around the fascinating variety in skull ornamentation and the puzzle of its possible function.

His points also demonstrate a commitment to open-ended speculation; a hope for transmitting experimental possibility to exhibit visitors by way of a layering of exhibit techniques that align with the palaeontological questions being asked. As will become clear in the chapters to follow, these matters present crucial points of both *conjuncture* and *disjuncture* between the curator's scientific conceptualizing and the ultimate finished outcome in the display, as well as its knowledge effects for visitors.

Henson[6] put forward the key ecological question being posed in the speculative and reconstructive work around duck-billed dinosaurs, which would in turn be posed in the exhibit: "Why are *Hadrosaurs* important?" For Henson, the importance of these creatures derives from the way they are connected with other life, in particular the flowering plants with which they have co-evolved, their sociality, their herding behaviour. All of this, in turn, he regards as generative of multiple stories or narratives:

> So, there are then all these different narratives that you tie together, but you give it at the same time the sense of "How is it that this has developed?" ... "How do you go from just bones in the ground to a neatly mounted skeleton in the gallery," *and beyond*, you know, putting flesh on it ... or "Can we make inferences on how the animal moved, what it looked like?" ... One of the dangers of putting a skeleton in an exhibit is that people appreciate its own aesthetics, its own aesthetic dynamic ... but at the same time, it's very difficult for non-anatomists to put flesh on them ... (Henson's emphasis)

The generation of a storied imaginary of connection and relation in these accounts is as important as Osborn's generation of an animated imaginary of force, size, ferocity, velocity, terror, and climax. Henson offers a proposition founded on articulation (beyond the bones themselves) with all kinds of indicators of the conditions of life for dinosaurs in prior times.

The statement that non-anatomists do not know how "to put flesh on" dinosaurs, as they are distracted by prior aesthetics, is a pointed one, considering that the visiting publics at this and other North American museums are extensively exposed to fully fleshed out dinosaur images through schooling, television, and other media. Henson is suggesting that there is important mediating work to be done in showing the technically correct procedures of "fleshing out" these skeletons. The work of palaeontology in this exhibit, therefore, was to make up the distance between the specimen and the spectacle – a case of actively performing suggestive connections that might resolve into the Mesozoic nexus when fossil specimens, scientific discourses, and techniques are put to work in imagining and materializing creatures, their worlds and world-moments. Henson presents two vectors in this work: (i) proceeding from fossils in the ground to reconstructed, mounted skeletons (emphasizing the artefact, the specimen); and (ii) continuing from the skeleton to a reconstructed, living species in a reconstructed, living world, a living "context," which is meant to signal ecology.

On the latter vector, Henson expressed caution regarding how far one can go with certainty into reconstructing the living animal in its life-world:

> So that's why we had the animation of the actual animal taking a drink or snorting or running around in the forest ... just to give a little bit of a feel of how this animal looked ... you know, we tell people, "We don't know the colour, but we know a lot of other things," for instance, the animals had scaly skin ... most dinosaur reconstructions make them look like the "Goodyear Blimp" ... just these pudgy creatures with apparently perfectly smooth skin, or in some cases with an elephant-like skin texture ... Yet every example of dinosaur skin impression ever found ... always is made up of these polygonal scales ... So basically all of the existing dinosaur reconstructions are incorrect ...

This comment on existing visions implies, notably, that there is considerable power and agency in preceding reconstructions. There is also a decisive stance on Henson's part to undo the force of those reconstructions not by presenting certain alternatives, but rather by suggesting that which leads to properly articulated, *in situ*, reconstructed dinosaur visions. These previous, smooth-skinned reconstructions would have sufficient force (through scientific and public representational and imagining practices) to limit alternative reconstructive visions developed with newer scientific data (e.g., "polygonous scales"). Henson points out that even such fossil-sourced shifts in reconstruction are subject to "imponderables" – which may be pondered nonetheless:

> [T]here are knowable things, and then there are imponderables ... unknowable things that we try to infer from various sorts of evidence ...

Here is a moment when uncertainty and experimental phantasizing are caringly admitted. The supposedly systematic and phantasmatic fuse, working with inference, from the visible to the "unknowable" to a different register of the visible, thought of as entire world moments, as ecosystems.

Nonetheless, in the case of the Maiasaur Project, such imponderables demanded a consideration of the resources that would allow such inference-making to be realized in exhibitionary form:

> Now ... the big trick was to do this all within a manageable budget ...

Indeed, budget limitation, the "big trick," was alternately a constraint, an opening, a challenge, and an uncertainty expressed to me not just by Henson, but by everyone involved in the making of the Maiasaur Project. It would have tremendous implications for the display's outcome.

Curiously, the administration did not support a previous proposal which, according to Henson, would have saved money for the ROM, and was a "safe bet" from a marketing standpoint in that it conceived of a different, more familiar potential centrepiece – a Tyrannosaurid dinosaur from central Asia:

> My initial idea had been to go out to Mongolia and collect a *Tarbosaurus* skeleton … because people have this fascination with the great meat-eaters … Well, this notion was thrown out as being too expensive … which retroactively I find very amusing, because I could have done this for less than what we spent on the Maiasaur Project … But that's an aside …

Apart from the fiscal point being made, this is actually a highly pertinent aside, though Henson does not acknowledge its particular relevance. Henson gestured here to the opposition between the potential of harnessing what he saw as public "fascination with great meat-eaters" and the opting instead for presentation of a passive plant-eater. That opposition – great meat-eater versus passive plant-eater – became a crucial resource in the course of the exhibit-making and its "reception." And this opposition was a major consistency in scientific, communicative, and public efforts at making sense in this exhibition – to launch a trajectory of difference in regard to dinosaurian possibility through *Maiasaura*, as against one of sameness through *Tarbosaurus*, a carnosaur not unlike *T. rex*. Expanding further on the development and its seeming serendipity, Henson continued:

> So, we had this offer of a *Maiasaura* specimen … and I tried to convince senior management to go ahead with this idea … So I went through our formal exhibit development process with a proposal … People in exhibits were quite interested in it … But there was this feeling that the preparation activity – since it unfolds in rather gradual steps and little increments over time – that this was not an exciting activity, that people would not be interested in this …

Seemingly, generations of exhibitionary skewing to drama, quick action, and affect brought forward a resistance to the potential of putting live technical work on display.

> … I must say, without any arrogance, that I knew this to be wrong, because I had seen a much smaller dino lab like this active at the Smithsonian Institution, and it was a huge hit with the public … even though the fossils were much less obvious … a whole bunch of skeletons jumbled in blocks collected from the *Coelophysis* quarry at Ghost Ranch … so the bones were much smaller and you had a much harder time making out things. Yet I was convinced if you had this facility with these bones gradually emerging, with technicians who are good communicators, who would come out a couple of times a day to talk with the public about what they were doing, that this was a success … for the very simple reason that people love activities … this is something that many museum professionals do not understand …

The curator trusted in the anticipated intra-action of technician and specimen, of technician and visitor, activating work-tools, fossils, blocks of unprepared stone, to generate potent connections that would intimate the sense of a dynamic process that he knew to be so important:

> People *like* to know how things are made … Like, it's exciting to see, say, a piece of Navajo pottery sitting somewhere, it's even more exciting to see a Navajo potter actually making one … People love this … People want to know how things come about …
>
> … This is that same dynamic … So, I knew that it was a very hard sell … and I knew that only by putting in an over-abundance of interactive technology which somehow made it hip and modern … was I able to sell that very simple idea …

Here was the human allure of practice. Henson's proposition of a dynamic process was thickened by this new approach – he was after not only the dynamics of life as it played out in past-worlds, but equally the dynamics of how this might possibly be brought forth as knowledge and how that process of knowing could be demonstrated. In short, it would have to be smuggled in with "an over-abundance of interactive technology."

The lab was truly the key component of the display for Henson, the site of actual technoscientific action, and the opportunity to present dinosaur fossils via aspects of the craft that make them comprehensible and visual. The other new media and interactive technology, as he notes, was needed to make it "hip and modern." These technologies of spectacle were precisely what Henson needed to "sell" managers at the ROM on what for him was the far more important element, the element which would expose the "dynamic" of science-in-the-making as the complex of action between technicians, specimens, curators, and all the instruments put to work in that process.

To say that Henson was able to "sell" the idea is a significant understatement of the complexity of the action which had to take place. What Henson did here, in the socio-technical terms of Michel Callon and John Law,[7] was to *enrol* the powerful agency of interactive media, and with that, engage spectators lured by this agency. Interactive media had become a rising fixation at the ROM, as it has for large-scale museums in many wealthy nations.[8]

Foresight

The enrolment of interactive media into the exhibit would, of course, have significant effects, beyond providing the conditions for the lab to be built and to operate publicly. Henson, as I learned, was highly literate not just in leading palaeontological techniques and theorization but also in museum practices, the history of science, mass media, monster and science fiction movies, and much more. He offered an ongoing assessment, narrating the life of the exhibit beyond its inception and into its operation:

> Now, it's interesting, once the project was underway, having a distinct museological interest myself … I went out and observed, time and again, how the visiting public would view things … It was interesting, people would go and push a few buttons, see the interactives, but even when the interactives were still on, they would often go back to the lab … I know from our technicians, some people revisited that lab over the life of the project thirty or forty times … So obviously, this "boring activity" which the administrators had identified was actually a viable one … people were actually interested, they lived with this one … this was like seeing a child grow up, or some other complicated process like that …
>
> … So, people were really taken by this dynamic activity … even though the increments were small … and coming on a day-to-day basis you wouldn't see much … but say if you came every other month, you saw a lot of things happen … We had the drawing of the skeleton, and filled in the bones that were being exposed and conserved … So you had a real sense of a dynamic process, something unfolding … the evidence being literally unearthed and exposed to the public …

Henson held to his trust in the "dynamics" of the lab preparation. In the process of development, as he noted, some of the exhibit specialists would be troubled. One specialist pointed out to me that watching real-time fossil preparation is "like watching paint dry." While it appears from my other interviews that almost everyone in the exhibit development process eventually came to embrace the lab as an experiment in display technique that worked extremely well, in this moment of idealized, almost memorial recollection, the curator appears compelled still to justify the laboratory, amplifying how very important and challenging this approach had been. His commitment was in attempting to articulate the specimen with public sensibilities, like "something unfolding," or in his more personalizing terms, "like seeing a child grow up."

What Henson identified here is something symptomatic of museums historically, including natural history museums: (1) they tend to show the specimen, black-boxed as an aesthetic or systematically identified object; or (2) they reach for off-the-shelf spectacle which wraps the specimen in an elaborate set of visions and narratives. They resist the exposition of the craft of scientific knowledge production. The difficulty in selling a publicly visible working lab was that it appeared to risk exposing all of that which museums had attained through a history of hiding the action of science. At least partially, the lab promised to articulate and so to reconnect the thing found with the resulting public vision of its recreation and transformation, assembling all of this together as process and outcome, and as elements of what we call "nature" – a politics of nature founded in engagement and articulation.

At the same time, this remained a relatively costly exhibit for the ROM. The specimen alone was one of the most expensive the museum had ever acquired, with a purchase price of over $200,000 Cdn.[9] In more than one sense, a lot would be at stake for those responsible. Justifiably, as curator, Henson accepted both credit and responsibility for raising the stakes by pressing for the lab. But the consistent agreement among staff about the lab's success led me to ask what I thought was an obvious question: why was the lab removed from the exhibit two years after it was launched, while the rest of the exhibit remained?

BN: Okay ... but now [the lab] is gone ... Why did the management not recognize the ongoing value?

AH: Well, the project was completed.

BN: Yes, but so what? ... don't people evaluate something and go, "Hey, this is one of the best things we've got going in the museum" ... It was that, wasn't it?

AH: Yes, I think so, yes ... No, this is not statistically significant, but I have had a lot of feedback ... that this was very good.

... We had the inherent limitations ... This was something outside normal museum activities ... The principal technician was a contract technician ... There was a finite amount of money there ... And it would have been inconsistent with the overall mission of the exhibit if we had suddenly put someone in there to break rocks with brachiopods ... that kind of thing ... It is a little unfortunate.

Budgetary constraints were cited yet again as the key here (as was apparent conflict with the official museum mission). Budgets to support staff in preparing the specimen, ROM #44770, into displayable and researchable skeletal form would only go so far. Perhaps unsurprisingly, Henson capitulates to the institutional imperatives around what was intended to be a temporary exhibit rather than one that fitted with longer-range exhibit hall plans. The dynamic process he sought was that mandated for this exhibit on *Maiasaura*, however important the ROM's brachiopods might be – including those from the renowned Burgess Shale.[10]

Henson's retrospective now sets out the terms for what I consider in the next several chapters, and I will bring forward more of his comments in these chapters. His retrospective touches upon many key actions required in producing the most efficacious life-world visions of the Mesozoic by way of a hadrosaur dinosaur – specifying the particular articulation work needed. The actions he proposed were ones that were open to possibility, that resisted being hemmed in by repetition, ones in which humans, usually experts, were actively participating. In Henson's account, in order to properly allow for the participation of these experts and that of the visitors, he also saw the necessity of allowing for the participation of the specimen. The intention, then, was less to *animate* the specimens in the way that Osborn did – imposing the contours of life possibility upon the specimens – than to *activate* them.

For this, he knew he needed a laboratory, technicians, and all the trappings that would allow the laboratory to work well, and to articulate well with the gallery visitors. In addition he offered a considered, respectful means to allow for the entry of phantasmatics – through inference and pondering the imponderable – acknowledging these as part and parcel of the technical work of palaeontology, the strategic moves which scientists make to advance their practices, to lead to plausible, emergent Mesozoic natures and ecologies.

Yet he also acknowledged the play of curatorial action with that of institutional action and wider circuits of public engagement – decisions made by those who approve budgets; the need to sell ideas, to understand and act upon "missions"; the requirements to design both the ideas and the exhibits; to choose specimens well and in a manner that anticipates the wants and conceptions of visiting publics; to understand what "hip and modern" might mean to exhibit promoters and developers and managers and what kind of technology might offer this best, and for whom; and how not to lose what he sensed to be most important in the process. In a way, he was searching for how *not to be captured by the dispositif*, the set-up in which these revised politics/natures might emerge, and how positively to ensure that his articulations with this particular specimen of *Maiasaura peeblesorum* would thrive.

It is from this last point, on how to possibly generate a productive marriage of Henson's "real sense of a dynamic process" with competitive marketing concerns, that I turn to a discussion of how the ROM would manoeuvre all of this into opportunities for achieving its institutional goal: "getting the visitors through the door."

10 A Really Big Jurassic Place: When Specimens and Chronotopes Meet

When the Royal Ontario Museum opens "The Maiasaur Project: The Life and Times of a Dinosaur" in mid-June, the public will witness the most ambitious use of computer graphics ever by a museum, and the first time multimedia installations have formed an integral part of a ROM exhibit.

1995 text from computer graphics industry magazine article on the Maiasaur Project[1]

Summer 1995 – Launching the Maiasaur Project

Keenly aware of the tenacity of recirculating, received notions of dinosaurs and their past-worlds, Andreas Henson sought to offer a more articulate means of bringing the public into an intimate relation with ROM specimen #44770. But doing so would require a trade-off. What this amounted to was a yielding to the impulse for spectacle so as to give ample service to the potentiality of the specimen. In this chapter, I tease out more of the action at the ROM as staged by the play between what counted as good public science and what counted as competitive public attraction. At the end of the chapter, I return to a discussion of the interplay of Mesozoic natures and chronotopes with the complex articulation work that the Maiasaur Project sought to enact.

It bears recalling how Henson had told me that, in order to advance his project of presenting a working lab and giving a "real sense of a dynamic process, of something unfolding," he would have to position the exhibit in a particular way:

> I knew that only by putting in an over-abundance of interactive technology which somehow made it hip and modern ... was I able to sell that very simple idea ...

This trade-off and tension would come to characterize so much of the work involved in realizing and even talking about this exhibit.

From the outset, the official marketing program for the Maiasaur Project would oscillate in its expression of commitment to the laboratory and the interactive multimedia components. The official invitation to the members' launch had placed equal emphasis on the lab and the interactive media, and had used the project's original name:

> Please join us for a Members' Preview of
>
> **the Maiasaur project**
>
> the life and times of a dinosaur
>
> Step back 80 million years in time to experience the life of a Maiasaur at the Royal Ontario Museum's new interactive exhibit. In a special working lab, technicians will work to unlock the skeletons of an adult and baby Maiasaur from the rock in which they have been embedded. Complementing the lab, state of-the-art interactive multimedia exhibits provide the latest research findings about the Maiasaur and the Cretaceous world it inhabited.[2]

However, while the lab was given equal billing in this opening entreaty, the major emphasis in the project's promotions – as I came to learn in reviewing the development of the project and its wider public extensions – would come to be the interactive media.

"The Maiasaur Project: The Life and Times of a Dinosaur" opened in the middle of June 1995 to a flurry of news media interest – interest that was carefully coordinated and extensively promoted by the museum's Marketing and Communications Department. The exhibition's launch and summer run resulted in "27 broadcast reports and 29 print pieces" – a level of interest deemed "excellent" by those in the department.[3] The news and feature reporters were given access and rights to the use of ample high-quality audio and video material from the exhibition's "state of the art interactive multimedia" displays.

As mentioned, the "skew" or "spin" of this exhibit for its marketing staff and consultants was to emphasize the multimedia. The marketing coordinator had noted to me that they developed a second project title used for much of the promotional materials:

> Now the actual title of the project, we actually deviated from that somewhat ... It was originally "The Maiasaur Project: The Life and Times of a Dinosaur" ... And we changed it to "The Maiasaur: An Interactive Exhibit" ... It actually says more ... "Interactive" was kind of a buzzword at the time ... everything had to be "interactive" in museums ...[4]

Yet all of this wondrous, interactive media technology was developed, ostensibly at least, to support something much more decidedly museological: the presentation of a notable fossil specimen acquired by the ROM. The news release for the exhibit would point out that this was not just any fossil specimen, but rather "the world's best skeleton of an 80 million year old Maiasaur dinosaur."[5] The media and visitors, as a rule, would not learn that this specimen was actually purchased from a private fossil dealer, Bearpaw Palaeontological, Inc. of Calgary, Alberta, whose principal market was public museums. Rather, the display content and the information supplied to the media attended solely to the person who discovered the specimen, Sherry Flamand of the Blackfeet Nation in northwestern Montana State. This was a minuscule but notable gesture in acknowledging both the discoverer and the peoples' collective to which she belonged.[6] As it turns out, Bearpaw Palaeontological hired the Flamand family to collect the entirety of the skeleton from Blackfeet lands through an agreement the company had established with the Blackfeet Tribal Council. In turn, the specimen was sold to the ROM in 1995 for a price somewhere between $200,000 and $250,000 Cdn, with a royalty returning to the Blackfeet Tribal Council.

These are points that received no play in the content of the exhibition or in its promotion. I mention this as an example of the selectivity and erasure that take place in the presenting, and more pointedly in the crafting, of a display. Many other elisions such as these would come to my attention as I continued to engage those involved in the making and programming of the Maiasaur Project, demonstrating again and again how easily the process of following and building certain articulations can cause many other interesting ones to recede, sequestered by accident or intention from expert and public scrutiny – this, despite the practical force these articulations have in constituting the resulting natures.

Returning to the exhibition conception, texts in the promotional material for the launch read in ways that would deflect impressions that the specimen-related work of museums and their researchers was in any way prosaic or dull. They emphasize that even the skeleton was to be presented in a special manner. Unlike previous dinosaur displays at the ROM, which consisted largely of mounted skeletons or full-scale replicas set against dioramic scenes, this skeleton was not to be presented ready-made, a *fait accompli*, a black box of scientific knowing. Rather, the media release for the exhibition launch invited visitors to "watch history unfold" as "researchers work to free the remains of an adult and a baby Maiasaur from the rock in which they have been embedded for over 80 million years."[7] These phrases imparted not just a sense of process but also the impression of salvation and miracle-working – at least in some secular sense.[8]

Some Unspoken Genealogy of *Maiasaura*

What these statements do not point out is that *Maiasaura peeblesorum* already had an established public and palaeontological history dating back to the 1970s and 1980s, a history grounded in the life of specimens. That was when Montana palaeontologist Jack Horner and his colleague, schoolteacher Bob Makela, made the first finds of this kind of dinosaur, including enormous bone-bed sites with the remains of thousands of these animals, plus some rather astonishing remains of "baby" Maiasaurs, of nests, eggs, and even eggs containing embryonic bones.[9] The details of these discoveries help to situate ROM #44770 and the Maiasaur Project as a marketable exhibit in a partial, palaeontological genealogy, one that has some clear gender politics already in-built but which also begins to suggest the larger shifts in politics-natures underway.

An important line of palaeontological analysis on *Maiasaura* focused on dental wear patterns in hatchling Maiasaurs, and on the growth patterns indicated in their bone structure (i.e., bone histology). As Andreas Henson explained to me:

> You have these little animals that have poorly ossified joint ends of the limb bones, yet show wear on the teeth. So obviously, in the standard model, they weren't capable of getting up on their lower legs and wandering about and foraging for themselves. It seemed more likely that they stayed in the nest, and then somebody brought them food.[10]

The interpretation which Horner and Makela drew from their review of the fossil material was that this kind of dinosaur almost certainly cared for its young. Their fixity on parental care was revolutionary in certain senses, insofar as it allowed for the raising of astonishingly different questions about dinosaur social behaviour from what had been contemplated prior to that time.

The first technical article on *Maiasaura peeblesorum* was published in 1979 (in that most preeminent of natural science journals, *Nature*) with the title "Nest of Juveniles Provides Evidence of Family Structure among Dinosaurs." The scientifically ground-breaking article is notably marked by languages of "families" and "babies": terms with as many human valences as animalian ones.

Certain rhetorical patterns in the article aid in strengthening particular familial affinities. One passage notes that it is not possible to assign the fossil material from the juveniles to a particular species:

> Because of the immaturity of the juvenile dinosaurs and the ontogenetic changes which probably occurred in the skulls during growth, a specific identification of the 15 individuals from the nest may not be possible until more specimens of older individuals are found. Characteristics of the skulls, however, suggest that they are generically similar to a large skull (Fig. 2) found within 100m of the nest. The large skull represents a new genus and species and, although it has some rather peculiar characteristics, is placed in the subfamily Hadrosaurinae.[11]

On further reading, it becomes clear that Horner and Makela find enough similarities to assign a common genus to these dinosaurs. The sex of the adult animal was indeterminate on the basis of anatomical characteristics in the skulls of the "little hadrosaurs." What Horner and Makela do attempt, however, as if to secure the association between the juveniles (or "babies," as they are called in some instances) and the skull of the adult found a hundred metres away – another moment of articulation work, though possibly a tenuous one on the face of the evidence – is to feminize and maternalize the new generic name: *Maiasaura*. The article notes precisely:

> Etymology: Maia from the Greek which means good mother, saura which means reptile (feminine)[12]

Out of a "jumble" of bones of juvenile dinosaurs and a dissociated skull we arrive at this particularly canonical, gendered family association, the baby-mother dyad, its veracity reinforced through an inscription in line with Linnaean rules, but also following the relational sensibilities of Horner and Makela in their historical-political moment.

It is not merely a coincidence that problems in human evolution in the 1970s and early 1980s began to unravel androcentric presumptions associated with "man the hunter" propositions, and there was a move to discussions of gynocentric possibilities through "woman the gatherer" models.[13] *Maiasaura* offered a parallel move from over-determined male carnivory to female herbivory, from *T. rex*, hunter of prey, to *Maiasaura*, forager of plants.

Having already noted that the juveniles could only be described down to their genus, *Maiasaura*, which the authors argued is shared with the adult, later in the article the figure referred to in the previous passage is presented (figure 10.1). Here it shows *both* the adult skull *and* a skeleton of a juvenile. The caption implies a common species identification of the juvenile material and the adult skull, even if, technically, the species name applied only to the skull in this moment of description.

By inscriptional accident or design, the specific affinity between these bones is bestowed by the journal *Nature*, the mother-and-baby dyad cemented all the more. In Austin's sense, this is very much a performative utterance, where the palaeontologist as authorized subject in this most authorial, technical venue announces the relation and effectively brings it into being.

This is also a moment of phantasmatic transference equal to that of Osborn's naming of his prized "king of the tyrant saurians." Today, Horner and Makela's new genus, *Maiasaura*, is commonly translated to the English phrase "good mother lizard."[14] For these men, parental care was sufficiently encoded in the Latin term, but this designation nonetheless sets limits on the gender alignment with female parenting, leaving male parenting – good or bad – out of the immediate meaning at the very least. A post-tyrant-king politics/natures, one that nonetheless borrowed from gender oppositions, was emerging through Horner and Makela's "discovery."

Even though Horner and Makela had done the lion's share of technical and scientific work on *Maiasaura*, and although those results informed so much of the Maiasaur Project, that point would not be presented in the final exhibition. The exhibit's content would be promoted and presented as an immediate experience of scientific knowledge in

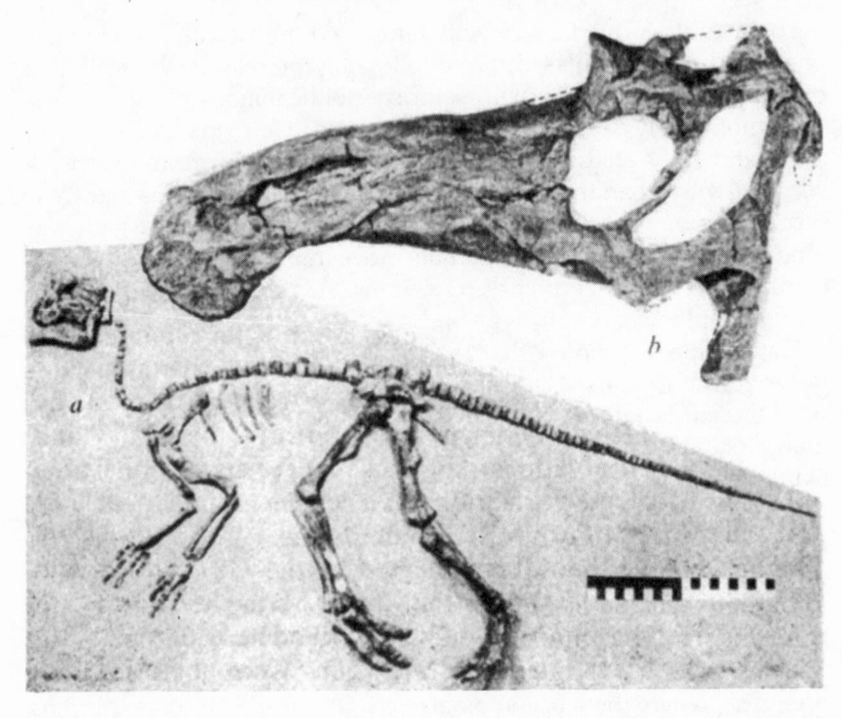

Figure 10.1. *Maiasaura peeblesorum*: a, incomplete reconstructed composite skeleton of individual from nest (PU 22400); b, holotype skull (PU 224405).Source: Horner and Makela 1979, 298, Fig. 2. Reprinted by permission from Macmillan Publishers: *Nature* 282: 296–8, copyright 1979, enhanced with images supplied by John Horner.

the making. The omission of this history would, undoubtedly, enhance that sense of immediacy.

In order to give the proper sense that scientific knowledge was unfolding in real time, the promotional materials indicated that an actual "working laboratory" had been installed in the exhibit halls, flanked by various other dynamic display media. One or more technicians – who were sometimes referred to as "researchers" – would work on-site in the gallery laboratory over a projected period of two years to prepare out the fossil material still embedded in the sandstone matrix (i.e., the encasing sediment). This fossil-bearing matrix had been removed *en bloc* from the original locality of the find. The laboratory was intended to be the "centrepiece" of the exhibition, and as such the display was meant, quite literally, to be a "living" case of science in action, intent on providing "the latest research findings about the Maiasaur and the Cretaceous world it inhabited."[15]

The media release text also attempted to ward off possibilities that visitors might find "plant-eating" dinosaurs uninteresting by quoting the curator in charge of the exhibition, Andreas Henson:

> We'll now have a much better idea of what these friendly plant-eating dinosaurs actually looked like. Once the adult is recreated I think people will find her quite adorable, with her big eyes and duck-like beak. I'm counting the days until our Maiasaur begins to literally rise out of her bed of rock, piece by fascinating piece.[16]

Again, the language of godlike re-creation and resurrection appears, as does the language of emotional connection and affect in the description of an "adorable" dinosaur. Upon reading Andreas Henson's vivid entreaty to the sentimental – however sincerely, strategically, or ironically intended – I found myself humming the American, Disney-propagated conservationist jingle: "Be kind to your fine-feathered friends, for a duck may be somebody's mother …" This text in the news release presented multiple senses of what was promised in the display experience: visitors could witness researchers effectively "liberating" the bones from their implied "confinement" in the rock of ages, resurrecting the creatures bit by bit, and as if with the virtual power of deities, coming to "re-create" better than ever what this "friendly" dinosaur "actually looked like."

Yet, from these statements, it was clear that some things were indeed already known of this dinosaur: "her" sex was already presumed in the language of the release and "her" visage anticipated. This was to be a resurrection of a very individual creature, through a process of skeletal extraction, articulation, and reanimation. Considered in this complex manner, the specimen as yet unrevealed by the lab workers was, in fact, pre-revealed in the imagination of the curator in charge and so many others on the exhibit team. In my experience, this is common in many museum exhibitions.[17]

The illustration used for the invitations and other promotional materials portrayed in a single view the entire drama which the exhibit appeared to promise. The image showed the entire creature "in transition," a transmorphic being – partly naked skeleton, partly fleshed out, and, more cryptically, partly cocooned in a strange, grid-like envelope, or wire frame, which turned out to be a graphic technical aid used in the visual work of producing computer animations.[18] The profile view set the animal into a posture that would be repeated throughout the actual exhibit. The stance of the animal on all four limbs would become, for all intents and purposes, the iconic posture for this exhibition's Maiasaur (figure 10.2) – much as Osborn's advancing *T. rex* became for so long the icon of large, carnivorous dinosaurs.

Of course, dramatization through gendering or imaginative reconstruction is arguably a feature of the performance techniques of exhibit-making and, more to the point, of media promotion. It is a means of "capturing" interest, of providing "hooks" to attract potential visitors to the museum. Such posturing does, however, have powerful consequences, not to mention underpinnings in human political concerns. These, in turn, have consequences in relation to the politics of science, as Western society's favoured, most relied upon vehicle for explaining and speaking on behalf of the natural, and museums, as the favoured, most relied upon medium for exercising this sort of advocacy. When publicly funded research museums present knowledge, it is usually taken seriously. When the natural matter encourages dramatizations along lines of this or any other sort, those too may be taken in public circulations as authoritative, legitimate, true.

As important here, both in Henson's release statement and in the illustration, is the conveyed sense that the skeleton would be the very object to be reconstructed, fleshed out, and transformed into the moving, animated being before the eyes of the public. But as the interpretive planner of the exhibition, Jennifer Ross, explained to me, the interpretation of the creature *preceded* the specimen's preparation. The actual preparation would only take place over the two years following the finished installation of the display itself:

> [I]t turned out we really did not have too much solid material to interpret from. The skeleton was not ready yet. So, it was a really weird exhibit in that way. It consisted mostly of material that we created, interpretive materials. Interpreting what?! This thing that had not happened yet!

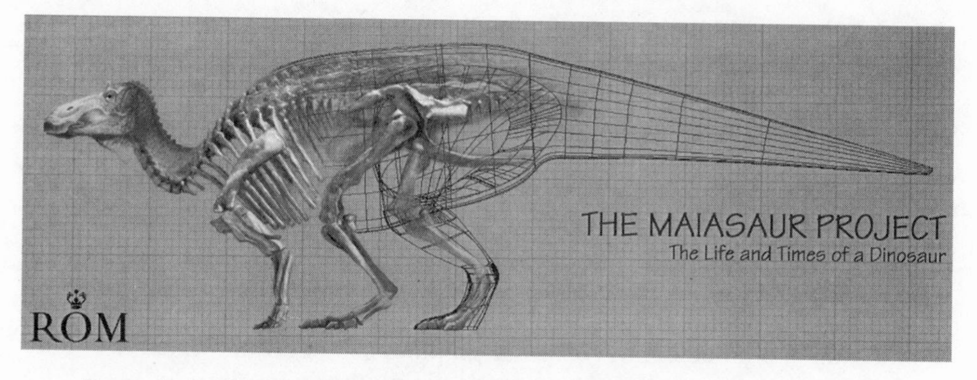

Figure 10.2. Transmorphing Maiasaur image. Source: Illustration reproduced with permission of the Royal Ontario Museum, © ROM.

The implication, of course, is that there was a hidden knowledge disjuncture between (a) the displayed preparation work which would lead to the mounted dinosaur skeleton, and (b) the interpretive content of the exhibition. The divide between the action of the laboratory and the rest of the display was widening, though all would be presented as though no such divide existed; as though the knowledge were derived from research on this very specimen.

In terms of the media promotion, expectably enough, this disjuncture and paradox was never an issue – if, indeed, anyone directing the promotion even took notice of it. The marketing and promotions task was to accept the legitimacy of the exhibition, and then to isolate those characteristics most likely to "appeal" to the intended audience. The interactive multimedia appeared to trump the lab in this regard. The news media coverage that followed from the promotions and the launching of the exhibition signalled that this was indeed the case. The news media immediately recognized that museums were going through some significant transformations in how they communicated with the public. In contrast to previous understandings of how public museum exhibitions were expected to operate, this was – for the ROM and for the news media as well – an exhibition with a difference. Rather than the robust articulation of the specimen sought by the curator, this approach to the exhibit was beginning to amplify articulations between visitor and the performed legitimacy of received scientific vision.

"A Living, Breathing, Environment"

Among the many broadcast reports resulting from the launch of the Maiasaur Project, the audio-visual production staff of the ROM felt that one in particular merited keeping. When, in 1999 (some four years after the launch of the exhibition), I approached audio-visual staff members David Ritter and Carl Dailey for a compilation of video material depicting the multimedia and computer graphics content of the exhibition, among the various clips they provided to me was a four-minute news story produced by the Toronto-based CityTV for its "MediaTelevision" feature program.[19]

The video feature oriented around four components: the interactive multimedia displays and interviews with their producer, Walter Tomasenko; the Maiasaur laboratory plus interviews with the "Curator-in-Charge," Dr Andreas Henson; the advertising "angle" for the exhibition, including interviews with advertising consultant Les Grantham of Roche MacAulay and Partners Advertising; and the response of a number of young visitors to the exhibition. Notable in the piece is how it profiled a crucial series of non-human/human articulations, drawing attention to a series of key dyadic connections between the physical exhibition and specific actors: the interactive multimedia with the media producer, the laboratory with the specimen and the curator, the television advertising with the advertising consultant, and the exhibition with the visitors. A further, hidden dyadic articulation can be added: that between the new CityTV program itself and its audience. Here were some very direct indications of the power of this exhibition to produce an extensive and socially potent network of connections in this particular location and moment.

The piece is introduced with CityTV reporter Sheila Cameron exclaiming how the makers of the exhibition were "bringing the 80-million-year-old Maiasaur to life."[20] From the outset of this feature, the temporal moments of the dinosaur were set – eighty million years ago, brought to the present – offering an almost supernatural sense of time suspension. The first images are of people moving through the active, busy exhibit space, then children crowded in front of a huge screen with moving dinosaurs in the background. This scene is followed by comments from Walter Tomasenko, the project's creative director:

> We don't try to teach as much in museums as we did in the past …
>
> What we do is we create an environment … a living, breathing environment, for people to enjoy, and to explore, and to understand, and to discover.

In juxtaposing the "people" visiting with what "museums" now do, the "environment" he refers to is the total exhibitionary environment, not simply the media-generated scenic content of one or another display component. Immediately, a second visual juxtaposition is presented in the video clip as Andreas Henson appears – another white male figure (though dressed more casually than Tomasenko). Though the report has given priority of narrative position to the multimedia components and their creative producer, the curator Henson states what is suggested to be the institutional priority, using the collective possessive form of "our" exhibit:

> The centrepiece of our new exhibit on the Maiasaur is the preparation laboratory, in which we're standing here … and where we're carefully cleaning up the remains of two specimens of this dinosaur called *Maiasaura*, where the bones are carefully removed from the rock, preserved with very expensive chemicals, and then ultimately made available for exhibits and scientific study … And we estimate that this will take two years …

Henson is seen standing within a technically sophisticated fossil preparation laboratory, reminiscent of a surgical theatre with great plexiglass viewing windows in the background. A technician is seen busying herself over a laboratory bench, then gazing at a specimen through a microscope. Visitors look on through the glass at the activities within this exclusive space of "careful" preparation work, of "specimens," and "expensive" chemicals. Immediately, a fast cut shows children leaning over a computer display of oddly shaped, duck-billed dinosaur skulls, with their highly varied crests. They groan as one intones, "Ughh, look at the gigantic tumour!" This tightly edited moment, standing in stark contrast to the seriousness of a technical laboratory, provides the reporter with her opportunity to put the thematic point of her story to the palaeontologist: "The museum has become now a place of both learning and of entertainment … ?" To which Henson responds:

> That's correct. Museums have become a leisure-time activity. So it's very important to compete with other leisure-time activities and to provide

entertainment. Plus it's also often easier for children to learn complex things if they're presented in an entertaining and enjoyable way.

Aside from his loaded admission that museums are "competing" in the leisure entertainment industry – and an apparent move away from the commitment to intra-active knowing – Henson's comments underline a longstanding tension encountered by museums as institutions understood to be at the interface of science and public, civil education.[21] The project of deploying museums as places for education was certainly underway in the English-language museum tradition in the mid- to late 1800s, as seen with the Kensington museums of London. In this instance, a highly accomplished scientist is welcoming entertainment into the presentation of scientific knowledge, as though it were a novelty. In concert with Henson's statements, Tomasenko's comments on the turn away from "teaching" towards "providing an environment" for various sorts of visitor experience suggest a general lessening of authoritative control over knowledge mediation and a granting of more choice to the visitor – with the visitor figured specifically as a consumer.

Superficially, at least, the report indicates that a working accord between researchers and exhibitors has productively taken place in the making of this particular exhibition. Underlying that accord is a commitment to certain sorts of exhibitionary practice, reorienting both to a more complete environmental experience and to an increasingly experiential and informational complexity. Tomasenko emphasizes that visitors want more choice. Henson emphasizes that they want more entertainment. Both anticipate an enhanced quality of experience for the visitor, which for Tomasenko is to "enjoy," "explore," and "discover" – and for Henson is "to learn complex things."

Against a historical backdrop of tension over the primacy of scholarship in opposition to public visitation objectives at the ROM – something recounted innumerable times to me by professional and non-professional staff alike – this apparent reconciliation suggests a significant adjustment in the positioning and priorities of a leading Canadian public museum. That adjustment accelerated substantially at the ROM after 1997 and the arrival of a new director, Sidney Lawson, who had been director of the Powerhouse Museum in Sydney, Australia. Lawson earned a reputation as an interactive display promoter, and for this was regularly criticized as sacrificing curatorial and collection interests all too readily.

In international terms, the Maiasaur Project exhibition can be seen to stand as a case of a larger shift in the democratic, civil functioning of museums taking place over at least the last two decades, towards what sociologist Andrew Barry refers to as "questions of consumer choice and visitor behaviour."[22] Concurring with the observations of sociologist Roger Silverstone,[23] Barry notes further:

> In recent years ... the liberal conception of culture as a means of individual improvement has had to run alongside – if not compete with – neo-liberal notions of culture as a consumer product. The traditional museum has been accused of being too paternalist, too dominated by the concerns of curators and the fetishism of the artifact, and too dependent upon public subsidy. What is said to be required is a new recognition of the competitive character of the visitor business in addition to the older preoccupations with scholarship and public education. The museum is "but a part of the leisure and tourist industries."[24]

Barry's observation rightly suggests the common understanding among museum professionals that there are only these two options: models of liberal improvement, or neoliberal models of pandering to visitors and consumers. Where does Henson's dynamic process begin to thrive in such a stark opposition, one wonders? With that noted, the MediaTelevision news story on the Maiasaur exhibition comes forward as an incremental move to the neoliberal impulse, a move into the leisure industry marketplace, by offering exceptional forms of virtual mediation for museums to draw in their visitors, now wrought as clientele.

But counter to these journalistic observations, here also was a curator whose commitments were being drawn – through a willing compromise with hip and modern techniques – into line with market interests. Yet, he also recognized that to ensure that interest in the specimen could be sustained – more than the conservative sense of "the fetishism of the artifact" as an aesthetic thing – much, much more had to be included in its mediation.

The plan, then, was for nothing less than a determined theatricalization of the regular technical preparation of the fossil specimen for the visiting audience. By theatricalization, here I mean the moving of "behind-the-scenes" technical work into public display areas in a form which amounts to a viewing theatre. These seemingly contradictory points of what was theatre, what was science in action, and what sort of theatrical science made for good substance in display became key matters that repeatedly pressed upon me throughout my ethnographic encounters at the ROM. Here, it was clear that the curator had complete faith that technical laboratory work would be quite entertaining, making for good performance, even though it was to be live, everyday, technical preparation, unfolding at a painstakingly slow pace. His dynamic process could be rescued by the laboratory's presence, and more importantly, by its activation with an important rare specimen of the moment – and by the work of technicians before the visiting public.

Before considering further the possible tensions or resolutions within the mixing of the techniques of working science and interactive media in the Maiasaur Project exhibit, it needs to be pointed out that the MediaTelevision story relied upon a public sensibility of what a museum ought to do. Certainly, the ROM's printed media briefing package had foregrounded the two major juxtaposed features of the exhibit – interactive multimedia and the "special working laboratory" – as well as the promised action from all this, which the reporter also passed on: that of "bringing the 80-million-year-old Maiasaur to life." Bringing a long-dead creature "back to life" has to be understood to mean bringing it "virtually back to life" – again, making up the distance between the dead fossil and the living re-presentation of the animated life-world of the dinosaur – that is, of course, its phantasmatic materialization.

In the public cultures of dinosaur *natures*, this same claim to virtual resurrection had been circulating widely since 1993 in the form of the virtual dinosaur animation work in Spielberg's *Jurassic Park* films. In addition to being a story of how technoscience could bring dinosaurs to life, *Jurassic Park* presented other apparatuses which had parallel elements in the Maiasaur exhibit. *Jurassic Park*, as book, film, and chronotopic story, also presented the island theme park with plexiglass-enclosed laboratories, computer interactive displays, mounted skeletons, excited children, theatrical presentation spaces – a blending of high-tech amusement park entertainment and technoscientific action. Here, then, was a museum exhibition that appeared to be offering something just short of that conjured cinematically by *Jurassic Park*, and yet in some ways considerably *more* than that of *Jurassic Park*, as it still continued to offer a "real" specimen, an actual (not merely represented) working technician, and technical instruments – not to mention the authority of its state-supported museum location.

The media releases had cued to the production association with the makers of *Jurassic Park*, noting that the project used the same computer animation technologies as the film, of which I will say more in later chapters. This was an exhibition behaving quite seriously like a Hollywood film, as well as a theme park, as well as a palaeontological laboratory, even going so far as to draw upon the same technology of production – and, to a large extent, the same sorts of animated visual outcomes, the same technical preparation activities and outputs. Here, then, for the media as well as the museum, was a story ripe for the telling – extreme, interactive animation technology meets live-action, technical dinosaur science in a tale of dinosaur resurrection, all located in what media-producer Tomasenko had said was a "living, breathing, environment."

"The Biggest, the Best, the Most"

The television advertisements produced for the exhibition – their focus also drifting away from the lab and the specimen, which curator Henson had emphasized – turned more towards the interactive and multimedia components in order to achieve the right marketing results in a world of competitive leisure attractions. The ROM's marketing coordinator for the exhibition, Brenda Mikelsen, described the situation:

> [W]e know that tourists stay between three and five days … they're interested in seeing "the biggest, the best, the most, the most outstanding" … you have to try to appeal to that … So, we wanted people to know it was different than what you normally get at the museum … it's interactive … it's not just standing there reading a label in a case … you come in and you can actually touch buttons … You could *do* … what you wanted … and the big screen up in the Maiasaur exhibit, where you could touch the buttons and you could make the Maiasaur run across the screen … You could make them do any number of things … We wanted people to go "Oh! That's kind of different for the museum!"[25]

Having conveyed what she understood as a generic sensibility of visitor interest – purified into a common fixation on "outstanding" attractions and an eagerness to actually "do" things interactively – she went on to describe the television ads used:

> You'll see we did two TV spots … one with a dog running out of the museum with a big bone … and some kind of crazy-looking curators or technicians running after him …

> Then another TV commercial with this guy, standing there, and the ground shakes and you hear a roar, he gets this look on his face, and then you see this big shadow coming at him as he's pressing different buttons on a console ... So you get the feeling that you can come in and touch a button and "oh you can actually maybe feel the tremors, hear the sounds"...

One ad, then, was a tongue-in-cheek play on scientific fixation on the object of study – the fetishism of the specimen. The other was an equally tongue-in-cheek play on funhouse thrills in interactive form – the fetishism of the spectacle, especially the spectacle of the terrifying and monstrous. Aside from the main title, "The Maiasaur Project," neither ad presented anything about this particular specimen or dinosaur. Instead, they simply represented two hyper-determined poles in the specimen-spectacle complex, as oppositions, either of which television viewers could presumably understand.[26] As mentioned already, the curator's intention was for the exhibition itself to provide some kind of mediation work between these purified poles, something more about the articulation practices that were familiar to his own modes of studying fossils. Put another way, what was encountered here was a commonplace late twentieth-century approach in museums and their media arms to the convention of deploying fetishism as a means of capturing and concentrating attention into things and productions. In later chapters I will speak to the power of the material outcomes wrought by such practices – the proposition of factishes, after Bruno Latour and Isabelle Stengers – but here I wish to hold to the conventional ways that fetishism was made to work, ostensibly to the benefit of the exhibit project and the ROM's institutional intentions.

Both of the ads fitted with the style established by the ROM in most of its media promotions of the past decade or longer – a clever, witty style aimed at savvy, well-educated folks. It was a style usually laced with a sense of irony about what a museum, in a canonical sense, ought to offer.[27] CityTV's "MediaTelevision" story presented an interview with the advertising professional who had consulted on these ads, Les Grantham of Roche MacAulay and Partners Advertising. He added:

> In the advertising, we wanted to focus more on the interactivity than the show in general, because that was the unique thing about it. So what we did was show this poor guy who steps up ... He's a part of the dinosaurs at that time period and really kind of gets thrown for a loop. This was also

> to appeal to a broad range of people. It wasn't just kids and it wasn't just adults. It's anyone who wants to be a part of something they could never really have experienced.

In classic lost-world, chronotopic practice, Grantham is attempting to suggest the power of virtualizing Mesozoic "experience" for anticipated audiences, the possibility of engaging with something otherwise remote, alien, exotic, and untouchable.

In addition, the default here is to a notion of unmarked, undifferentiated "mass" audiences, and equally to the potential of being practically consumed by the spectacle, to be "part of the dinosaurs at that time period." I asked Brenda Mikelsen about how the media promotion for the exhibition did not give any cues to the thematic of either *Maiasaura*, "the life and times of a dinosaur," or the point that this was a herding, nest-building, duck-billed dinosaur. These themes were the planned interpretive account. Mikelsen pointed out to me:

> I guess what we were trying to do was appeal to the masses … everybody knows the ROM has dinosaurs, but we were trying to put a unique spin on it … it's our "newest collection" … here we have "the Maiasaur, an interactive exhibit" … which to us meant … You come in and you actually watch them pick away in the lab, which is really neat … it was great …

Great as the "newest collection" and the potential of watching them "pick away" was for those in the Marketing and Communications Department, this was still about pitching the virtual experience, paying only peripheral attention to the technical components of science on display. This virtual interaction was something that they believed would appeal to "the masses," here meaning the widest scope of their traditional audience demographic – largely middle-income, central Canadian "families," or tourists in the summer, typically "families" again. This intended appeal translated into a media campaign that reduced meanings to elemental forms – the "newest bones," "an interactive exhibit." The pivot point for these political-natures was shifting to such elemental forms, and away from the curator's dynamics and practices. The ostensible reason: to allow for competition in a marketplace already crowded with interactive spectacle, with "the biggest, the best, the most."

I asked Mikelsen how well coordinated the various sections of the museum were when communicating with the public on the exhibit. She told me that the management of production and diffusion of ROM's media materials was inconsistent:

> [T]hat's sort of a tough question … we do find that the communication materials that go out are kind of fragmented … they don't all look the same … Membership may take a totally different spin on something than Marketing would, or shops would … It's getting a lot better … But at the time, the Maiasaur stuff, it was pretty much doing your own thing …

Despite these multiple approaches concerning what was important or interesting about the exhibit, those in the Marketing and Communications Department were not fully blind to curatorial interests. For instance, it was understood within the Marketing and Communications Department that the use of "Project" in the titling of the exhibit referred to the technical project of getting the fossils prepared out of the sandstone matrix in which they were embedded. Commenting some three years after the exhibit opened, Mikelsen added:

> For the *Maiasaura*, the plan was two years uncovering, one year display … but upstairs, they still have the big screen and the cast … But the lab is no longer necessary …

The lab was understood as a necessity for technical work on this specimen, but also as a temporary fixture that would not carry the exhibit – it was dispensable. It was not intended to be an ongoing display, even though it was paid for through display budgets. Indeed, the bearing of these costs was a point of considerable internal consternation for those in charge of display planning.

As the MediaTelevision report closed, the reporter persisted in her line of questioning about what the exhibit meant for museums as places of science, education, and objects in cases – questioning that followed quite logically from all the claims of the museum's media releases and advertisements. When interviewed by the reporter, curator Henson found himself underwriting the advertising campaign as much as the exhibit itself:

> People have sort of a threshold anxiety about museums because museums are dusty old places, and I hope that this advertising campaign, along with the exhibit, will dispel that notion and make museums an attractive option for people to spend their leisure time in.

What a complicated position to be in as a scientist, navigating all these concerns, even to the point of being enlisted as an advocate for "leisure time"! All of this was done in order to make palaeontology relevant, to win the public's participation in the scientific practice to which the curator was so deeply committed. This was the major risk taken by Andreas Henson: the interactive media with which he was able to "sell" the Working Lab to management was also what would have to be emphasized to bring in the crowds, parlayed now as a bid to supposedly save museums (and vertebrate palaeontology) from being consigned to the dustbins of public lack of interest in an environment of leisure-time competition. In this sense, marketing and marketability were *performing* the display concept, shifting and reconfiguring the material outcome in the direction of the environmental, interactive experiences. The very definition of "museum" was coming under closer scrutiny, and with it a nervous spin towards media attraction was starting to accelerate.

As far as the marketing and communications staffers were concerned, the curator had, more or less, given his endorsement, as it was their practice to ensure that exhibition curators had ongoing input to the process:

> The whole campaign, the curator would have seen all these pieces, and given his approval that it was accurate, that it well represented what he wanted to do with the gallery before any of this stuff hit the streets ... The curator has a big hand in everything that goes out ... We don't operate in our own little bubble ... We try and bring in as many people as possible ... It's really important in an institution of this size where there's so many people working on so many different elements ...

My sense was that Mikelsen had identified the relevant points – yes, the curator had been consulted and had the chance to give his input. But, as if to compound matters, she pointed out the more encumbering feature of the process, the importance of keeping people involved "in an institution of this size where there's so many people working on so many different elements." This was where control over media content and display configuration overflowed any single individual's interests, including the curator's, or those of any single administrative unit of the ROM. In the interactions, the alliances made, the tolerances for the contingency of multiple players and approaches applied to multiple elements (e.g., display design, multimedia development, marketing, fundraising, collections management, interpretive planning, project management, curation, etc.), and in the struggles to produce meaning and coherence in the display outcome – in all of these places, certain features would be more foregrounded, others backgrounded, in ways that no one could fully predict. What was getting foregrounded more and more was the orienting of the exhibit towards an experience akin to modern theme park attractions, and – with the interactive animation components – akin to a particularly popular Hollywood film, *Jurassic Park*.

Ultimately, the evidence confirming the parallel *Jurassic Park* identification was also provided in the MediaTelevision report, as the reporter canvassed the responses of children in the midst of their interactions with the exhibition. One boy, touching the interactive, morphing screen (which morphs continuously between multiple duck-billed dinosaur skull forms, with an array of crests), remarks, "Well, we're just pressing the screen, and it just evolves!"

"Evolution," then, becomes a technological reality with the ease of a touch. The next child speaks, fully cognizant of the regular mediation techniques of museums and what is new here:

> Usually at museums, they don't have much virtual stuff … but this … it's like really neat and everything with the dinosaurs walking, it shows you how they walk and everything … it's not really like a museum … it's more like, umm … yeah, like a video game, sort of …

And finally, a young girl brings the point home:

> It's not at all like a museum … it's like a really big Jurassic place!

Her choice of "Jurassic" to signify the living time-space for this "Cretaceous" dinosaur points to the *Jurassic Park* invented by Crichton and Spielberg. That movie also generated a "place" that was "alive" with dinosaurs; a tour, a site for interaction with life-size, running saurian creations – and even, in one scene, a glass-enclosed laboratory as the site for the "recreation" of living beings out of fragmentary source material.

For this young girl, Tomasenko's wish for producing an environment had been achieved. Here she was, made small yet given power in this "really big" place of moving, push-button and joy-stick activated, full-size and animated dinosaurs to virtually "make the Maiasaur run across the screen," as Brenda Mikelsen had aptly put it. Here she was, in a living, breathing environment of real fossils, live researchers, television shows, animated power tools, and imaginary worlds enacted. The living, breathing environment had incorporated and enrolled her into its performative replay of the *Jurassic Park* adventure, which would lead her, in Penelope Harvey's words, to become "inscribed into the exhibition" – that is, articulated in an embodied, performative part of the practical action.[28] The shift to environmental interactivity and virtual world-making appeared to have provided the conditions that the curator needed to sustain the relevance of scientific practice taking place in the palaeontological laboratory. The question remained, however: had something else of value been lost or diminished in the process? In cultivating *Jurassic Park* articulations, might other articulations have been lost – ones more responsive to the fossils themselves?

Working through Oppositions

In the first section of this volume, I traced how the Mesozoic time-space apparatus evolved as a malleable, recirculating nexus of fossils, knowledge, phantasy, and technical work, and how this formulation of palaeontology-made-public has been constitutive of political natures over much of the twentieth century and remains so in the twenty-first. The unabated oppositional play between the capacity to materialize phantasmatic worlds past and the action of engaging with dinosaurian remains and their reconstruction would repeatedly rear itself in the development of the Maiasaur Project. What is striking about the enrolment of the *Jurassic Park* imaginary into the Maiasaur Project – and, reciprocally, the gallery visitor into the *Jurassic Park* imaginary – is how "learning" and communicating ideas about this dinosaur was already buried in a highly fused set of imaginings and imaginative sensibilities within these children: the power to morph skulls, the experience and cadences of playing a video game, the imagining of being in a "really big Jurassic place."

Chronotopic – that is, time-space – worlds and ways of engaging them are yet circulating here, undergoing smaller or larger transformations at each moment of translation and engagement. Moreover, both the Maiasaur Project and *Jurassic Park* are visible expressions of the specimen-spectacle complex, a bifurcate dispositif in operation. The former is located in an institution meant to enhance civic engagement in what counted as nature's history; the latter in an institution of commercial consumer entertainment. Yet the two locales, one public and one private, become increasingly blurred as the media and narrative content of each are increasingly in trade – and as they are pressed into the ever-accelerating competition for market share and, with that, for accelerated "return on investment." The visitor or viewer is drawn into the play of articulations and articulators that ensures the autonomy of the specimen will be preserved enough, in the very action of making spectacle of its existence.

In the Maiasaur Project, however, the attention cleaves to the act of mediation, in the most McLuhanistic sense. This attention to acts of mediation is signalled, in the case of the media producer, with his transforming of the museum exhibit into "living, breathing environment"; the curator transforming it into a site for performing the "dynamic process" of palaeobiological techniques and reconstruction; for the visiting children transforming it into a "video game" or a "Jurassic place"; and indeed for the reporter herself, who profiles these transformations in a story about the museum as, quite literally, a place of interactive multimedia.

Now, summoning forward references to Karen Barad's propositions, if we approach each of these layered-up mediations not as *interactions* (between mutually exclusive autonomous things or agents), but rather as *intra-actions* (between relational things or agents that affect and shape each other), the effect is to enliven the apparatus, affirming its being, while at the same time allowing politics-natures of the moment to be assembled, shaped, and reshaped – however gradually – by the incursion of potentially novel or unexpected concerns of each *intra-acting* agent. One resilient precipitate, recognizable in all of these mediations within the nexus, is the oppositional struggle between the action of specimens and the action of spectacles.

The thumbnail MediaTelevision story certainly relayed the opposition of techniques of science against technologies of interactive animation, which were contained in the carefully crafted media information package. The information package, in turn, had taken its cues from the actual planned opposition between these two sorts of experiences in the exhibition. This raises many questions about what sort of knowledge is actually constituted when visitors of many sorts engage contemporary exhibitions such as this, when the comprehensibility of how to engage the exhibition is, at least in part, acquired by the visitor *elsewhere* in the wider landscape of entertainment and educational media experience – this elsewhere including the Hollywood film industry.

Resisting While Being Captured by That Which Circulates

I will attempt now to gather up some of the kinds of redirections that were emerging as a consequence of curator Henson's trade-off in the promotion and marketing associated with the Maiasaur Project. With that, I aim to bring forward and elaborate on matters addressed to politics/natures, as discussed in the first part of this book, notably in chapters 5 and 6. As noted, so much in the project turned on the play and oscillation between action that allies to specimens and action that allies to spectacles – a kind of *duplex* operation, to borrow Marilyn Strathern's term.[29] By means of this duplex, we can begin, as well, to detect what it is that is circulating and precipitating in these institutional actions and diversions, and the challenge this poses to articulation work. Not only was the curator contending with how to articulate his technical palaeontology of ROM #44770 with institutional new-media concerns, but he was also having to contend with the reassertion of the Mesozoic phantasmatics brought forward by a media-steeped visiting public, as well as a marketing and digital media production staff who sought to capitalize on this.

So, politics/natures are at play here, once again. This is a case of the back-and-forth trading between museum science, as an active producer of natural knowledge, and more distributed, public circuits of science. The knowledge carried in by the visitor combines and fuses intra-actively with the knowledge produced by the museum and translated into exhibitionary form. Where one child uttering "It's a really big Jurassic place" draws upon all the cultural (i.e., phantasmatic, material) cues that the phrase might entail, she also properly links and articulates her museum dinosaur exhibition experience to her Hollywood-, theme-park-, or television-produced experiences. Those are experiences anticipated, if only in the most general of terms, by the makers of the exhibition. Some resilient, shareable materialities/phantasies permit this linkage to occur, circulating readily at this and several other moments of connection.

In particular, through a chain of dyadic connections (media and media producer, curator and laboratory, advertisement and advertising consultants, child and interactive display, etc.), the revisable phantasy of a living world of the past, experienced in the lived world of the present, has been a common proposition as a performative resource. The shiftable, individually particular material-phantasy of the Mesozoic is distributed across these various dyads, emerging in utterances; in technical action in the laboratory; and in animated reconstructions – imaginings of dinosaurs, of the fossils themselves. As I have said before, it is this chronotopic phantasy that is the animating tissue of each dyad, allowing a relatively shared performative engagement. This is something more than what Bruno Latour refers to when he speaks of the "thread of networks or practices and instruments, of documents and translations" which can connect and so collapse the most local and the most global.[30] These various Mesozoic natures as they are performed, imagined, and materialized also have to be taken into account, for they too have agency in the "thread of networks" to prompt certain articulations rather than others.

A case illustrating the articulating effects of the performative Mesozoic encountered in material practices was offered directly by Andreas Henson, who told me how, since he was a child, he would imagine worlds and scenarios when reading palaeontological texts (and *any* text, for that matter):

> When I was four years old, I got my hands on a copy of Zdenek Burian's *Prehistoric Animals*,[31] and I thought this was just the coolest thing … thinking of these completely different worlds and these times far back in the past … I realized this was all based on these very fragmentary remains and that intrigued me … this was something where you had these rare and precious objects that then helped you in your mind to reconstruct these ancient worlds …
>
> … When I read books then, as I do now, it's basically like a film moving before my eyes … I actually visualize the texts in visual images …

With Andreas Henson's cinematic conjuring in mind, the circulations by means of phantasy and materiality can be rendered equally well in the dialogical terms of Bakhtin – where the *reader* is in dialogue with the *word*, encountered as text.[32] Images, narratives, and phantasies become palpable as the reader exchanges with the text, at the moment of her/his engagement. Here, what Bakhtin refers to as "the word" is replaced by the reading material, images, an exhibition, which also simultaneously constrain and enable what may or may not take place. In thinking of that dialogue historically – as many dialogues taking place over time – Bakhtin writes:

> There is neither a first nor a last word and there are no limits to the dialogic context (it extends into the boundless past and the boundless future).[33]

In the case of the Maiasaur Project, all those involved with its articulation – the visitor in the exhibit, the television producer making a media story, the scientist conceiving of an in-gallery laboratory, the digital media producer generating the apparition of a running-dinosaur computer graphic, and the child experiencing an interactive display – are each engaged in moments of dialogical interchange. As a proposition, that interchange may be thought of as a zone of exchange, or perhaps a zone of *intra-action*,[34] between these various persons and the various machineries of communication, exhibition, and technical practice along with the litany of meanings they entail. At each interchange, a transaction and a measure of translation take place. Still, what both precipitates and circulates in that zone are the current iterations of Mesozoic politics-natures – resolving as a figure, a picture, a relationality, a sensibility of "nature," a "world," and the means by which one encounters it. This materialized phantasy continues to be transmitted and circulated robustly across the multiplicity of intra-action zones. From the transitory moment of its enunciation by this child – " a really big Jurassic place" – multiple registers of differentially articulated assemblages are brought into some degree of coordination, and into further articulation. Indeed, the Maiasaur Project becomes one such assemblage, linked in turn to ever-wider networks by that now widely distributed performative nexus, the Mesozoic, along with the chronotopic narratives it inaugurates – and has done since at least the early part of the nineteenth century when it was first imagined and named by a geologist.

Negotiating with Competing Fetishes

The second summary point is on how the performative entity of the Mesozoic underwrites the emphasis of the high-tech multimedia experience, and simultaneously backgrounds or obscures other experiences and articulation possibilities. The extraordinary, recurrent attention given to the materialized environment or world where dinosaurs come to life, and where humans get to experience their resurrection virtually, immediately places emphasis on world-making, a principle constituent of modern spectacle. The difficulty for the curator was in providing a technical spectacle equal to the world-making, multimedia spectacle which was so reinforced by Mesozoic performativity. The multimedia components came to be so over-determined in the marketing campaign, in the press coverage, and in the exhibition itself as to make that challenge all the more difficult. In effect, the "dynamic process" of the curator was at continual risk of being subordinated by this thoroughgoing and highly deliberate multimedia fetishism. A tension between what could be called *the fetishism of spectacle* and *the fetishism of specimen* was set in motion – or continued, when considered within the longer historical frames already contoured in this book. To compete in the leisure attraction sector, the Maiasaur exhibition was developed and positioned as a new multimedia spectacle, and as a move beyond the traditional fetishism of the museum, the fetishism of the specimen. The specimen ROM #44770 was gradually being obscured amid all these efforts to produce the right combination of spectacles.

I use the term "fetishism" here in the sense of both the Marxian commodity (what Marx called *Fetischcharakter*)[35] and the Freudian fetish, for that is precisely how other marketing campaigns at the ROM, as well as that for the Maiasaur Project, have been conceived and managed. In other words, fetishism was not a means of denunciation, but rather was being put to positive work here: mobilizing desire to make the exhibit *more* of a commodity. The aim of such campaigns was to identify an "audience market," imagine the tastes of that market, and highlight those features or items within the exhibition which exemplify that connection between imagined desire and identified audience market. As an example, in the ROM's Victoria and Albert (V&A) show, "A Grand Design" (which profiled selected artefacts of *manufacture* from Britain's Victoria and Albert Museum), even the object itself was obscured by the fetishist identification of recognizable human figures: the media campaign used the names of a famed artist and two contemporary haute couturiers (Leonardo da Vinci, Christian Lacroix, and Vivienne Westwood) to stand in for the objects, which didn't even have to be identified.[36] In the case of the V&A show, the emphasis was that of out-and-out, high culture artefact and commodity fetishism. In contrast, the Maiasaur campaign worked with technological, multimedia fetishism, *Jurassic Park* fetishism, and fetishism of the new in the form of interactivity which would displace "outmoded," didactic, and object-oriented museum display practices. In the case of both exhibits, the marketing staff and exhibit developers would select and embellish the fetishistic element or entity that cut across and strengthened the marketing flow.

This, then, leads us to our next question: how much do these fetishistic amplifications of the Maiasaur exhibit – from the promotional visions of the Marketing and Communications Department to the feature program of CityTV – correspond with the exhibition as it was planned or as it ultimately came to be executed? Well, to leap slightly ahead, I can say they corresponded closely in relation to the exhibitionary media deployed, and hardly at all in relation to the planned "story line" of the exhibit – which, as I discuss in the next chapter, keyed in on the "good mother" thematic of *Maiasaura*. The marketing coordinator also told me that this theme was not where they felt their efforts to capture an audience would be best invested:

> [W]e were focusing more on the active nature of it ... the "good mother" is a nice story ... but I don't know if it's as appealing as "Come in and punch buttons and roar and watch this thing run across the screen and make it do these things"...

The marketing and promotions now worked to background not one, but two elements of the exhibition: the interactive multimedia appeared to be superseding the laboratory and its dynamic process, while the experience of interactivity and virtuality appeared to be superseding the "nice story" to be told of the "good mother."

But the marketing coordinator's purpose is clear enough: to get visitors through the door. What these visitors would encounter once in the gallery itself is yet another story, and one that enacted still different articulations.

To this point, I have presented only a very partial introduction to the Maiasaur Project exhibition. Here, the questions begin to multiply. How might connections to the story of the "life and times of a dinosaur" be achieved when what is being conveyed is as much about the media experience itself? What would the role of the specimen be in rescuing these connections? Might Horner and Makela's "good mother lizard" make it back into the dialogues, and how important would that figure be in the outcome – and for whom? Did the laboratory actually work as "the centrepiece" for others beside the curator? What sort of articulation was to be made between the lab and the animations? How much did the concerns of the many co-developers, designers, writers, technical staff, managers, and visitors come to reside in or interplay with the outcome?

While I return repeatedly to these questions and others in the coming chapters, let's turn now to those whose job it was to ensure that the right articulations were made in the exhibit in its relations with anticipated visitors. Now that we know what the curator was up against in his trade-off with market forces, forces that might simultaneously threaten and enable his dynamic process, it would be the planners and the tools of their trade that could ally to the concerns of both the specimen-wielders and the spectacle-makers to generate a more seamless continuity across the oppositional arrangement faced in the Maiasaur Project.

11 Need to Say, Need to Know: Planning to Articulate Specimen and Spectacle

There were two phrases in this department that we picked up a few years ago from the plain language movement in government ..."need to say" and "need to know" ... I thought, "That's it!" Sometimes there is perceived here to be a huge gap on the scientific, curatorial side of this. My side of the museum is concerned with both. Are we driven by the curator's "need to say," or are we driven by the "need to know," which means the audience, what *they* need to know? With a lot of the curators, they know they are working in a museum ... they are well aware of it and will tell me *why* they are fascinated with this stuff ... about copper, which is a recent example ... Some of them are just naturals. If they were all naturals, I would not have a job. No, I would still have a job because it is a lot of work ... relating "need to say" and "need to know"...

Jennifer Ross, interpretive planner for the Maiasaur Exhibit (March 1998)[1]

"Let Us Talk about Soup!"

This chapter and the next consider dimensions of the specific planning process for the Maiasaur Project, while at the same time beginning to draw in discussion of the exhibit as encountered by visitors in the ROM galleries. I introduce aspects of the planning work with some comments from the interpretive planner for the Maiasaur Project, Jennifer Ross. Ross was charged with consulting closely with the curator, researching the content, developing the storyline, coming up with an overall exhibit coherence, and ultimately preparing the textual elements for the various display components. What follows is the text of a conversation between myself and Ross in March of 1998, where she offered a sense of the chasm of translation that often confronts folks involved with exhibit planning situations. This translation is emblematic of the conditions of public exhibitionary practice in those late twentieth-century public museums in cities and nations of privilege, where curators and exhibit developers must find the means of collaborating in the development of exhibitions of nature, culture, and otherwise. I began with a question related to the Maiasaur Project, but Ross quickly turned to another matter – soup:

BN: It strikes me that the lab was the location where you could see the real specimen and the real equipment, a live technician … this stuff was being revealed at the moment, you could experience it. Is that a guiding philosophy in development of exhibits? To present, to give that sense of the authentic? Or is it to tell stories? Or what else is it?

JR: Sometimes it is both. Sometimes they are at war with each other. Particularly, I am thinking of the decorative arts, which is a whole other type of work. Its specimens, its objects, its artefacts – frequently, the classic conflict is the curator wants to just put the stuff out there so that people, their kind of people, like collectors, can come in and say, "Wow, that was made by So-and-so …" … You can get a real thrill out of that if you are into, you know, pottery or something … On the other hand, people in my section tend to say, "There has got to be more to it than that."

… My favourite story is about an exhibit of soup tureens. This was a travelling exhibit. Campbell's Soup owned an incredible collection of soup tureens. Each of them required a case *this* big [*gesturing with hands*]. These were ultra-lavish things. The curator had organized them … into medieval – you know, pre-medieval, and early stuff … Baroque, Rococo, and neo-classical and Regency. He organized them … and that was *it*. And he was going to say who had made them – a very important silversmith, or a very important ceramics factory, Limoges, or something … what kind of decorations were used, whether it was gold chassis or cul-de-bleu, and that was it. I said, "We should say something about what they were eating out of them, what soup, what was put in these things." It just stopped him in his tracks. He said, "I don't know anything about what they were eating!"

… I thought, "Woah!"… this huge gulf here, where I saw something – was it actually served at the table, was it used at the side, did they actually use these things? You know, the way people have fancy vases. Did they actually use them at all, and what were they eating out of them?

Why were they so fancy? How *did* they use them? And he was looking at them strictly as a connoisseur, and I thought, "How do you bring this together in a harmonious fashion?"

... So sometimes the object's authenticity is seen to be the draw. And sometimes you think – that is not enough! That is too limiting, people deserve to know more. The general sense in my whole section, and certainly with [the managers] Lydia and Wendy, the view is, "Let us talk about soup!" I kind of leaned on him. I was nice, but – soup! That is kind of a classic break sometimes.

BN: So, you guys here tend to be animators, you are animateurs?... to bring things, in a sense ... to revive them?

JR: Yes, and make them relevant. That kind of sense – "Why should they care?" What about your twenty-year-olds? ... It's fine if they're here for a date. What about these people who are *not* connoisseurs? They still come here. They deserve more – we could tell them more ... make this kind of neat without lecturing to them.

The mission statement I heard consistently in discussing the planning of the Maiasaur exhibit with many staff in the Interpretive Planning section of the ROM – notably an all-female staff, in contrast with the predominantly male curatorial and exhibit production staff – was that of making museum objects *relevant*. In the case of the tureen exhibit, relevance was created by animating the tureens with the missing soup. In this way, relevance and animation were intimately related. The object had to be fitted or located into a larger context, and it was best if that context had something to do with the experience of the visitor, to make it "kind of neat without lecturing." Twenty-year-olds eat soup, too, even if not from ornate, Baroque soup tureens. By bringing soup back into the action, the interpretive planners activated the bowl from which it was eaten, *lowered* it from precious thing to active thing. The bowls which twenty-year-olds did eat from would be conjured – indeed, perhaps even the Campbell's soup they ate from those bowls – as these contemporary equivalents contrasted with those in the exhibit. The Campbell's Soup company benefited from sponsoring this exhibit, not simply because a competent ceramic artist made beautiful vessels, but more powerfully, because people then and people now have to eat soup out of something: this was the marketing advantage to be had. The logic of Ross's project was clear: *animation* is an important route leading to *relevance*, and *relevance* connects visitors to the things in exhibits (and to sponsors). Fill the bowls with soup! – here, then, was the rallying cry to practise articulation well.

In the case of the Maiasaur Project, the curator quite clearly came with a sense not merely of how to fit the fossil specimens into a systematic typology or a natural historical trajectory, but also of how the bones could be articulated, how they could be fleshed out, how to revive them, to imagine them in an entire world moment. The curator had told me that he could visualize the creature as though it were living and interacting, even when reading a book or when examining a diagnostic fragment. However, as I have pointed out in previous chapters, recall that this exhibit was, in more ways than one, a *pre*-animated thing. Much of the story came from achievements of palaeontologist Jack Horner, and Andreas Henson was a curator who had some clear ideas about how to mobilize those achievements.

Official Statements: Unruly Circulations

In this chapter, I refer to the official documents and planning tools that were developed through communication between the curator and the exhibit-planning team of which he was a member. These are the documents which change hands between the curator, the technicians, and all the exhibit planners, designers, managers, builders, etc. – all of it in a sometimes coordinated, sometimes seemingly arbitrary way. I present these "official statements" as particularly forceful steps in translation, contouring how the dinosaurian natures of the Maiasaur Project came to be configured. They led to so many other actions that redirected the project's configuration, both in the making of the exhibition and in its audiences' engaging of it. The enrolment of technical, administrative, and display instruments, plus the networks of other palaeontologists, is also signalled through this and subsequent descriptions. The temporal period concentrated upon in this discussion is that from 1993, when the project was first conceived, through to early 1995, when the plans guiding physical production were completed.[2] Some limited reference to subsequent actions – such as production, installation, visitor engagement, physical outcome of the exhibit – help to anticipate some of the effects of these earlier translations. In addition, I include some specific retrospective commentaries on the display's "Working Lab" in operation, which aid in demonstrating its crucial effect in the ultimate exhibitionary outcome.

The approximate process of developing an exhibition was described to me by research and planning staff in the Exhibit Programming department. Jennifer Ross described how the department was involved in one stage of this process, the development of the first key document for an exhibit, the *exhibit brief*:

> Within our department, once an exhibit is approved and is going to go ahead, they produce – this changes all the time, of course – an exhibit brief. It answers lots of questions ... "Who will produce it? Do we have the staff? When do you want this thing to open? How long do you think it will take to do it?" It's not so much, "How much money do we have?" but rather, "Where does it fit into the schedule?" So – it only comes to me – when I get the brief, it says: "This many square feet, here is the budget, here is when we expect it to open." Then all the other departments start kicking in. The media relations, they need a long lead time to start building up a campaign and letting the papers and the TV stations know. They need time to plan their material.

The exhibit brief would outline, in a general way, the content, interpretation, design parameters, marketing plans, and costs. Documents such as this become potent tools in making connections work, bringing ideas and practices from disparate places into proximity, and also as instruments that connect with other players, other documents.The brief was typically developed in advance of the actual interpretation and design work, and in this instance it was developed on the basis of a "curatorial proposal" where, as Ross noted, the "curator will submit a two-page proposal that says, 'Here is what I am thinking of doing, and here is why.'" The brief for the Maiasaur Project apparently had its own idiosyncrasies:

> With this one, I don't know the politics of it that well. By the time it got to the exhibit brief stage, certain decisions – and this is very unusual – about how we were going to present information were already decided. In this case, it was as if some of the high-tech stuff that was highly promoted by the advertising department – some of that had already been decided by someone who is not here anymore. His specialty was real high-tech computer interactive stuff, and he was really keen on that. I think that he had been talking – I am sure he had been talking to Andreas beforehand, and saying, "If you were to do this exhibit, it would be really cool if we

> had this gigantic dinosaur, and these other things where we could punch buttons on it." So it was kind of weird for me to get what is called an exhibit brief where they had already made so many decisions!

The implication that the multimedia components were not simply of marketing and promotional value but had been pushing this exhibit forward from the start was becoming increasingly apparent. Indeed, it seemed that this former media-specialist staffer and the curator had formed a prior strategic alliance to move all of this forward.

Once the brief was in place the interpretive planner could begin her work:

> The brief came with Andreas's curatorial proposal ... he is a very organized guy, very thorough ... I was the exhibit interpretive planner. My responsibility on any project like this is to go over the material that is being proposed to be in it. The curator might be saying they are going to put all these skulls in and they want to have a skeleton, I want to have this and that. I would usually say, "I think you might want to have a few more skulls," or "I think it might be better not to start out with that stuff. Why are you doing this?" Ask questions.

The interpretive planner, then, had this particularly crucial trading role with the curator, and with the entire sequencing of the exhibit development process for that matter. The flowchart (figure 11.1) recreates the document flow in the development process from the time the curator supplies a proposal to the initiation of design and production.

I asked Ross directly, "Where are you in relation to all this ... are you with the curators or the design team?" She responded,

> I am in between. I go back and forth, like the messenger. And the design team cannot start working – they will say, "Where is the plan?" They need it before they can start making those decisions.

From this position, the interpretive planner was also well backed by her exhibit programming colleagues, and a great deal of moving of documents back and forth takes place in the revising process. Notably, this group was charged, more than any other in this project, with attending to the matters of audience definition and responsiveness. Much of this, as I found out, was based on some generalized museum visitation statistics, combined with an anecdotal sense of who the probable

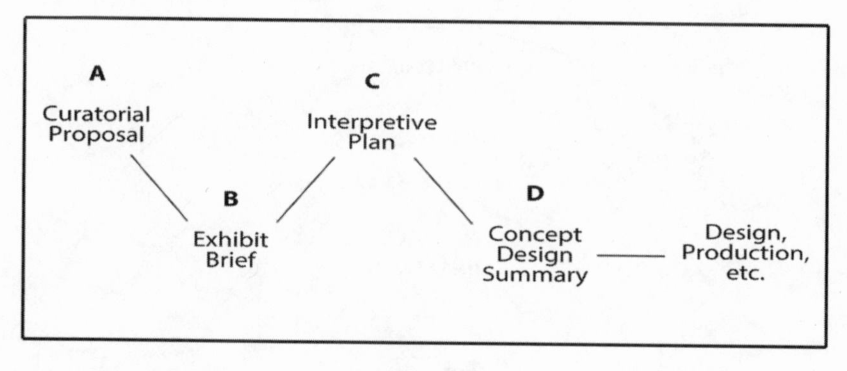

Figure 11.1. Reconstructed flow of official planning documents for the Maiasaur Project, as outlined by those interviewed, schematic.

visitors were and how they might react to different sorts of media and content.

Leashing Down Henrietta

Jennifer Ross remarked on her own approach:

> I sometimes say I am an "audience advocate." My responsibility is making sure the scientific information that is being proposed for the content is organized in a way that is best matched to the way the public might take it in.

When asked how she would achieve this goal of *organizing scientific information* into the *content* which visitors could *take in*, Ross presented me with a diagram (figure 11.2). The depiction showed her pivotal translation of the curator's original statement, preserving his emphasis on the centrality of the Maiasaur specimen, which she now refers to as Henrietta:[3]

> I will show you what I eventually ended up with, as a result of looking at all this stuff that we were given. I came up with this little chart. This was a draft of an interpretive plan … the central idea is this particular dinosaur, this particular specimen.

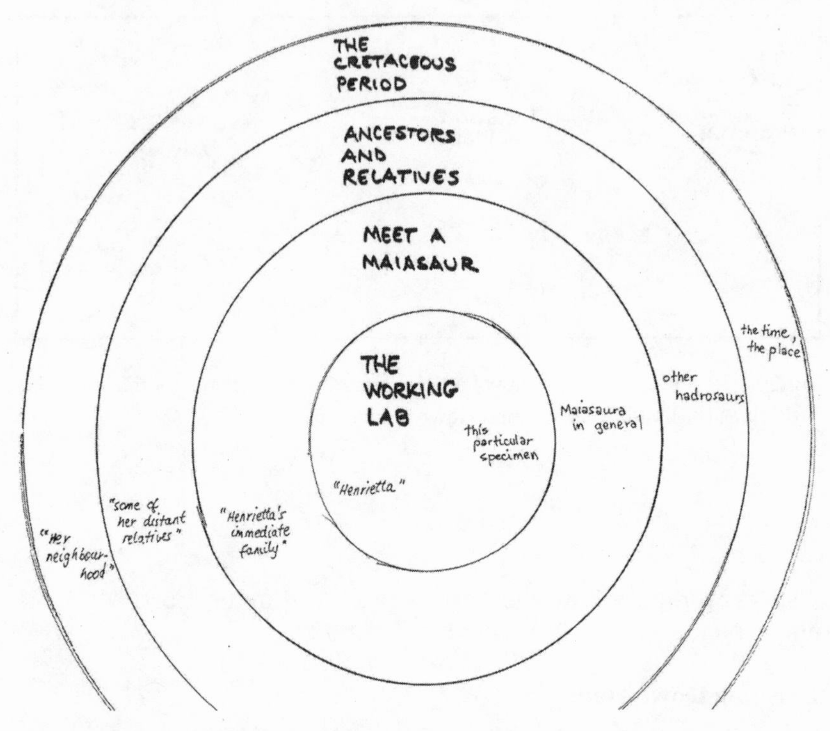

Figure 11.2. Jennifer Ross's concentric "leash-down" "chart" for exhibit planning (articulates time-space, vernacular/technical conceptions, specimen/spectacle). Source: Illustration reproduced with permission of the Royal Ontario Museum, © ROM.

… This was, I said, "Everything that is proposed to go into this exhibit – it was like we have a post in the middle, and there is a leash on it, and there is a limit to how long I will let this leash go." This specimen and that, you can only see it in the lab. [*pointing to diagram now*] Then there is a section – these are actually sections of the exhibit … – but the focus [of each section] had to be on this particular specimen, all the time, trying to relate it back to why are we interested in this particular specimen. So it was not going to be an exhibit about *Maiasaura*. It was an exhibit about this – the finding of it, all this stuff.

> ... The most distant one was information about the Cretaceous period. I just said it was her neighbourhood. It was to help visualize it, and so on.

For interpretive planner Ross, everything had to have a tie-in to the specimen, Henrietta – this one diagram was the working document that provided the fullest, simplest sense of the conceptual continuity for the exhibit. Even the worst-behaved dog would be kept close to home base with this leash – or, at least, that was the sort of power that Ross hoped this depiction would have. All of this was part of her effort to keep participants well connected to the specimen: visual and sound designers, construction teams, curators, technicians, visitors. In this sense, the term "centrepiece" – which Henson would come to use in media interviews to describe the specimen at the time of the exhibit launch, and in speaking to me four years after this diagram was drawn – had a very particular and important history. Moreover, here was a sketch that captured much that I have posited throughout this work: the collapsing of time and space, vernacular and technical conceptions, and the linking of specimen to spectacle, scientist to audience, phantasmatically selected matter to phantasmatic rematerialization.

It is also notable that, while on one hand Ross had preserved the curator's central priority on the lab and specimen, she had also begun to translate matters effectively from technical and impersonal terms to those very personal ones – terms which, presumably, she felt the gallery visitors might find most captivating, most relevant. The specimen (ROM #44770) had also obtained the name "Henrietta" from Bearpaw Palaeontological, the company which collected the skeleton. Mirroring meteorologists and their hurricane-naming practices, Bearpaw Palaeontological assigned an alphabetical name to each of their dinosaur specimens, in the order of the find over time. Henrietta, then, was the eighth dinosaur they collected; their intention was to shift to masculine names once the first alphabetical sequence was finished. The connection to the gendered name *Maiasaura* is coincidental, but that coincidence, all the same, only serves to reinforce the gender identification here.[4]

The name Henrietta was used regularly in office-planning documents up until the January Concept Design Summary in 1995, by which time the nickname had faded out of usage – to the relief of several people in Exhibit Planning who were concerned about being "scientific." By then, however, the specimen was being gendered, not just through its scientific name, but indeed through its everyday, playful identification, which had been well established. This gendering did not have grounding in palaeontological analysis. The curator had noted to me that it is very difficult to identify the sex of dinosaur skeletons, and that no such identification had been made with this one.

Following Ross's plan to relate all of the exhibit back to the specimen, the dinosaur genus *Maiasaura* was entailed as "Henrietta's immediate family." Other kinds of hadrosaurs would be "her distant relatives." The time and place of the Cretaceous was "her neighbourhood." Aligned with each of these translations was the display component which would contain the elements and the translations. This depiction of concentric circles illustrated, in as direct a fashion as one could ask for, the process of translating bones in the ground from specimen to spectacle, technical to vernacular, content to audience, a singular form of dinosaur life to a living world – and a very familial one – in time and space.[5] In creating this well-integrated diagram, Jennifer Ross launched a set of translations which corresponded at each step with many of the curator's intended translations. This articulation was what would ensure that the otherwise inanimate fossil of a potentially "boring" plant-eater would be animated, and so made relevant – soup.

It would seem from this that planning processes surrounding the specimen, the scientific practices, the lab, and the curatorial intent in general were more or less secured. However, once again, Ross raised the point that this exhibit upset expected planning and development roles:

> It is like an editing job, in a way ... it is an intellectual editing ... sometimes it is organizing the intellectual content of it ... [except] this show was atypical. By the time Maiasaur came to me, it had all already been decided. We were going to have this video and that video, and this. I went, "Umm? ... oh!? ... okay." Things are getting weird these days. We are doing things so fast that we don't produce documents like this anymore.

So, Ross laments the limits being placed on the power of articulating documents, witnessing how the smooth process of planning the content of the exhibit had been disrupted, a point confirmed by many others with whom I spoke. As in the Marketing and Communications department, the enacting of the curator's intentions and visions was considered to be a serious objective in the planning process. These, in turn, were translated through the exhibit brief, via diagrams like that prepared by Ross; then through the interpretive plan; next, through the design; and finally in the production management phase. In this phase the exhibit would be transformed from imaginings on paper into two-dimensional, three-dimensional, spatial, visual, and electronic forms. What I gathered had become most disruptive to the regularity of this process, however, was the degree to which the new interactive multimedia was garnering increasing amounts of attention, especially in relation to the budgeting for the project. One of the Maiasaur exhibit designers pointed out:

> The interactive component was a done deal before we even began our design ... A lot of people had their backs up over the funneling of so much money to the multimedia ... whole components were cancelled ... A major graphic component on the research was just dropped ... It was basically about field research, how and why the [fossil] matrix was prepared ... the budget wouldn't allow it ...[6]

The fissures between the multimedia and the scientific-technical components were beginning to show in the planning of the project, and continued into its design phases. Some pointed to the "'just get it done' attitude" of the project manager (who was in charge of production schedules and budgets), or to lofty arguments occurring at the level of senior management. Prevalent here was the idea that, somewhere in the decisions made by management, the alignment of resources with the multimedia components had taken increasing precedence, even to the point of eliminating components which would have complemented the curator's initial hopes of presenting a "dynamic process." I am reminded, however, that the curator's exact words were a "sense of a dynamic process." So long as the lab was retained, that sense could presumably be achieved – but many of the planned articulations were rapidly or slowly being cast off.

The official exhibition content statement was eventually summed up and presented in a subsequent series of drafted and redrafted documents ultimately presented under the provisional title of "Concept Design Summary," and was meant to guide the entire exhibit development team.[7] The role of this document was to hold all team members together in a common effort towards a somewhat understood, if still contingent, palaeontological nature. The package I work through here was the final version (January 20, 1995). The document was provided to me on March 3, 1998, along with several earlier drafts of the "Exhibit Brief" and Ross's diagram of concentric components. Attesting to the power of documents, the Concept Design Summary had been distributed by the associate director for public programs to no less than twenty-four individuals ranging from the museum director through to staff and managers in eleven different departments or sections of the museum.[8]

The "Interpretation Statement," which was drafted by Jennifer Ross – though it had been modified by the project managers before its final form – describes the projected experience as a total, including the storyline, media used, major sections, and communication goals. Other sections were drafted by respective design and media specialists. In the discussion below, I draw upon this document, the earlier "Exhibit Brief" documents, and drafts of the "Curatorial Proposal" (first draft December 3, 1993; revised draft March 1, 1994)[9] written by exhibit curator Henson. My main source, however, remains the Interpretation Statement, as it would prove to be a key nexus-defining statement for the entire program – the statement that could do the most to rescue Henson's dynamic process from the mess of ostensibly careful planning aimed at suturing specimen to spectacle to visiting publics.[10]

Enrolling *Maiasaura*'s World: "Summary of Interpretive Approach"

Aside from a "way-finding" system to orient visitors to the exhibition within the museum, the exhibit was to have an introduction and three major "sections," each with several display "components." The most spare, concise summary of the exhibit's material and intellectual content was the framework that was presented in the Interpretation Statement of the Concept Design Summary. It read as follows:

ORIENTATION SCHEME
- Name and Subtitle, Brief Concept, Dates
- Where to Find It

INTRODUCTION TO EXHIBIT
- What the Exhibit Is About *(graphics; scale model)*

THE CRETACEOUS PERIOD
- 1a, A Visit to the Cretaceous *(animated video)*
- 1b, Plants and Animals *(exhibits – skulls & fossils)*

MEET A MAIASAUR
- 2a, Impressions of Maiasaur *(huge interactive video)*

THE WORKING LAB

- 3a, The Palaeontology Process *(lab; video; exhibits; hands-on specimens)*
- 3b, The Story of This Specimen *(video)*
- 3c, Progress of the Work *(diagram)*
- 3d, Skull Specialization *(interactive morphing video)*[11]

In the most general sense, at least, all of these sections and their noted components would be completed in the final exhibit. Also contained in the Concept Design Summary was a technical drawing of the floor plan for the exhibit (figure 11.3).

The Interpretation Statement noted three goals for the exhibition: (1) communicating "a wealth of existing knowledge about Maiasaur and the times it lived in"; (2) stimulating "interest in Maiasaur by showing an actual specimen in the process of being cleaned and assembled"; and (3) "dramatically" evoking "a sense of the once living animal." Notably, any necessary set of steps articulating the active preparation of the specimen on display, and the other knowledge and dramatizations of the animal being displayed, were left unstated in the Interpretation Statement.

The Statement identified "the audience" very roughly as "family groups containing children between the ages of 4 and 13," with no other specific reference to what a "family group" was. Some expectation of how visitors would "behave" was noted more lucidly in the Interpretation Statement:

> [W]e can expect visitors to be in small groups, either family or friends, and to use the exhibits in a conversational way. They will point out things of interest to each other, read exhibit text and explain things to each other, ask each other questions about what they see and hear. Constant interruptions will be normal, as will random sampling of different exhibit elements.

Gender, ethnicity, national origin, socio-economic status, and any other indicators of individual identity were fully unmarked in this homogenizing, universalizing description of who would attend the exhibition.

That said, the film *Jurassic Park* was used as an experiential standard for considering how to present *Maiasaura* to the visitors/audience, cueing to anticipations of contending with a Hollywood movie-going audience and its expectations. The Crichton and Spielberg productions had impacted deeply on the planning process. Playing forward the recirculating matter discussed in chapter 6, the Interpretation Statement noted how the film had the effect of

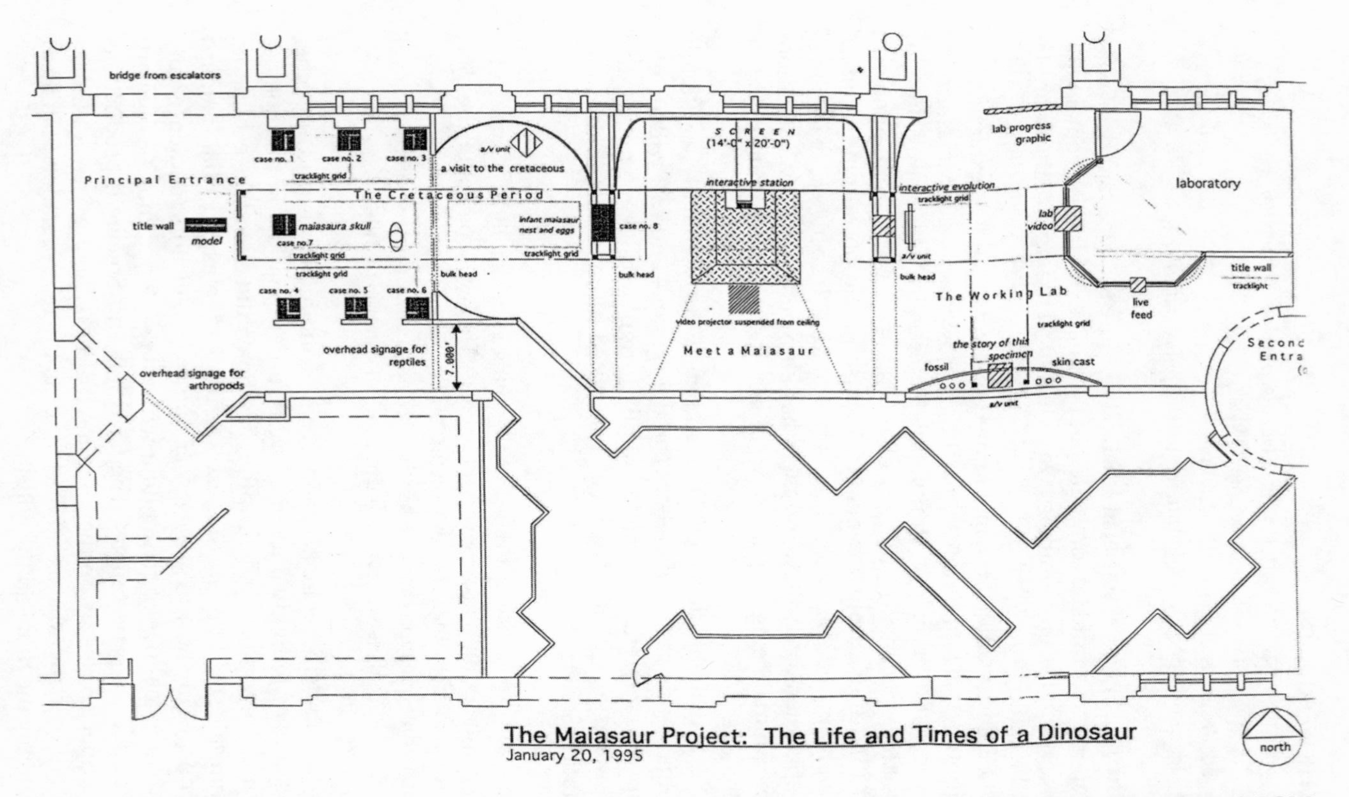

Figure 11.3. Exhibition floor plan. Source: Illustration reproduced with permission of the Royal Ontario Museum, © ROM.

> raising the public's expectations of special effects and dinosaur animation. Because *Maiasaura* was not a fearsome bloodthirsty hunter – it was a herd-living browsing plant-eater noted for the care it gave to its young – it is not appropriate to overdramatize the presentation.

The contrast here with a "fearsome bloodthirsty hunter" – highly charged language right out of the discursive lexicon of frontier dinosaur-collecting and the phantasies of Henry Fairfield Osborn – is quite telling. The killer dinosaur was being used as the benchmark for considering how *not* to present "a herd-living browsing plant-eater noted for the care it gave to its young." Quite obviously, the concerns were directed at the power of technologically sophisticated animation media to generate something which could easily "overdramatize" the presumably less – but nonetheless somewhat – "dramatic" life and behaviour of this particular dinosaur.

The implicit challenge, drawn from the curator's concerns, was how to counteract the dimension of excessive spectacle in the dinosaur-as-killer that had been generated historically through the mass configuring and replaying of dinosaurs like *Tyrannosaurus rex* and *Velociraptor* as "fearsome bloodthirsty hunters" in the dioramic, theatrical spaces of museums and popular movies. As noted in earlier chapters, Osborn, Knight, Willis O'Brien, and Hollywood had long ago deployed, popularized, and stabilized such images. This left a predicament for future museum curators and exhibit developers. Finding alternatives to the fearsome dinosaur that could still draw an audience was one of the recurrent challenges in this project. A distinction can be drawn here between two performative apparatuses, representable by the equations [Hollywood + imperial science + colonial adventurism + dramatization + bloodthirsty killer dinosaurs] versus [ROM + authorized science + nurturant, plant-eating dinosaur]. Part of Henson's "dynamic," one could argue, was in the movement across these two set-ups of politics/natures.

Henson had stated the communication goals soberly, with similar interpretive restraint, in his first proposal to develop the exhibition by means of a set of simple yet highly loaded questions which the exhibit would answer – or at least pose.

On comparison, it becomes fairly clear that the interpretive goals and the physical exhibit sections had been shaped over the following year into responses to each of Henson's questions. I have summarized these select yet crucial translations in the following table:

Table 11.1. Planning translations of the curator's intent

Questions from the curator (Dec '93)	Response – interpretive goals (Jan '95)	Response – interpretive sections (Jan '95)
What were they like? Dinosaurs as living animals (behaviour, feeding, locomotion, growth, social structure)	Evoking a sense of the once-living animal	"Meet a Maiasaur" displays
Where did they live? Interpreting the world in which the dinosaurs lived	Communicating knowledge about Maiasaur and the times it lived in	"The Cretaceous Period" displays
How do we know? Retrieving information about the history and past diversity of life on this planet from the record of the rocks	Showing an actual specimen in the process of being cleaned and assembled	"The Working Lab" displays

While there is an apparently striking correspondence here, even a seemingly elegant coherence, these are only general, paradigmatic associations. Each translation, even here, already signals significant transformations: "how do we [i.e., museum scientists] know" is far more complex than a specimen "being cleaned and assembled," and that in turn is but one limited set of actions that takes place in a working lab. Some of the more elaborate translations that would be enacted were already being motioned to in Ross's circle diagram, with its "families," "relatives," "neighbours," and "Henrietta." That diagram was the first draft "interpretive statement" to follow the curatorial statement, and it preceded the official Interpretation Statement in the Concept Design Summary. Even by the time the latter appeared, no physical display existed *per se*, and between January and June of 1995 an additional set of phantasmatic transformations from textual statement to physical display would yet have to be achieved. As I will be discussing in subsequent chapters, a great deal more *and* less came to reside in the outcome, and even more so in the ways that the exhibition would be engaged and interpreted by visitors.

The Curator's Public/Scientific Strategy

Notwithstanding the multiple translations which would ensue, the curator's central motivation hung on the stressed point of mobilizing around the expected predominance of the "killer dinosaur" in the public imaginary. Henson would later explain to me how the purpose of the exhibit was not merely to counter that expectation with the opposing example of plant-eating dinosaurs, but also, in presenting a creature which displayed nesting and other sorts of social behaviour, to promote a new appreciation of the complexity of dinosaurian life:

> It becomes useful because you start thinking in a sophisticated fashion about dinosaurs as once-living, complex animals that did a variety of things. You similarly start thinking of nature as a complex thing ... You know, nature's not black and white ... There are not "bad" animals and "good" animals ... There are not primitive brutes, and you know, a bunch of sophisticates ... It's the first impetus to think of the world around you in a more differentiated manner ...

In this hopeful, responsive strategy, curator Henson was acknowledging the conventional nexus, yet seeking an alternative to it. He continued:

> I think that what came through was that you have to think about dinosaurs in a different way – dinosaurs are not just these lumbering vicious brutes; dinosaurs were part of an environment, an ecosystem; this ecosystem existed in a world very different from today ... and all of this has a historical component – witness the exhibit where you can change an Iguanodon skull into all kinds of other saurian sculptures ...

Dinosaurs, in this new, historically shifting, "ecosystemic" scenarization of the Mesozoic, were becoming much more highly "differentiated," plural, and able to coexist in conditions of biotic differentiation. This was a significant turn from the "survival of the mightiest" logics of Osborn's eugenics imaginary. Osborn had enlisted the fossil specimen of *Tyrannosaurus* into his Mesozoic world conception, and in doing so provided a case for his contemporary, eugenic, human-world logics – his "agenda for antiquity," as Ronald Rainger called it.[12] Correspondingly, Henson was speaking in terms familiar to 1990s action in regard to rising concerns for sustainable biodiversity – not an "agenda," per se, but certainly a contemporary concern of the moment.

While the concerns shifted, as did the nature-politics they mobilized, here the Mesozoic had offered up an effective machinery to Henson and the ROM for capturing new resources – including human narratives and interests – in symmetry with its acquisition of newly discovered, non-human fossils. Henson explained to me how his colleague Jack Horner had, in effect, enlisted the first finds of Maiasaur fossil material into newer conceptions of the Mesozoic world, clearly for different reasons from Osborn, as one would expect:

> Well, to him [Horner], I think the significance of the find was that he realized, based on this find, that dinosaurs were behaviourally much more complex than people had been giving them credit for … and this came at a very auspicious time, because this was the time when the great debate was raging about "are dinosaurs hot-blooded versus are dinosaurs just big, overgrown reptiles," and of course, an advanced mode of thermo-regulation was seen to be intimately tied to more sophisticated behaviours, anatomical adaptations, that could not be understood in a traditional, orthodox reptilian model, and so on … so, to have evidence that seemingly puts duck-billed dinosaurs – this specific duck-billed dinosaur, *Maiasaura* – beyond what we thought was typical for reptiles was then seen as evidence favouring this new, revolutionary view that dinosaurs were these very advanced, very active animals … So to him, that was the main agenda, and that's why his discovery resonated as much as it did …[13]

The resonance at that time with this new agenda in dinosaur bioenergetics and behaviour was so powerful for Horner that innumerable international news stories followed his discovery, including articles in *National Geographic* and *Time Magazine* – as did publications in the journals *Nature* and *Science*, large grants, and, in due course, a MacArthur Fellowship.[14] Horner had captured into the Mesozoic not just some new fossil material; he had unearthed, in symmetry with the fossils, certain narrative material of a contemporary human sort: the possibilities of parental care, family living, peaceful coexistence – and an entrée into the leading-edge bioevolutionary debates of the 1970s and 1980s. This was the very sort of matter which could advance him in the mutual arenas of publicity (where interest in gynocentric sociality was gaining ground on androcentric sociality) and disciplinary palaeobiology and ethology (where newly complex, animalian social dynamics were gaining ground on narrower models of competitive sexual selection).[15]

In effect, by choosing to obtain this "large, spectacular, and scientifically invaluable"[16] skeleton of *Maiasaura* as the topical key to this exhibition, Andreas Henson was allying himself with the revised ecosystemic principles amplified by Horner's successes in order to alter public fixations on dinosaurs as bloodthirsty monsters. Through the mutual aid of phantasizing and exhibitionary materialization, the Mesozoic diversification was extended to the public, and the familial phantasies of the public were incorporated in turn into the Mesozoic, attached as they were to this new kind of dinosaur. By cautioning against the overdramatization of the multimedia animations, possibilities of over-comparison with *Jurassic Park* (which had bloodthirsty dinosaurs aplenty) would be diminished, and, it was hoped, the scientific authority of the display would be increased. The entire Mesozoic world of *Maiasaura* was mobilized as Henson's ally in this way, simultaneously offering plausible demonstrations that the ROM was involved in advanced palaeontological research on dinosaurs.[17]

Demonstrating that he made a clear connection between public interest, the new specimen, Horner's *Maiasaura*, and the revised palaeontological knowledges, Henson also felt that *Maiasaura* would be construed as a "friendly" dinosaur by most children. He pointed out:

> [A]s far as kids are concerned – and in this case we played this very consciously to the family audience, who, at that time, demographics identified as our key audience, and I think they still are – children think of dinosaurs in two categories ... there are "friendly" dinosaurs, the kind of dinosaur that has all the dinosaurian attributes, except it's not fierce, therefore it's a safe dinosaur ... and then the vicious killers. So, it's like "friendly dinosaur versus vicious killer"... Maiasaur clearly falls into the first category – it's a "friendly dinosaur," it's a plant-eater, and so by implication, as far as children are concerned a "friendly dinosaur," a dinosaur you could possibly keep as a pet, sort of like Dino on *The Flintstones*.

So, while there was a sense that the ferocious dinosaur had to be displaced from the imaginary, it was done with a recognition that there was a pre-existing binary aspect of the imaginary which could be reduced to "friendly dinosaur versus vicious killer." In this way, the existence in the imaginary of the vicious killer presupposed the effectiveness of this curatorial communication goal to extend a revolutionary way of thinking to the public. Was Henson not already, and all along, doing what planner Ross had called for: advancing a technique that articulated what he and his palaeontology cohort *needed to say* with what the visiting publics *needed to know*? Curiously, an important side note to this is that the curator continued to have an interest in carnivorous dinosaurs, an interest which reached far back into his childhood:

> Actually, I always had more interest in meat-eaters [*chuckling*] … the reason being was that, when you look at the diversity of the natural world … the meat-eaters seem intrinsically more interesting … like there's a lot more involved in being a meat-eater than being a plant-eater … because basically when you're a plant-eater, you're just this huge digestive device … you forage … the plants don't run away … you just walk up to them, you eat them … the chief problem, biologically speaking, is how to digest them … but with a meat-eater, you know, most of your food is still on the hoof … it takes complex behaviours … it takes a certain amount of planning … and it takes special adaptations to track down, subdue, and process prey … so all these extra steps … whereas for a plant-eater, there's just nothing to track and subdue, it's just process …

I puzzled over the apparent contrariness of it all – despite this preference for carnivores, here was an exhibit about a complex plant-eater, *Maiasaura*, the "good mother lizard." It was beginning to appear that the Maiasaur was a strategic default, one to which Henson could in due course turn his research activities. At the same time, this reaffirmed my sense that the binary lives in the phantasmatic terrain of palaeontological practice, where many contemporary dinosaur scientists were quite likely to concur with Henson's point that meat-eating is "intrinsically more interesting" while plant-eating "just process." Maiasaur, then, was the next best thing to a meat-eater, both for Henson's palaeontology and for the ROM's display ambitions.

Nonetheless, this all demonstrated how highly cognizant museum-located palaeontologists can be in imagining the predilections of the "public." In this way, the science itself was being calibrated to the predicted audience. Henson had told me many times that he would go into the galleries to watch visitors interacting with exhibits, and Jennifer Ross confirmed that many curators did such things:

> A lot of them are out there, talking with the families, like on "Family Sundays." They get roped into it, but some of them are keen, they love it … They know that they are working in a museum, they know that they are not working in some isolated lab somewhere.

Based on their rather commingled observation and imagining of visitors, the scientist's and planner's conceptions of the audience were now performatively realigning dinosaur life in the exhibition with that of the scientific work undertaken by others on *Maiasaura*, the "friendly dinosaur." Much as the *market* had performative force in shaping the advertising campaign, now the *audience* was enlisted into the performative shaping of this exhibition of the natural world. In contrast to Osborn and his *T. rex* exhibitionism, this was a radical departure in assembling nature, one that took seriously the place of the public in nature's constitution. This could only take place *because* the research objects enlisted into the process (fossil eggs, nests, altricial bones, and this associated ROM specimen) presented an opportunity for it. In this manner, Mesozoic dinosaur kinds could more easily be divided into "friendly dinosaurs" and "vicious killers." The materialization and the corresponding agency of that binary were intensified through this complex, historical flow of public-scientific action.

Detailing the "Interpretive Approach"

But I have digressed. Continuing with the Interpretation Statement, it offered guidelines for general "Exhibit Characteristics." While *overdramatizing* should be avoided, the Statement suggested nonetheless that the exhibit elements should be "dramatic, streamlined, easy to comprehend quickly, and they should stay close to the central theme." In addition, it noted, "People should be able to 'get into' an experience without preamble, but feel satisfied 'getting out' fairly quickly." The sense garnered here is about the temporal conditions of embodied experience in the exhibition space; it is described as a continuously fluid experience, similar to a "ride" in an amusement park, where the drama produces "highs" and "lows" of experience, where it is over before the visitor knows it, where they need no introduction but have been prepared only to be surprised, if not thrilled. It was noted in an earlier "Exhibit Brief" document that the expected rapidity of information absorption by the visitor would suit "family" audiences, "so that adults can pick out information easily and respond to children's questions." Again, the anticipation of what the audience expected would put limits on how much of the "dynamic process" of scientific techniques – in their methodical slowness – would be conveyed. The Statement continued with details for each section. Here, I discuss them in order.

"Orientation Scheme." As mentioned earlier, in addition to the physical, spatial components of the exhibition proper, the Interpretation Statement started with an "Orientation Scheme," broadly conceived to include all those actions "outside" the exhibition space that aided in "getting the word out about the existence of the exhibit and the concept behind it." This would take in publicity, advertising, and posters as the means of attracting visitors. The publicity would be complemented inside the museum with "Graphics Signage" meant "to guide people to the entrance(s) and confirm they are on the right path while getting there," such as an image of a hadrosaur's head, or dinosaur footprints painted on the floor. A secondary set of connections was to have helped visitors connect to other galleries, like the permanent Palaeontology and Life Science galleries, noting, "A cast of *Archaeopteryx*, illustrating link between dinosaurs and birds, could signal the way to the Birds Gallery for example." These secondary linking illustrations were never actually produced.

"Introduction to Exhibit." Confronting a spatial arrangement that gave visitors five different ways to enter the exhibit (all of which would have entry signs), the "Introduction to Exhibit" emphasized two "main" entrances which would concisely convey an "introductory message." One of those would include a "scale-model cast" of the Maiasaur to establish a "powerful visual image of *Maiasaura*," an "appealing landmark" that would "'slow down' the visitor long enough to comfortably take in the introductory messages." Once more, the planners were predicting not just what the messages would be, but the very manner in which visitors were expected to behave. The exhibit-shaping presumed to know the visitors as much as it did the Maiasaur that it was meant to be displaying. The introductory component centred on (1) reiterating the exhibit's concept; (2) providing a visual image of *Maiasaura*; and (3) communicating that this particular species was the subject of current work in the ROM.

Although the Statement noted the multiple ways in and through the exhibit, depending on which entrance the visitors used, it proceeded all the same to describe the exhibit in an ideal, linear, narrative plan: from the scale-model "Introduction" (of which I will be saying much in coming chapters) through to the "Cretaceous Period," to the "Meet a Maiasaur" section, ending with "the Working Lab" – the supposed "centrepiece," according to curator Henson and the ROM media releases. In actual spatial terms, the physical centrepiece in this plan of the exhibit was the "Meet a Maiasaur" section with its "huge interactive video."

"Section 1, The Cretaceous Period." After the Introduction section, what I would term the *chronotopical work* of the exhibit was to take place in "Section 1, The Cretaceous Period," the primary purpose of which was "to create a picture of the time in which this Maiasaur lived." The mixing of time and space by translating them as abstractions into pictorial vision – creating a "picture of the time" – would take place here. In the terms I have already presented, this would establish the proper performative relation of the visitor in the contemporary moment with *Maiasaura* in the Mesozoic moment of the Late Cretaceous. Display components, described very generally, were ordered under the headings "A Visit to the Cretaceous" and "Plants and Animals."

The "Visit" components were "interactive animations" which established a timeline by presenting: (1) the changing shapes and placement of the modern continents of the Earth over eighty million years; (2) a video animation of "the Cretaceous environment" using existing animation footage; and (3) an "essential package" placing *Maiasaura* in the Cretaceous world and presenting a coherent story of nesting, herds,[18] migration, care for young, the proliferation of flowering and seed-bearing plants, and questions of adaptive connections between such plants and *Maiasaura*. Complementing these interactive animations, the "Plants and Animals" components would entail an array of showcases with skulls of "Cretaceous neighbours, two friendly and one not" (signalling the curator's proposition), and another array with skulls of other hadrosaurs (i.e., Maiasaur is a hadrosaur, or duck-billed dinosaur) for comparison of "size and common elements."

It is worthwhile to interject here with some points on changes in the plan that took place between the distribution of the Concept Design Summary in January and the final installation of the exhibition five months later. According to the Statement, an empty case was to head the "Cretaceous Period" space – the eventual receptacle for the skull of the specimen being prepared in the laboratory at the opposite end of the exhibition. The literal "absence" here was meant to mark the ongoing work of the Maiasaur *Project* – an *in vivo* project, happening in real time with the visitor present. Curiously enough, this empty case element ultimately would not be produced and installed in the exhibit. Was it that this *presentation of an absence* turned out to be antithetical to the positive, visible, material presentation work expected of museums, an exhibitionary *horror vacui*? In its place, a number of smaller specimens including some skin impressions and small limb bones would be displayed in this showcase, ensuring that the materiality of the Maiasaur specimen was indexed – at least minimally. Also, an additional – and, as it turns out, crucial – element was incorporated into this hall of showcases: a "static" graphic, showing a timeline cueing "80 million years ago" as the moment when Maiasaurs lived, was mounted on a pedestal at the most central point of this first major section.

Somewhere in the production phase, a decision had been made to provide this extra redundancy for the time-orientation message – time being a potentially confounding point in a display with a specimen from the "past" being prepared in the "now," and yet being presented as a remarkably well known creature. The effort to allay visitor confusion over when this dinosaur lived – in the Cretaceous (nearer to 80 million years ago) and *not* the Jurassic Period (nearer to 140 million years ago), which had been signalled massively as an almost default age of the dinosaurs in *Jurassic Park*[19] – was expressed in the first of two draft renderings of the timeline. The beginning of the Jurassic was prominently crossed out in this layout. The final artwork would begin with the time of the Maiasaur, 80 million years ago, late in the Cretaceous Period (figure 11.4).

But there was also a major omission of an element that had been part of the curatorial statement: the matter of Maiasaurs as herding dinosaurs, and the possibility of such large-scale gregariousness arising as a co-evolutionary outcome of floral diversification at the end of the Cretaceous. In true articulator's form, Jennifer Ross drew a direct connection between these omissions and difficulties associated with the cost and technical demands of designing and producing multimedia components:

> Where are we showing these ideas? Text. [*incredulously*] That was it. [*begins referring to document*] "Lived in herds of several thousand" – but because we are doing it with digital animation, which is really expensive, and every action costs so much money – we could not show it. This idea that they lived in herds of several thousand was nowhere. It was only in text. Major weakness, I thought. "They migrated" – not depicted anywhere. "They built large nurseries" – there's something in the Tippett film, the little stop-motion animation, which we bought … sixty seconds for five thousand dollars … a bargain. We edited it down to sixty seconds and paid for the rights to that sixty seconds. Five thousand dollars. Fabulous. It seemed expensive, but it was *not* expensive. You *see* the nests. "They cared for the babies"… that shows up in the film … but it only shows one nest … again, like the animator is building these little things … he can't build a thousand of them all moving …

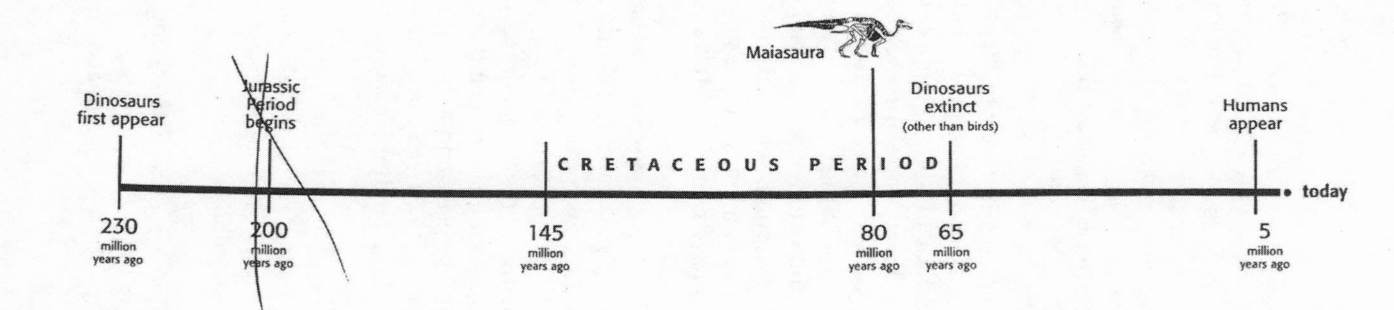

Figure 11.4. Timeline artwork for Maiasaur's Cretaceous world moment compared with human world moment (with designer's editorial cross-out of "Jurassic Period"). Source: Illustration reproduced with permission of the Royal Ontario Museum, © ROM.

> ... Now this didn't affect in any way the popularity of the exhibit ... But, as long as we're doing this exhibit, what are we trying to say? Apart from "make it run, make it jump, listen to the noises" ... What are we trying to say about this animal? The thing about the gigantic herds would have been a really cool idea to get across ... Just the thought of these tens of thousands of animals this size, ranging up and down what really was a pretty narrow strip of land here ... and *eating* ... and they weren't the only hadrosaurs, either. There were many, many. A good dozen species of hadrosaurs. We were unable to get that across and I was disappointed with that, too ...

I include Ross's lengthy comments in order to highlight the concern about scientific intent being undermined as an effect of multimedia, as well as the potency of the technical outputs to more or less rescue these deficiencies, at least in a limited way. Popularity was not sacrificed, but Henson's interest in getting folks to "start thinking in a sophisticated fashion about dinosaurs as once-living, complex animals that did a variety of things," about complex dimensions of biology, was getting increasingly diluted. More to the point, set against the visualization of the single Maiasaur nesting scene, there was a risk now of the scenario of dinosaurs as generic, "family-living" creatures to become all the more foregrounded.

"Meet a Maiasaur."[20] With the "time-travel," chronotopic action well accomplished, and with the tales of the world and lifestyle of the Maiasaur broadly contoured, this next section, "the most dramatic sequence in the exhibit," was intended to bring the visitor face to face with a Maiasaur, to illustrate "the *Maiasaura* as living, breathing animals." This was the "large-scale realistic computer animation" using a "20 foot video projected screen" and a "touch panel interactive videodisk" along with a selection of twelve twenty-second animated "behaviours." A direct quote from curator Henson was included in the Interpretation Statement:

> Get a sense of the movement and scale of the animal; avoid details that we have no way of knowing; skin texture is more reptilian than elephant-like, necks are generally poorly muscled ...

Striving to not exceed positive evidence, and staying close to "knowable" details, meant accepting the curator's authority on what the creature would have looked like – based, presumably, on his reading of secondary literature, not on primary research on the specimen itself. Henson continued:

> for models, better off using large birds like ostrich; elephants are too straight-limbed ... in moving there is a kind of slow swaying, especially in the tail.

The technical and imaginative translation required to move casually from specimen to appropriate analogue to "a kind of slow swaying" is not signalled. It appears, for the most part, to be arbitrary – to be taken as an article of faith. Again, loose statements like this appear to have been intended as a means of giving *a sense of the curator's sense* of what it was that the animation could be allowed to dramatically evoke. A notably conservative tone was being struck, and yet it still required considerable imaginative intervention. In this instance, the document was working to ensure that the authority of the scientist was not overtaken by unauthorized dramatization on the part of the multimedia animators, as though their interpretations would undermine the curator's proper inferences and so would need to be held in check. The artist, Manfred Tolman, who worked closely with digital media producer Walter Tomasenko, and who produced the sculpted model on which the animations were based, told me that the curator did indeed hold him to highly static and conservative poses for the creature.[21]

There were further indications of a practical and conceptual divide developing between the curator and the digital media producer, which I encountered repeatedly in interviews and discussions at the museum – a relatively minor contest which, nonetheless, was already showing indications of changing the outcome of what the "life and times" of *Maiasaura* would become in this exhibition. The locus for the contestation, dividing, and trading of interests was *the imaginary*, something in which both the scientist and the digital media producer knew they had a very high stake – and something which the planners were working continually to resolve.

A notable *inclusion* in the Interpretation Statement was a brief discussion of an "alternate medium": the use of 3D stereovision for the projected animations. While this was not attempted in the end, perhaps due to costs of production (and of providing 3D lenses or viewers for visitors), the impetus for considering this medium tellingly notes the competitive sensibility of those involved in the project, vis-à-vis the entertainment industry. The Statement reads:

Payoff: no one else has done this yet so we will be ahead of a certain famous movie.

That "certain famous movie," *Jurassic Park* – as much or more than ROM specimen #44770 – was once more suggesting standards for the shaping of this public scientific exhibition. Indeed, it would not be until 2007 that 3D animated movie-making became a mainstay of Hollywood cinema. The ROM, in this thinking, was ahead of its time, even though the 3D work was never actually undertaken.

Other indications of how this large screen space should operate suggested the advantage of a "stationary viewpoint: more like the effect of being able to hide somewhere and watch this large beast, like an explorer or a naturalist." Such an approach, centring around "scopic vision,"[22] borrowed upon historically circulated travel and big game hunting tropes as a guide for producing the visitor experience in this exhibit, echoing back to Doyle's and Osborn's productions of the Mesozoic/lost world. Another suggestion, which cued into the curator's wish to give a sense of scale, was the painting of a full-sized Maiasaur silhouette on the wall adjacent to the big screen (though, ultimately, this was not actually produced). As with other components of the display, reading requirements were to be minimized; visual cues to scale, like these proposed silhouettes, were favoured "as opposed to stating it in metres."

Furthermore, the Statement suggests that offering only a "small number of choices [of dinosaur animations] (4–6) would encourage visitor turnover." The effect of "getting in" to an experience and then "getting out fairly quickly" would presumably be enhanced by such approaches. I later learned from one of the interpretive planning staff members that there had been fears that the animation theatre would hold people up, creating a bottleneck. There was a great deal of anxiety expressed over this component – just as there would be with the following component, the Working Lab. Indeed, as I discuss in the next two chapters, there were significant disparities between the descriptions in the Interpretation Statement drafted by the interpretive planners and those in the Multimedia Statement drafted by the digital media producer. These disparities suggest that the debate over how these exhibit components would work for visitors was a perennial, if not definitive, matter in the political natures that were rendered and in which the visiting public would come to participate.

"The Working Lab." While each of the first three exhibit sections took a page or less to be described in the Interpretation Statement, it took three full pages to describe the Working Lab. Measured by the work of transforming a laboratory into a display – that is to say, the amount of effort that the museum had to apply to make this component work as a display – the "Working Lab" could indeed be reckoned as the "centrepiece" of the exhibit (even though its location at one end of the space, rather than centrally, did not attend to this priority). Moreover, the lab consisted of enormous numbers of technical elements including pneumatic tools, ventilation equipment, binocular microscopes, toxic chemicals, and so on. The details of how to incorporate these elements required far more description than was usually needed in preparing an interpretive statement.

Having now coursed through the Interpretation Statement to arrive at the Working Lab, I will begin to turn beyond the planning elements to an extended discussion of the lab as it actually operated. Given its pivotal importance in the plans and the project, as well as the controversy it generated, this feature of the exhibit plan deserves its own chapter.

12 The Difference a Lab Can Make

Give me a laboratory and I will raise the world.

Bruno Latour, on the power of Pasteur's laboratory science to recompose society[1]

There was a real political battle in advance of the design really happening ... in both a grab for whatever percentage of the budget to put into the multimedia, versus the lab ... There was always this kind of rivalry ...

Sam Enright, designer, the Maiasaur Project (July 1998)[2]

The problem of how to be faithful to the Maiasaur specimen and the dynamic process of reconstructing and learning with it and from it was continually challenged by marketing impulses that favoured interactive multimedia techniques as a means of engaging publics. Countering this impulse, there were two particularly potent actions I would encounter that assiduously worked at articulating visitors more directly with the specimen (and its palaeontological interest and genealogies): Jennifer Ross's Interpretation Statement work; and, in pragmatic ways, the actions and intra-actions of the Working Lab itself.

The caring work of Jennifer Ross revealed itself as a genuine and generous effort to marry, and to generate a seamless interchange between, the radical elements of the specimen-spectacle complex that weighed on the Maiasaur Project collective throughout its work. And she was effective at this, straining against the forces that might otherwise tear things apart. Her task was to make peace among these forces. Her leash-down diagram would do for the museum what Andreas Henson sought from his technical preparation laboratory. These were committed gestures to articulation. The challenge and worry was whether they could draw the exhibit development actors together, and so draw the visitors together, while retaining a high premium on living palaeontological practice – and on the specimen.

As noted previously, the *Maiasaura* preparation lab was in operation only for the first two years after the exhibit opened – that is, from May 1995 to May 1997, prior to my own arrival at the ROM. Although this timing prevented me from conducting direct study of the Working Lab in operation, I was able to conduct some remarkably telling interviews with the technicians who worked in the lab, most notably Phil Thomm, the preparator who undertook the lion's share of preparation work and direct public interpretation. Thomm's remarks are pivotal in what follows, especially as counterposed alongside comments from interviews with interpreter Ross and curator Henson.

Finally, the detailing of the lab that follows will also give greater relief to the complexity of the planning work for this exhibition in relation to the translations of the fossil skeleton, ROM #44770, *Maiasaura*, "Henrietta." I continue to pose the description against planning documents, notably the Interpretation Statement. In addition, and perhaps most importantly, this description gives a clear impression of the intricate, detailed action of *articulation* to which I have been referring over many of the chapters in this work.

"To Share in the Research and the Specimen, in the Whole Thing"

According to the Interpretation Statement, the Working Lab itself would "offer an accurate picture of part of the process of palaeontology" (see figure 12.1). The planners knew this was only a part of the process of scientific techniques, and that it was only a representation, "an accurate picture" of palaeontology in action. At the same time, the Statement made the emphatic point that it was a "real lab with real scientists!" That the "real scientists" were actually real technicians – as opposed to the curator – was apparently not an issue. The effectiveness of this component was by no means considered a *fait accompli* during the planning, as Jennifer Ross pointed out:

> I actually had my doubts about the lab. I had seen a lab, I think at the Smithsonian or the Tyrrell ... which was all yellow and horseshoe-shaped

Figure 12.1. The Working Lab in action. Source: ROM Photo, reproduced with permission of the Royal Ontario Museum, © ROM.

> and so boring! You could sort of look, but there was nothing to see. The work consisted of people going "scratch, scratch, scratch" for years! This isn't very riveting.

For the exhibit programmers, displaying a laboratory was a complex and confusing matter: how to *rivet* visitors to the action, how to get the laboratory to behave more like a representational display (i.e., an "accurate picture"), how to bring attention to the specimen, how to make the otherwise alien activity taking place inside intelligible.

The apprehensions and uncertainty were well justified. As it turned out, the very act of placing a very real, "behind the scenes" lab in a museum gallery automatically transformed it into a display, a picture, a piece of theatre. Technician Phil Thomm (see figure 12.2), who worked in the lab over most of its two-year operation, pointed out to me how it

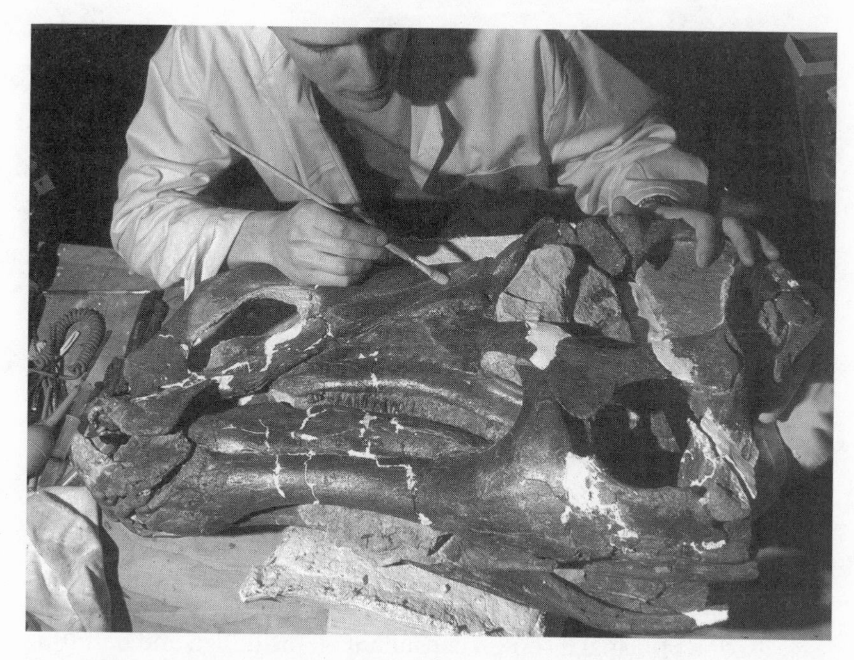

Figure 12.2. Phil Thomm preparing ROM #44770. Source: ROM Photo, reproduced with permission of the Royal Ontario Museum, © ROM.

was the expectation of a theatre-like "set-up" which conditioned visitor responses:

> [S]ome visitors would do a complete walk-by and they would think it was just a fake exhibit and that I wasn't real, just some mannequin … I think most people had the perception initially that it wasn't real … because every other exhibit in the museum is a representation, it's not the real thing. You know there's stuffed animals, there's old English rooms that are kept as if somebody was about to walk in the room, but you can't go in them … everything's a set-up … so when they first come up to it, their mind is trained to think, "This is just a set-up," and "That is a really realistic-looking dummy," and "Boy, the museum went to a lot of trouble to make it look so real"… when you moved, you'd sometimes see and hear people screaming and jumping as if you'd given them quite a fright … after that, you could hear them laughing … they would actually be quite embarrassed …[3]

The actual issue, then, should not have been "how to give an accurate picture," but rather, "*how to overcome the expectation of accurate pictures* and how to make the set-up come alive for the visitors." Instead of a highly composed, scenic reconstruction of a habitat diorama or a historical room, the visitor was presented with what would most likely be a puzzling array of technical apparatuses and materials: work benches, microscopes, pneumatically driven tools, air hoses, suction ventilation, magnifiers, plastic phials, steel trays and shelves, paint brushes, bottles of chemical solutions, bones, scientific papers, power supplies, tubing and piping, huge white-plaster-coated blocks of stone with bone-like shapes in them, photographs, protective gloves and goggles, mechanical pencils, graph paper, laboratory coats, and of course, the technicians themselves.[4] This museum diorama was, after all, a working laboratory.

Another major challenge – though more so for the curatorial staff, and specifically the technicians who would work in the exhibit – was determining how to keep this transformation of work-space into display-space from impeding the actual work that needed to be undertaken inside. The senior vertebrate palaeontology technician, Paul Anderson, was also the principal technical consultant on the design and outfitting of the Working Lab. He expressed to me the issues which tended to be overlooked by the exhibit planners, and which caused him considerable dismay at the time. Central for Anderson were concerns that a theatre-like lab did not meet protective labour and workplace standards, and that visitor safety could be jeopardized, given the presence of toxic chemical preservatives as well as dust and shards from the removal of the matrix in which the fossils were embedded. Anderson pointed out that it was crucial for the technicians to have a comfortable, productive, and – more to the point – safe working environment while literally being put "on display."[5] Presenting live-action technical work was something alien not only to ROM visitors but also to the entire staff of the institution – and to natural history museums in general.[6]

The Interpretation Statement noted that visitors could see "technicians/scientists" at work "at certain guaranteed times of the day, every day of the week." They would be able to ask questions of the workers behind the plexiglass barrier by means of a telephone handset. This interactive telephone feature was not actually developed, as palaeontology technicians felt it would be too distracting from the work. In due course, the compromise between the seemingly incommensurable actions of conducting science *and* producing theatre in the form of a laboratory was finally arrived at: the on-duty technicians would come out of the lab at two posted times during the day to answer questions. The rest of the time, the technicians would attend to the preparation-related work on the specimen. Work was planned to start with the Maiasaur skull, and proceed strategically through the remaining skeletal elements.[7] As it turned out, the skull – the element of most palaeontological interest to curator Henson – was the last element to be prepared, fully taking up the last seven months of the two-year preparation phase.

To bring visitor attention to the preparation work without interrupting the technician, visitors would be brought in close to the action by watching the detailed preparation on a video screen, which showed the view through the technician's microscope as he or she worked. Jennifer Ross recognized the potential this offered during a visit she made to the palaeobiology labs of the ROM during the planning stages. She found one technician at work preparing a specimen:

> I looked through this microscope and thought, "Holy Geez, that's fantastic! … we *have* to attach a TV camera to this." So we managed to link up that little TV monitor to the microscope. When he worked through the microscope, you could see it. It suddenly became really neat – it was such a clear picture, so vivid! You could see this little scratch, scratch, these little chips going, and suddenly the tedium of it became interesting.

It was critical to produce pictures, especially pictures with action. With the right instrument of translation, "tedium" could become "interesting" – so interesting, in fact, that some visitors would find it difficult to pull themselves away from the lab, as Phil Thomm told me:

> [I]f they didn't think we were real, they would just walk by quickly … but once they saw we were real, the average stay, I would say, would be ten to fifteen minutes … if there was a microscope being used they'd stand there for half an hour or so … some would spend a couple hours looking at the lab and then go away into the other Maiasaur area, and come back again …

The lab was proving to be the most effective theatre of all – and all because it simultaneously was and was not theatre! Equally importantly, it was knitting together the action involving the specimen with the action in the other displays in rich, even dynamic ways. The dynamic process was being signalled by the practical work of articulating, and making intimate, the work of specimen preparation.

Among members of the Maiasaur Project design group, the lab was ultimately seen as highly successful, largely because it drew public attention to the central matter of museums: their collected objects, their specimens. Designer Sam Enright described it this way: "People come here to see objects ... that's the principal reason for coming here, to see the real McCoy." He also elaborated his position on the internal "battle" over which section of the exhibition should be given primacy:

> I think people came principally to see the lab ... that real-time thing is invaluable ...
>
> There was a real political battle in advance of the design really happening ... in both a grab for whatever percentage of the budget to put into the multimedia, versus the lab ... There was always this kind of rivalry ... And in spatial terms they kind of take up the same amount of space ... I was quite encouraged that the lab – at least, based on my personal experience in wandering around the exhibit – won out in that. Because people seemed to like it much more ...

"There was always this kind of rivalry" – I was reminded yet again that much of this museum's organizational "culture" turned on protecting the primacy of the collections' objects in the communicative work of display, yielding at every turn to projections of the visiting public's "need to know." At the same time, this was consistently set against an anxiety that high-tech media spectacle was the source of greatest threat to that primacy – but also a boon to the public/political mandate of the museum and its managers.

The lab was complemented by other components (see schematic layout, figure 12.3).[8] These included a pre-recorded, explanatory video describing the lab equipment and process (which was later replaced by an interactive computer-animation display); graphics with questions and answers on the topic of fossils; and a real (i.e., actual) piece of fossil to touch, "to induce a sense of connection to lab activity" – as suggested by the Interpretation Statement. This sort of action, meant to "connect" the visitor in many different ways with the entirety of the display as an experiential total, was directly expressed in the Interpretation Statement:

> A cast of Maiasaur skin that can be touched. The emotive experience will be enhanced by the realistic animation of *Maiasaura*.

On some occasions, when emerging from the lab to answer questions, the technician would even permit visitors to handle some of the more

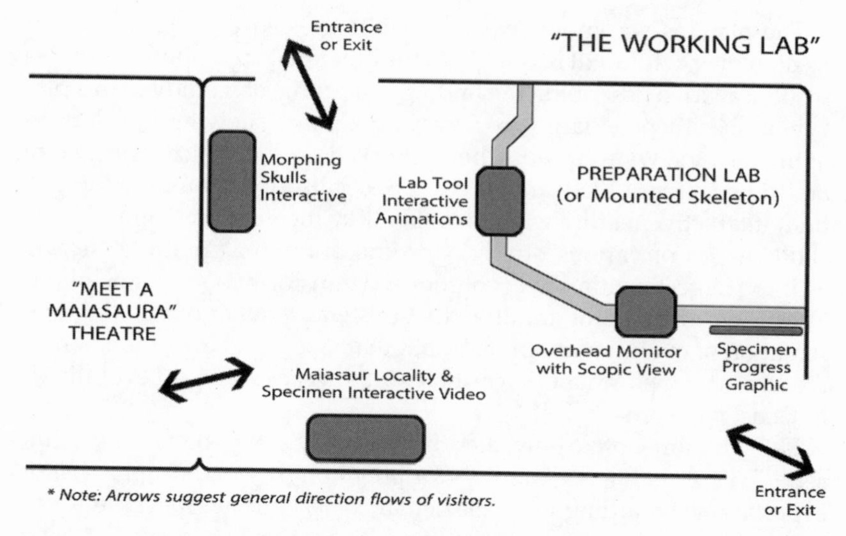

Figure 12.3. The Working Lab – schematic of the presentation space.

robust fossil elements of ROM #44770 that he or she was preparing, amplifying the synecdochic relation: how the singular, handheld item could stand for the larger project. Thomm explained:

> That was a really amazing situation, when you can explain, "Here's what I'm doing in here, here's the specimen, and it's eighty million years old," you can see you've got their full attention … and then you put it in their hand … [*pauses, as if to recreate the moment of rapt wonder*] … and then you just stop talking and let them think about what they have in their hand … And that's the first time, probably, they've been in the museum, that something that was behind glass, protected from them, and then they're allowed to interact with it. That's basically what the whole Maiasaur Project was about … allowing them to share in the research and the specimen, in the whole thing …

It was as if the visitors, in the moment of tactile contact, became aware that they and the fossil were both part of the sweep of biological, geological, planetary history. This, of course, was an unplanned moment, but arguably one that in the process had found something of the dynamic and the imponderables that Henson had hoped to elicit.

The plan, as we know, was to attempt to keep the technical work itself in play, while still relating it to the other major components of this section: a video describing the finding, collecting, and study of this particular specimen; a "large progressive diagram" showing which parts of the skeleton were currently being worked on and which remained to be studied (figure 12.4); and a "Skull Specialization" computer animation/interactive feature which would allow the visitor to "morph" the skull shapes of various ornithopod dinosaurs on a computer screen, as if to suggest continuous evolutionary transformation. In one of our interviews, the curator admitted that this was a rather misleading presentation of evolutionary process, but that it suggested, in the abstract, the idea of transformation over time – "the very essence of evolution," in Henson's words.

Phil Thomm noted how the "lab" display accompanied by "museum" displays impacted on visitor attention. It worked upon visitor expectations regarding what the actual "object" of the spectacle was – such as the specimen, or Thomm himself, the preparator:

> [T]he lab situation [with] the video background and the computer explanation tools, and toys for – some of it was toys, like the unit that made the different skulls morph – like it was educational but they could also interact with it … all those things really took the focus away from the people working in the lab, and focused it on the material being very important … I think that was an important move in the exhibit … The rarity of them seeing – normally it's a stuffed lion or whatever is in there – so they would initially be focused on "it's a live person," but with the other material that's there, they became focused on the specimen itself and the tools of the lab, and the processes of the lab, rather than the people of the lab … there was never any mention outside the lab of who was working in there, and what is their background, and where they come from … which I think in hindsight was a good thing …

Interestingly, once visitors saw that there was a human being working inside, doing something very serious and technical, it appeared that communicative effect was articulated and distributed to these other elements of the display – and, in turn, to the specimen – in new and potent ways. This helped to limit the attention paid to Thomm, the person actually performing the work:

> [I]t helped in not feeling like you were so much on display; you just happened to be the one working on the specimen … you were just a player

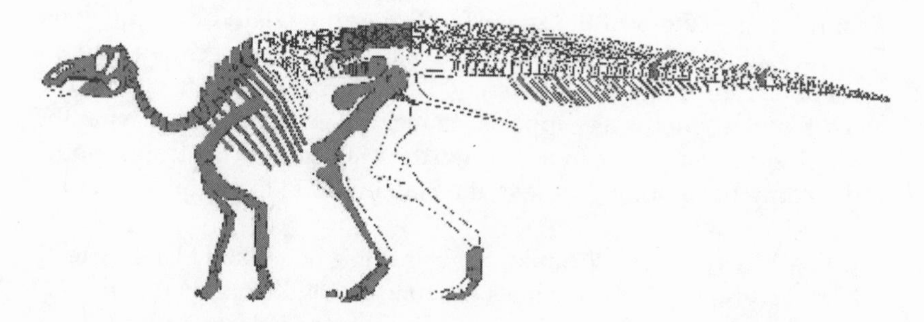

Figure 12.4. Specimen "Progressive Diagram." Shaded areas indicate parts of the specimen already prepared.Source: ROM illustration reproduced with permission of the Royal Ontario Museum © ROM.

> in the thing. Otherwise it may have made the situation working in the lab, that you're the one that's on exhibit, and you know, the dinosaur is secondary … So, those extra exhibits there to help people understand the specimen turned out to be very important on the working atmosphere inside the lab. Without them there would have been too much focus on the live people working there …

The concern for the specimen would be supported through this intricate play of foregrounding, backgrounding, distracting, and grabbing attention. Of the other exhibit elements, the pre-recorded video on the specimen was to include location footage showing how the specimen was found – as well as by whom (Sherry Flamand) – and to create a "sense of excitement and luck," as noted in the Interpretation Statement. In her continuing quest to find relevance for the expected family audiences, Jennifer Ross explained to me how she felt this to be an extremely important story to include in the display:

> I jumped on it when I heard the skeleton had been discovered by a kid. She was sixteen at the time. I said, "That's great, let's make something of that!"

Ross was hoping that visitors, especially young visitors, would identify with Flamand and her find, much as they might identify with the live technician working in real time on the specimen. This would produce yet another possible set of linkages – what Ross called "relevance"

– connecting visitors to the specimen through a chain of associations. The track of associations would then lead from the specimen into aspects of its morphology and biology. The Interpretation Statement also noted that the video was supposed to include explanations of what the research should produce in relation to the Maiasaur's "beak and snout" and feeding behaviour. A note in the Statement on the latter suggests:

> This is important since it emphasizes something new about Maiasaur research; it also ties *Maiasaura* to its environment, and hints at an ongoing "mystery" in the relationship between dinosaurs and plants.

In the outcome, however, this was yet another point that received very limited display support. Henson's curatorial statement explained broadly the connections he intended. He wrote:

> Particularly important … is the horny beak covering the front of the snout; this rarely fossilized structure will provide new insights into hadrosaurid foraging. As such, it is of great interest to an ongoing research program by Henson and his associates on the evolutionary acquisition of herbivory in different lineages of terrestrial vertebrates. Duck-billed dinosaurs, along with horned dinosaurs, have recently been implicated in scenarios to explain the tremendous evolutionary radiation of flowering plants (angiosperms) during the Cretaceous period. An understanding of their mode of feeding is obviously central to this debate. The assumptions of the aforementioned scenarios are that hadrosaurs were very efficient herbivores that required large amounts of rapidly growing vegetation and that early angiosperms, unlike other vascular plants, were weed-like in their habits and could quickly recover from an onslaught by herds of these large plant-eaters.[9]

With a few small moves, and in a very limited locale, it was planned that the exhibit video would effectively carry the experience of the visitors in the gallery space, bringing them to the specimen in field and lab, and ultimately to the intricate behavioural possibilities of the living world of *Maiasaura* millions of years in the past.

How much of this would be achieved? Arguably very little, at least for visitors. The knowledge that the skull and beak morphology was crucial to establishing a proposition about the co-evolution of flowering plants and Cretaceous herbivores could be offered as a gloss, but as the project involved drawing publics into the intricacy of action in how lab work supports the examination of such a proposition, its conversion to an effective "fact" was a remote possibility. Isabelle Stengers pointed to how laboratories have conventionally been deployed in the service of delivering facts:

> A "fact" in the experimental sense of the term, only exists when a laboratory apparatus succeeds in overcoming the objections that aim to put this fact to the test, to verify that it really does have the power of creating an agreement between the protagonists.[10]

The presence of protagonists, the demonstration of the lab work as applied to the testing of a particular proposition, the resolution of the argument towards an agreement – this was only roughly alluded to in a few graphics, and in an accompanying video with Henson speaking. The active engagement with the process remained beyond the reach of visitors – at most, they would get a "sense" of the process, but not the "process" itself.

The point here is that great effort in the planning of the exhibit and the Working Lab was being placed on making connections and articulations to the action of scientific practitioners and their empirical matter, the fossils and specimens, and the assembled natures/worlds in which they could be imagined to be living. One set of connections was being built between this section of the exhibit and the ones previously described in the Statement. The second and even more notable set of connections was that attempting to link a unique characteristic of this specimen – the horny beak of ROM #44770 – to a scientific problem, and ultimately to the larger palaeontological process of fashioning plausible Mesozoic worlds. All of this would somehow aid in reinforcing the otherwise tenuous connection between the specimen of the display and the spectacle of the display. The remedy was sought by multiplying the mediating actions and instruments between the complexity of the object itself and the even more complex animation display media, seeking what Latour refers to as "well-vascularized" relations among practices.[11] What this remedy did not fully address was how planning and marketing considerations had reduced the sense of audience complexity to a loose and embracing notion of "families," a matter to which I will return in coming chapters.

Once installed and operating, therefore, the lab did activate a chain of mediated connections, though not always the expected ones. One brief mention of the beak and snout was made in the didactic video about the finding and significance of the specimen, and, while the articulation of these fossil elements with other display elements may have been an aim of the project, this single connection was rather weak. While the principal preparator was relieved that the other display elements minimized direct attention on him as the object of the display, it was abundantly clear that the connections intended by Henson would not have happened without the technician's active engagement with the instruments, the specimen, the curator, *and* the visitors. In short, the technician – that is, this one human – had been crucial in activating the entire "chain" of connections, if and when he was able to speak with the visitors directly. Phil Thomm summed this up in several commentaries. First he pointed to how, in his responses to visitor questions, he could direct them beyond the simpler points they would raise about the specimen:

> [T]hey were definitely interested in the specimen, "How old was it? Where was it from? How did you know how old it was? What are you working on?" and, "How do you do it?"... it's like I'd almost get bored answering the question in a certain way ... and, over a year, say, I'd try to answer it differently by bringing in some other information so we could take the discussion somewhere else ...

Reciprocally, the interaction with the visitors had practical effects on Thomm as well, even redirecting his work on the specimen:

> [T]heir questions were 90 to 95 per cent of the time on the specimen, and I would point out, within the block that I was working on, what was exposed and I would be very up front with them, what difficulty I had, what my next stage was ... and I'd find it very helpful to be able to verbalize with people, who didn't really have any say in the decisions I was going to make, but in verbalizing them, allowed me to figure out the problems I was facing with the specimen ... It was kind of neat ... it helped me to put the focus on what was needed to do, and what wasn't all that important ...

Here, as intra-actions with visitors were multiplied, a much more interested and interesting engagement with the specimen was effected. Here, the public was brought intimately into the work that might recompose, even in small ways, the Mesozoic natures that would circulate thenceforward, and reciprocally into the recomposing of society as a participant in this action. Here, Thomm had become the most keen, pragmatically immersed witness to the difference a lab can make.

Tools of Revelation

Apart from direct interaction with the technician, however, the exhibit's articulation work was established by use of the computer-animated, interactive tool display. Several sorts of machines were demonstrated in this way: the "air scribe," the "pin-vise tool," the "filter system," the "dental tool," the "glove box," the "cast-cutter," the "sand table." Once one of these was selected from an array of lit-up buttons, a computer graphics animation of the particular tool would appear on a black screen. The tools would float across the monitor space, with texts indicating the name of the tool, the "matrix," and the "bone." The tools moved on their own, as if levitating and operating without human aid. Thomm explained what would happen as visitors glanced back and forth between the digitally smooth, *disembodied* animations, and his own live and most definitely *embodied* activity in the lab:

> [T]hey had the computer-animated thing about a dentist's drill and an air scribe, and that allowed them to look in the lab and go, "Oh, there's the dentist's drill, and here's what they do with it"… Otherwise they'd have really no way to make any sense of it all … those exhibits helped to take things out of the lab context, and say, "Well, this is what they do," and put them back in the lab context and then "There it is, and oh, he's reaching for it, and maybe that's what he's going to do with it …" I'd be hearing adults … trying to figure it out, glancing into the room and if the dentist drill was right in front of them, they'd say, "Oh, it's right there," and they'd hit it again and go "Oh, look at that."

Thomm had outlined, quite simply and precisely, how articulation takes place: the computer representation of the disembodied tool outside the lab was what indexed the technician-held tool inside. Following the technician using the tool, either with the video image through the microscope or by watching through the plexiglass, led eventually to the specimen, and from the specimen to the diagrams, to the videos about its find and significance, and so on. This rhizome-like complex of juxtaposed representations, of *partial connections*,[12] of tools, texts, and humans worked to bring the imaginary of the visitors into line with the palaeontological interpretation of the specimen. By their creative, narrative in-filling, they were indeed apprehending and participating in the epistemic work of palaeontology, as political participants in Henson's "dynamic process" of technical and reconstructive palaeontology.

In effect, the visitor would be enrolled into this network so that the scientific imaginary could be activated, and so that scientific imaginings could be circulated.

Here were cases of complex interactivity working as intra-action in the Working Lab as a theatre-like display. But the articulations carried out by the technician acting/working in the lab didn't simply create connections between the visitors, other display elements, and the specimen. They also provided the occasion for rich exchanges with the curator and his activities in ways that might never have occurred otherwise:

> Andreas would bring down a couple hadrosaur papers ... and I would take those things that I read and saw out into the talks with me ... not the papers ... but explain how many specimens had been found in Canada, sort of dabble into the science of it a bit ... those papers were very instrumental ... Andreas's own interests typically focused around the skull, which didn't really surprise me too much ... when it started to come out, he was down there more frequently, and I appreciated that, and he always answered questions I had ... and from that I started to get an idea of where he was wanting to go ... and it was mostly around the skull ... the dewlap was very interesting ... the neck-skin impression, sort of like a turkey ... and the beak on the front of the Maiasaur, this is the first one found in situ, I believe, on any hadrosaur ...

As a consequence, the curator's honest interests, resolving here around the skull and its anatomy, were repeatedly brought into play via the genuine interest of the technician, who would then be in a position to redistribute those interests through his interactions – both with the specimen in preparation and with the visitors. So much could come into coordination: the specimen in the lab, the visitor in the gallery, the technician, the equipment, the history of *Maiasaura*, Henson's project to revise how people think about dinosaurs, the Late Cretaceous of northwestern Montana, the experience of a teenager finding a dinosaur – indeed, the very rationalization of this major research museum as an institution of civil participation in the acquisition and production of knowledge. Thomm had summed it up already in talking about the moment of handing a fossil from the lab to visitors:

> That's basically what the whole Maiasaur Project was about ... allowing them to share in the research and the specimen, in the whole thing ...

Though expressed in new terms, here again was the sentiment in Ross's cry to "let them eat soup," Henson's similar recognition of people's excitement to "know how things come about" – this was the emergent commitment to responsible articulation of publics, and to the actions of pondering the mysteries of past life.

Decentring the Specimen

Before appearing to celebrate the accomplishments of the Working Lab too much, I must emphasize that such potentially powerful connections between the nuances of scientific practice and visitor experience could only take place when the technician emerged to engage visitors' questions. Much of the effort to produce an articulate set of connections would be diverted in the process of exhibition production, budget-management decisions, and emphases on the multimedia spectacle. Now, even more than before, the significance of the omission of the planned computer animations of Maiasaur herds and hadrosaur diversity becomes clear, as does Jennifer Ross's dismay over this. A vital articulating device had been abandoned. Along with other desired features mentioned by designers and planners, such omissions significantly upset the chain of connections between, for instance, the unusual beak and snout of ROM #44770 and the scenario of thousand-strong herds of Maiasaurs, which the curator and interpretive planner had so carefully outlined in their official planning statements.

Even more general connections between the specimen and the surrounding displays would stand to be lost when the lab was removed, perhaps above all other losses. Once more, Phil Thomm made the point unambiguously:

> The electronic exhibit stuff provided an interesting background for the lab material … it took it to another stage, to an educational and an entertainment stage, like with the computer graphics stuff … now that the lab is gone, I wonder if the exhibit stagnates a bit … I can only speculate on this … Whereas for two years the lab was a centrepiece of new stuff going on all the time … the lab environment was always changing … now people come back and they really can't see anything new …

Beyond just losing the possibility of seeing something "new," it appeared as well that the intricate, sensitive, cared-for complex of connections between the specimen and the media spectacle might also be eroded, if not erased entirely, by the lab's removal. The risk was that an over-reliance on received chronotopic phantasmatics might recur in this project.

Before turning to the other display components of the Maiasaur Project, some final comments should be made in relation to the place of specimens in this exhibit. It has to be recalled that this exhibit was meant to be about and centred upon one single, outstanding specimen – ROM #44770. As noted already, the rigorous study of the specimen had not actually taken place; only its preparation had been done. Second, the specimen came to be located not in the centre but, as it turned out, at the "end" of the exhibit. The incompatibility of the two physical entities, "Working Laboratory" and "Exhibition Hall," conditioned this eventual spatial relegation, as I learned in another exchange with Jennifer Ross:

JR: Because we had these solid masonry walls here, this became the best place to put the lab … hugging these walls. This is an original, old, brick wall from the museum built in the '30s, so it is really solid. The idea was that because it was predicted to be kind of heavy, that structurally it needed this.

BN: So you were physically constrained, in putting that centrepiece right there.

JR: Yes. It would be great if that was the first thing you hit, as it was central. However, just because of traffic flow, and because it is not available to everybody – people with strollers, disabled, and so on – they can't use this staircase that brings them up close to the lab. Then we had to designate this other end as the main entrance, instead of the lab …

The "best-laid plans" of the curator, the interpretive planner, and the designer had been diverted further – now by the physical structure of the building itself. Such architectural characteristics were rooted in museum histories which had not anticipated a technically equipped laboratory – a *behind-the-scenes* facility – working as a *display front of science*.[13] The term "centrepiece" had increasingly become an artefact of the commitments assigned to it during the planning process. While unquestionably a major component, the term was gradually losing both its physical and conceptual centrality, and with it the centrality of the specimen. Jennifer Ross – like the curator – made many efforts to make the specimen paramount in this exhibition. This was, after all, the *project* of the Maiasaur Project.

Paradox: On Translating Action without Translation

Despite Ross's efforts in preparing the Interpretation Statement, which was key to the development of the exhibit, what is striking is how little presence this supposedly crucial actor – the fossil specimen – had in the document itself. Of course, she did not predict the surprising way the technician and curator and journals and exposed horny beaks would act together, or how the visitors might be brought into this play of actors. The skeleton-in-preparation was undoubtedly the dominant specimen to be included in this display – supplemented with a handful of skulls and jaws from the ROM collection, a couple of small limb elements and plant fossils, and then a number of casts. The simple quantity of the official "stuff" of museums – their collections – was remarkably small. The accounts of how the key specimen was found, what made it unique as a specimen of scientific interest, and how it would be prepared would ultimately be relegated primarily to one video display. The overwhelming material content of this display would be the multimedia components and the technical and physical apparatus of the lab itself – as opposed to the specimen to which these techniques were applied.

To a great extent, the "leash-down" effect of Ross's concentric diagram had begun to unravel. What ultimately became the physical, spatial centre of the Maiasaur Project was the "Meet a Maiasaur" interactive multimedia theatre – the budget sensitivity of the institutional arrangement was beginning to revert this exhibit to the spectacle-marketing impulse. Over and over, recognizing the lab – let alone the specimen – as the central element of the project was difficult for programmers. As 3D designer Mark Alton's statement in the Concept Design Summary had noted, two of the exhibit's components – one being the lab – would display "context," while the third would display "current understanding":

> The exhibit is divided into three sections: the primary introductory section (The Cretaceous Period) provides the environmental context for the Maiasaur; the secondary introductory section (the Working Lab) provides the modern context for the present study of the Maiasaur ... Between these two introductory sections is the interactive theatre. This section showcases the most current understanding of the Maiasaur's behaviour and movement rendered in the most realistic manner allowed by present computer animation technology.

Here, again, the exhibit producers appeared to have conflicting senses of what was the "centrepiece" of the exhibit. For curator Henson it had been the lab. For Alton, the interactive theatre now had primacy; the lab would provide context for that section, being relegated to the status of "the secondary introductory section." In the planning process, the allure of the contemporary technological spectacle continued to challenge both the action of scientific techniques and, as importantly, the specific agency of the specimen itself, the purported point of the entire project.

In laying out and commenting on the "official" documents behind the exhibit's development, a number of notable contradictions came to be highlighted – especially on the question of what the actual centrepiece of the exhibit was or ought to be: the specimen, the lab, the theatre, and so on. The effect of those contradictions in the "finished" exhibit has yet to be considered – but the ability, or inability, to find the means of articulating and connecting the specimen to the spectacle and the visitor had been clearly identified as the critical matter for all.

Articulation and totality of imagining were central to this Statement as a work tool, but matters were getting visibly more complicated as the translation work proceeded – and as production of the multiple elements progressed, with new and unexpected turns in articulating arising. The curator's proposed lab remained. However, there were unanticipated (or unseen) struggles over where to allocate the resources, over the difficulty of putting a working laboratory in the galleries, over the newness and cachet of developing interactive, high-tech multimedia displays, and over what should be included to produce the optimum chain of connections – all of these conflicts came into play. The centrality and the articulation power of the lab were being eroded. The newness both of the lab as theatre and of computer-based interactive media was pushing the work network into unforeseen terrain.

It also became clear that the continuity of connections was limited between this specimen, ROM #44770, and the larger story of *Maiasaura* to be told. This is not to impute any error in the *attempt* made to represent the specimen as a reasonable index of the finished Mesozoic world put on display. *Rather, it is to suggest that the connections between them were complex enough, and the resources limited enough, that to animate those connections in a consistent way became impracticable.* The displays in the immediate area of the Working Lab space (including the videos, lab tools interactive, morphing skulls interactive), may have been quite well articulated through to the specimen. Now the divide that appeared to be developing was between the lab space and the two other main spaces of the exhibit: the "Meet a Maiasaur" theatre and "The Cretaceous Period" displays. To some extent, one would have to rely upon the power of juxtaposition to generate some sort of cohesion between these sections.

Recalling the table of translations (table 11.1), elements associated with the curatorial question of "How do we know?" were poorly articulated with the elements associated with the other two questions, "What were they like?" and "Where did they live?" With the removal of the Working Lab as a response to the exigency of institutional market pressures, the curator's "How do we know?" question, which literally anchored his "sense of a dynamic process," would fade out of the exhibitionary apparatus. But even prior to that removal, the gap between specimen and spectacle – though promisingly filled in by the lab and the actions surrounding it – returned once more in this new divide between multimedia theatrics and ROM #44770.

Having coursed through the promising play of articulating and disarticulating forces in the making of the Maiasaur Project exhibition, I am able now to take up Phil Thomm's very practical question:

> I think *now* would be the time to answer what role the lab played, now that it's gone, what sort of state the exhibit is in … that's something I can't answer … I left when the lab was done … I haven't seen how people react to the whole exhibit today, whether they spend time there, or what they do …

To rephrase Latour, if the articulating work of the lab had been attenuated by its decommissioning, what else might come to occupy *the position left empty by the dichotomy between the specimen and the spectacle*, or put another way, *between the physical presence of* Maiasaura *in this exhibit and the intra-active capacity of those who would visit the exhibit*? While continuing to consider the many contingencies and unexpected relations I have been outlining, the chapters that follow address the effects of the Maiasaur Project in its latter incarnation – that is, after the two-year project of specimen preparation, and once it became an exhibit devoid of its official "centrepiece." With a sense, now, of the difference a lab can make, we can properly consider the difference that having no lab at all will make.

13 A Perfect Time for Raising a Family: Kinship as New Syntax for Dinosaurian Natures?

Maiasaura: Revising the Mesozoic Nexus

In the previous chapter, I offered several interview commentaries that contoured dimensions of how the "Working Lab" – one of the three major spatial components of the Maiasaur Project exhibition – operated as a finished, materialized display. In effect, it showed the linking together of the curator's phantasies with those of other key museum staff and the exhibit's audience. What follows in this chapter (and the next) are accounts of the physical materialization of the remaining two-thirds of the exhibition. These accounts are based significantly on my participatory observation within the exhibit space itself, in addition to interviews and other documentary sources. The two components remaining to be discussed are, respectively, the combined "Introduction to Exhibit" and "Cretaceous Period" displays as well as the "Meet a Maiasaur" interactive theatre. The accounts in both chapters offer an engaged tracing of the exhibition and its knowledge outcomes for a particular Mesozoic nature/politics, drawn in part out of Jack Horner's scientific-technical work. They also discuss the articulations that are increased, upset, or suppressed as a consequence. The accounts extend the dialogues from the making of the exhibit into the exhibit itself, bringing the action of its developers and visitors into play.

My subtitle for this chapter gestures to the way that kinship, usually ascribed to human sociality, is articulated in reordering dinosaurian natures, generating a dinosaur sociality/nature that has increasing affinities to a *particular* human sociality/nature. This also gestures to the shift in focus that began at the end of the last chapter with a consideration of the lab in operation, where the prominent *nexus* of the exhibit had to that point hinged upon the relation of the specimen (via the lab) with the multimedia theatre and other elements of the exhibit. At the same time, the previous chapter and this one move from thinking about the networks acting upon the *conceiving* and *promoting* of the exhibition to the *materialized* outcome of the exhibition, and what actually transpired in its communicative, phantasmatic outcome, including a differently composed and differently inhabited Mesozoic time-space.

In terms of my original proposition, then, whereas the first section of the book emphasizes the translations and circulations of *the animating performative nexus* into and through *the network* – that is, by way of *a priori* phantasies and materializations insinuating themselves in the network of production – the discussions ahead emphasize how that procedure gets turned around. Now I bear witness to how the materialized outcome, *the animated performative nexus, is precipitated* out of the action *of the network*. Put another way, we recognize the particular materialized phantasies that the specific network (of so many actors and things and precedents) would produce as a nexus in the course of its activities. Along with a reconstituted Mesozoic world, particular forms and relations of life emerge in the exhibitionary outcome. Moreover, as was the case with the Doyle/Osborn nexus, the forms and relations in that outcome signal a discernible, altered, and reconfigured political nature – an always recomposable cosmopolitics, to use terms from Isabelle Stengers and Bruno Latour. It is here that the gender and kin relations, among other potentialities, become all the more distinct in the politics/natures of the Mesozoic, in late twentieth-century times of *Maiasaura peeblesorum*.

As noted, during the period when I was at the museum and conducting interviews, the Working Lab had been removed from the exhibit. My discussion now turns to a number of questions following from considerations of the lab in the preceding chapter. If the lab display, in its multiplex ways, connected visitors to the specimen, what sorts of connections were made in the remainder of the exhibition? If the curator's actual wishes to convey the dynamic processes of palaeontological reconstruction and of alternative propositions of the proceedings of dinosaur life were most activated through the Working Lab, what propositions and wishes were activated in the other two major components? If a widening divide was being created between the technical work of the lab with its specimen and the multimedia work of the theatre's Maiasaur world, to what extent would visitors obtain curator Andreas Henson's "sense of a dynamic process" and with that "start thinking in a sophisticated fashion about dinosaurs as once-living, complex animals"? What was the predominant material/phantasmatic world of past life that came through the exhibition and how did visitors engage with that world? In short, what politics/natures emerged in the exhibition, especially now that the animating nexus within the Working Lab had been removed?

Altered natures are cued in each of these two chapters. I will say more of the Maiasaur multimedia theatre in the next chapter. In this chapter, however, some dominant emerging politics relate to the naturalistic ordering of family life – both human and saurian. The components of the exhibit that I describe in this chapter were developed along the lines of more traditional canons of museum practices: presenting comparative collections and fashioning a didactic narrative to enhance their intelligibility.

I will begin my descriptions of the exhibit, quite literally, in the middle of things. This opening scenario took place in May of 1998, just under a year after the removal of the laboratory from the exhibit space.[1] At this stage, in place of the lab now stood the mounted skeleton of the specimen, frozen on the spot, in a standing pose – the upright posture, ready to move, signalling that this was not just a well-dead creature before me, but rather something already on its way to being virtually reanimated, virtually brought to life.

1: In the Virtual Space of the Maiasaur Project

Having positioned myself between the "Cretaceous Period" and "Meet a Maiasaur" spaces of the Maiasaur gallery, I immediately became aware of the overall soundscape of the space. To my left was the very loud "Visit to the Cretaceous" interactive module/console with its overhead video monitor. Visitors could touch one of the buttons to start videos on one of two topics. One selection was for "A Cretaceous Neighbourhood," presenting an overview of the Cretaceous period during which *Maiasaura* lived as a time of flourishing plant life – cue curator Henson, whose plans stressed this point of understanding concerning Late Cretaceous palaeoecology. The second selection was for "A Maiasaur Family," presenting *Maiasaura* nesting and parental care – now cue exhibit planner Jennifer Ross, who fought hard to leash together the life of *Maiasaura* with the lives of likely visitors to the ROM.

At this moment, as I stood in the midst of the exhibit, I recall that someone had chosen the former of the two selections, "A Cretaceous

Neighbourhood." To my right was the much-touted computer graphics animation theatre – the "Meet a Maiasaur Theatre," as it was called during the development phases – with its big screen, across which a variety of lifelike animated dinosaurs were seen to be walking and moving, the deep drumbeat of their footfalls resounding with each step – the same sort of visceral pounding which Spielberg had used to signal the footfall of an approaching *Tyrannosaurus* in his *Jurassic Park* films. Children clustered around the console, vying to be the first to hit a button after each projected sequence was completed.

Both interactive modules had been activated at that moment in the space, but the volume on the Cretaceous Period display was higher. At one moment, rising distinctly above the surrounding sounds, the resonating yet soft-toned woman's voice spoke out:

> "It was a time when food was everywhere ... it was a perfect time for raising a family."

A time of no scarcity, I thought to myself. Here, in the fossil record, was an economic prescription for raising a family – a kind of naturalistic moral about abundance and fecundity. I wondered whether and how this prescription played out for the visitors. Some were rushing through as if to simply catch a glimpse of the displays, as if to make sure they would at least be able to say, "I saw it all." Others were lingering, gazing, scanning, reading, touching, more or less quietly and gently. Still others – mostly young children – were pounding, bounding, and screaming their way through the space, very much bodily engaged with the surroundings. How and whether they could background or foreground the competing sounds, which easily mingled and bled into each other, was not clear to me. In a few moments, the video with its declaration played again – someone having pressed the console button for this segment once more, the words lifting loud and clear, "it was a perfect time for raising a family." There was no escaping this point.

I also wondered how much of a story, or what sort of a story-phantasy, these visitors were drawing and composing from their experience in this space. There were a number of approaches to this hall on the second floor of the building, so no unitary plan of experience could be predicted.[2] People could have made their way into this gallery having just come from any number of ecological, zoological, or cultural exhibits, or else from the main orientation hall downstairs. The hall did little more than orient the visitor to the various floors of the museum by means of some general floor plans and some showcases and videos sampling the sorts of objects to be found in the museum. Conjuring what John Law has referred to as the "mess of practices" characterizing the best-laid work of experts,[3] this hall made it difficult to imagine an idea of consistent, coherent narratives of natural history, or evolution, or life, or of the relationship of humans to other natural phenomena operating here – unless of course, visitors brought or imposed their own order upon this array of natural/cultural topics available within the public galleries of the ROM.

It was the video's "family" point that continued to resonate around me, both aurally and socially. Here were people moving through in small clusters that appeared to be what were said to be the target audience for the exhibitions: "families." As noted previously, the museum's understanding of that audience seemed very general. Exhibit designer Sam Enright noted the vagaries in aligning with this audience during the Maiasaur Project development:

> How the content is written, how we design is supposed to tie closely to the audience definitions … it may well be the very first or second question after budget questions are asked, "Who is the audience?"… but that is often the most vague sort of criteria … the current director has a strong intention of having this as a "family place"… that audience may have been better defined had he been here then … historically it's not been at all clear … there's a lot of educated guesswork …

However, in the modal North American sense spoken of by several people in the process – and stated quite clearly by the marketing coordinator of the exhibition, Brenda Mikelsen – "a family is an adult and a child."[4] I took this to mean some combination of a small number of adults and a small number of children visiting the museum together, but with the notion of "familial" relationship often read back into such clusters. The crucial point was the dyad – a child accompanied by an adult – but that notion still demanded Enright's "educated guesswork," leaving enormous room for imagining what it meant to be a "family" visiting a museum exhibition. The exhibit, then, would provide some cues to the assumptions made about what constitutes a "family."[5]

The "time for raising a family" phrase started making more sense as an indicator of the dynamic political nature at play. This exhibition for families very explicitly played up the familial thematic – the particular one of mother and babies at the nest (with no mention of a father).

Again, this had been aligned with the well-circulated scientific interpretations of this kind of dinosaur, most notably those proposed by dinosaur researcher Jack Horner. Recalling Jennifer Ross's circle diagram with "Henrietta's immediate family," "neighbours," and "relatives," this phantasy of "family" and the "familial" was clearly a key point of articulation being worked upon in the exhibition. For Ross, the dedicated diplomat-technician, this was the shareable, phantasmatic point of translation and articulation used to connect the knowledge-experience of scientific practitioners with the knowledge-experience of visitors – even though, in Isabelle Stengers's terms, this would carry the risk of betraying the specificity of translation in a way that the curator, Henson, might not have approved, but which he would probably tolerate under the circumstances.[6]

In the work of grasping for whatever narrative coherence and visitor relevance (i.e., articulation) the exhibition might provide, I returned on a later day to continue my close review, now following the visitor track intended by the exhibition developers[7] – but, adding further to my vexations, the gallery had gone through some spatial revisions in the month since my previous visit, introducing yet another disruption to any intended communicative strategy. I entered it from what I knew the planners had intended to be the primary entrance, the beginning of the idealized one-way walk-through. Where the exhibit had been designed and opened with four distinct entry points, the exhibit layout had now been made into a cul-de-sac with the "intended" entrance being the only way in or out. Moving through brought me to a point requiring me to return back through the space the same way I had walked in. Strangely enough, now that there was only one way in, the exhibit's flow design was actually training visitors to move through it in the preconceived manner that had earlier been abandoned in the planning stage, whereas the multi-entry potential of the original layout meant that the experience could permit the story flow – such as it was – to operate in reversible ways, allowing it to follow several permutations depending upon visitor choice or downright accidents of distraction. Indeed, many visitors had previously moved through this exhibit in what appeared to be a highly distracted manner – looking randomly at this or that element, suddenly turning around to look at something else, or even just marching through with a few sideways looks.

The other passageways into the space had been blocked when the summer's much-anticipated temporary exhibition, "A Grand Design" from London's Victoria and Albert Museum (the "V&A"), had arrived during the previous month. A shop with merchandise specialized for that show had been installed in the space which I had previously noted as being occupied by the mounted Maiasaur skeleton, which the year before had likewise replaced the preparation laboratory (see the next chapter, figure 14.1). Most visitors would not notice the transformations, but I was very aware of how (a) completion of the skeletal reconstruction had taken primacy over the display of the technical process in the lab; and (b) now the merchandising potential of an unrelated exhibit took primacy over the communicative effect of both of these. This surprised me somewhat, given the substantial investment in labour, creative planning, story development, communication objectives, and concern for visitor learning competencies for the Maiasaur Project. How could such experience-altering revisions to the workings of the exhibit be made so readily and with such a flippancy and disregard for nature's outcomes as put on display before an eager public?

This willingness to stray from design rigour and discipline meant something other than a relaxed attitude towards the planning and communication processes. It also suggested to me the difficulties of meeting the needs of operating a complex organization with plenty of internal differences on exhibitionary practices that could confound most any curatorial scientist, including Henson – something that, you will recall, Henry Fairfield Osborn obviated by a more direct dictation of dinosaurian nature, of king tyrant saurians, rather than by a dialogic or participatory emergence of such natures. I recalled how the manager of exhibit programming, Wendy Madsen, had informed me of the continual pressures to "rationalize" the museum that those at the ROM faced daily in managing an institution with a cumbersome infrastructure:

> You have to remember ... while the ROM currently has six curatorial departments, those six resulted from the amalgamation of nineteen curatorial departments two years ago ... And within each of those nineteen departments there were a whole lot of different collections and different interests ... So you have a huge number of competing interests ... for limited space, limited resources, attention, etcetera, etcetera ...

Shifting management priorities, contingency and complexity in interests, collections, and agendas had to be "balanced" against scarcity of resources – the purported "reality" of running any and all components of the ROM. As well, the V&A exhibit was expected to be the big summer audience draw. Additional admission was being charged for the

V&A show, which had to recover costs, and in general it was hoped that the exhibit would pull increased numbers of visitors into the museum. The coherence of one exhibit and the nature story it embraced could be put aside for these sorts of managerial complexities and financial imperatives – in other words, care for the natures emerging from study of fossils was being displaced by bottom lines.

The contrast between the layout of the exhibit with and without the lab is particularly striking.

To help give a stronger sense of the exhibitionary spaces, I have prepared my own translation devices in the form of several schematic floor plans.[8] Figures 13.1 and 13.2 are two generalized plans of the exhibit spaces as they were laid out over two different spans of time, indicating the various display elements they contained and which are included in my discussion here.

2: Entry Model and Fossil Showcases: A Strangely Familiar Order

At the time of my second visit, the spaces of the gallery had been reorganized in what might seem a familiar exhibit hall pattern: an entry area with a marquee or title banner, followed by glass cases, culminating in a space for the reconstruction of the featured dinosaur skeleton. Now these "traditional text- and object-based"[9] exhibit components were followed and "complemented" by the technologically sophisticated interactive computer graphics of the "Meet a Maiasaur Theatre." In the original layout, the theatre would have been followed by the lab/mounted specimen space, but, as mentioned, that had been removed.

A Fetish of the Familial

The signed entry area displayed two large chrome-like mesh panels with the name of the exhibit, *The Maiasaura Project: The Life and Times of a Dinosaur*, indicating the sort of chronotopic vision that was to be presented. The suggestion was that visitors would enter the everyday world of the Maiasaur – or, more precisely, of this particular Maiasaur. The first display object to be encountered upon entering the exhibit was an ultra-modern, shining metallic scale model of the reconstructed *Maiasaura* dinosaur (figures 13.3a, b). It was a futuristic figure of exquisite detail and preciousness. Little did I know at that moment what a potent materialization this otherwise diminutive reconstruction would constitute – small as a baby Maiasaur, yet composed to model an adult.

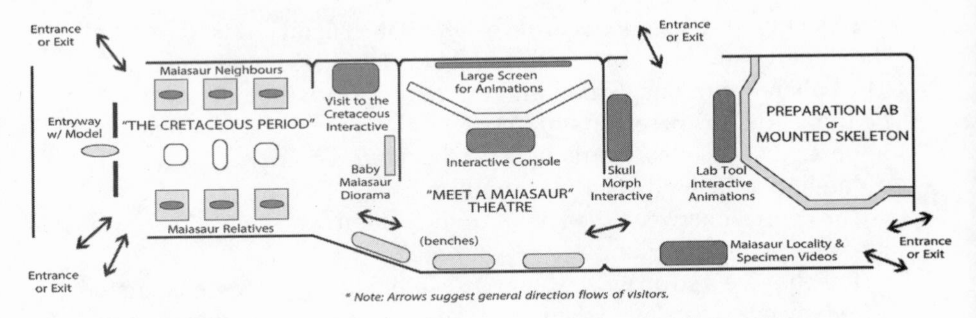

Figure 13.1. Original layout of exhibition (May 1995–May 1998), schematic.

The figure stood in a fairly static standing position – the sort of pose characteristic of a silhouette in encyclopaedic dinosaur identification books, to which many children growing up in the comfort of Euro-American education and commodified entertainment milieus are often exposed. The model's four-footed stance was somewhat unusual for this sort of dinosaur, which had for most of the history of palaeontological representation been shown standing on two legs. This revised position placed the creature in a more horizontal stance, somewhat reminiscent of contemporary stances rendered for *Tyrannosaurus*, a point noted to me by many of the museum visitors with whom I spoke.

The metre-long model was mounted on a wooden pedestal, its head positioned about four feet from the ground (about eye level for most ten-year old children). It was three-dimensional, free-standing, accessible for viewing from the front and sides, something to be admired for its craftsmanship – and, not being enclosed in a display case, it was free to be touched for its pewter, sterling-like finish, displaying in low relief the fine dermal architecture of the creature in life, based on Hadrosaurian skin impressions collected by palaeontologists. It is as if this precious object was drawn, in its totality, directly from the earth, polished to brilliance, and placed on display before us.

This was a finely crafted model with exceptional detailing in the skull, the folds of the skin around the neck, and the backbone with its row of tiny bumps. Even the subtlety of its pose – the creature stopped in the midst of walking, its left foreleg slightly raised – was striking. At the time, I did not appreciate the significance of these details. Some of these features were cues that I understood only much later to correspond with

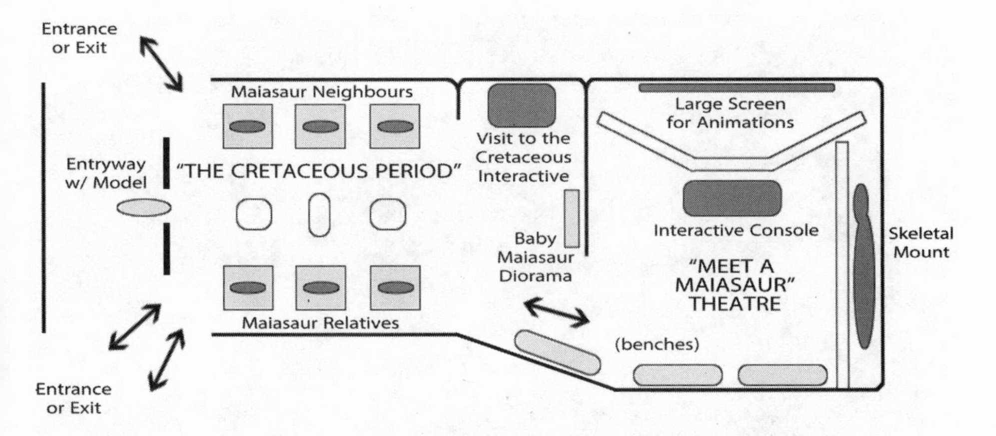

Figure 13.2. Modified layout of exhibition (May–September 1998), schematic.

Andreas Henson's inferences about the anatomy of this kind of dinosaur. Most notably, the model's interesting dewlap folds had been recognized from skin impressions associated with the ROM Maiasaur specimen, while the stance appeared to follow interpretations from Henson's previous comparative studies of ornithopod dinosaurs and from studies by Osborn's mentor, Edward Drinker Cope, a century earlier.[10] At the same time, however, there were no explicit display references here to signal the connection to the translation work undertaken by the curator in regard to the dewlap. While promising and compelling features, these articulations to the technical were not made explicit. Visitors could not know the process that culminated in the detailing of this model. This then misses out on the potency of *articulation* that Bruno Latour pointed to when he wrote of how such articulation "stresses the independence of the thing; reveals the two planes [of knowledge and belief] at once; maintains the character of historical event; ties reality to the amount of work."[11] At best, the dynamic process of articulating scientific practices would be left to inference in this initial encounter with this highly believable rendering.

On another visit to the gallery during the museum's "March Break" activities (when adult-accompanied children arrive in huge numbers during their annual spring week off from school studies), I sat on a bench across from the pewter Maiasaur. Child after child – from three

Figure 13.3a

years of age and older, and often tugging on the hand of their parent or adult companion – would reach out to touch, pet, caress, or otherwise sense the contours of the lamb-sized replica. The possibility of tactile contact with something precious, here a fetishistic reconstruction, was immediately offered and, in the case of most children and indeed many adults, was consistently taken up.

What took place here is worth recounting in some detail, and I draw this directly from my journal:

> As children would approach [the pewter Maiasaur], they would do a double-take, or march straight up, hands stretching out, to touch the chrome-like model. The accompanying adult – mother, father, grandparent, or some other guardian or friend – would turn and follow the lead. The child would invariably look over her/his shoulder up to the adult with a look of satisfaction or even awe. The adult might then read the text, as one did, remarking: "Oh, this is called the Maiasaur … that means Good Mother Lizard … that means that this kind of dinosaur took care of its babies. It was a good mother, just like people … did you know that?"
>
> The child would look at the creature again, as if the message was slowly sinking in. But on the other hand the creature was not like a big dinosaur.

Figure 13.3b. Figures 13.3a and 13.3b are two views of the pewter Maiasaur. Source: Photos by E. Siegrist (13.3a), B. Noble (13.3b), used with permission of the Royal Ontario Museum.

It was actually child-sized, about the size of a terrier dog or a lamb perhaps. The children's hands would reach most often for the snout of the animal, or run along the ridged spine (I remember that many kids would tell me what they learned from this exhibit was that the creature had spikes along its back, which of course is not typically the way duck-bills would be described in descriptive popular books, nor in the majority of technical descriptions). The next-most-touched bit was the foot, but that often corresponded with smaller children who could reach the toes most readily.

I watched for about fifteen minutes. I must have seen a couple hundred people walking by, in parties of two to five at a time. And certainly every second party had someone who had to reach out and touch the model. Adults generally didn't reach out, except in one or two instances where the child with them had reached first. Again, the adult took on the interpreter-teacher role, either reading and interpreting with, "Look, it's the good mother lizard" (very typically), or else asking the child, "Did you see this dinosaur before? Do you recognize it?" and the like.

As to what was going on in the difference between adult and child relations, a few things were pretty clear. Children were into experiencing and sensing the exhibit. They were delighted to find something that was not behind glass or cordoned off. Not only that, here was something their size, at their eye level. Most adults – excepting the rare few in wheelchairs, for example – would look above this, but the child's interaction would bring their gaze down. The text then gave the parent the means to relate with the child in an articulate way. But almost every child would dart out a hand, even if they were running by – a quick head turn, a flash of recognition, the quick brush of a hand against the nose, the possible one-second halt, gaze, and touch. To touch this dinosaur was almost a reflex for most children. It was testimony to what happens when museums ease their restrictions against touch. The metal shone in the spots of most repeated contact – toes, nose, the ridge of the back – leaving traces of the history of interaction.

The power of the model was astounding because I knew that, to some extent, it was a default element that survived the ravages of budgetary struggles. Jennifer Ross explained:

> This was something you could at least touch, although I noticed a lot of people say "No, no, don't touch it." But it was meant to be touched. It is about being at a museum, right? I think eventually we put up a little sign saying it is okay to touch it. Because you did not get to see it, I needed something showing the scale. The graphic designer had planned to have a full-scale projection on the mesh in the background, so you could get a sense of how big it was. He had tons of ideas – like, you could climb through and look through its eyes and see what it was like for a Maiasaur. There were pages of inventive ideas. We didn't have the money or it just didn't get done. So, instead, we just ended up with this little tiny model.[12]

The graphic (or "2D") designer for the exhibition, Sam Enright, had conceived of the pewter Maiasaur, noting how it was "to be viewed from all angles, and mounted on a low pedestal so that it is accessible to visitors who wish to 'pet' it." He continued:

> I was inspired principally by a display in New York on the Statue of Liberty ... They had portions of the statue recreated full scale and they were all hands-on touchable things that kids really like ... And I thought

> that was an opportunity to get kids to engage with this thing … that they might even approach the thing as a "pet"… And hold them there, because we were expecting to present a certain amount of didactic material there, that might not be engaging in itself to kids, but if the kids were somewhat entertained, the parents would help … in showing things like scale …[13]

I recalled Andreas Henson's remark about presenting *Maiasaura* as an animal that children might view as a pet, like the character Dino from the 1960s American television cartoon *The Flintstones*. But the precious metallic quality of the thing brought in other valences. The modern, sleek-lined feel of the exhibition extended well beyond the little pewter Maiasaur into the organically shaped, almost butterfly-like, but otherwise nondescript wire mesh forms hanging from the ceiling, giving a sense of the exhibit's new design elements against the older architectural structure of the building surrounding them. Sam Enright had also told me that the exhibit's ultra-modernist character had been conceived by the 3D designer Mark Alton, whose proposals reflected this point in the "Concept Design Summary":

> The aesthetic pursued will be one of modernity, clarity, and elemental simplicity …
>
> The materials specified will stress a constructed or component nature, sympathetic to *the exhibit character which assembles facts and information to construct the best ideological and experiential model of the Maiasaur.* The materials will provide a contemporary context that stresses the currency and continuing nature of the research being done on the specimen … The refinement of the architectural detail should stand in contrast to and further amplify the experience of the organic nature of the specimen. (Emphasis added)[14]

Here was the explicit nod to the intellectual project being put to work, which "assembles facts and information to construct the best ideological and experiential model," arguably the longstanding, even quotidian project of all natural science exhibit development, and very much akin to the modelling work typical of palaeontological reconstruction (see chapter 4). Here also was a precise case of producing the sense that things human and technological were utterly different from the fossil, which was non-human, organic, natural, unmodified. Later, seemingly contrary to this advice, the multimedia producer would instead attempt to draw everything into a seamless total, "a living, breathing environment," of which the visitor would be an interactive (if not intra-active) part. Nonetheless, the very abstract formal features built into this space had been meant to cue the visitor to the contemporary, aligning modern clarity with the ongoing virtual assembly of "facts and information" in the construction of "the best ideological and experiential model" research could offer. Gleaming shininess and technologically sleek elemental forms signified the cutting-edge re-creation work taking place here. This sensation was translated into the soundscape of the exhibit when few or no people were there. When "at rest," and although this would be less noticeable in the din of visitor traffic, the exhibit played an ethereal, fluid sound, the sound of unfolding, an ascending chord, then a glissando and gentle cymbal crash, rising then falling. It was gentle, almost akin to breathing, the rise and fall of waves – like New Age "nature" and "relaxation" recordings – but here it was naturalistic, organic sound-motion produced by digital, electronic means.

The Performativity of "Good Mothering"

The paired chromium screens flanking the metal Maiasaur showed the exhibit's title, once in English (left), once in French (right), subtly privileging English readers in this left-to-right reading/writing convention. With this titling, visitors were cued from the outset that this would be a bilingual exhibit:[15]

the Maiasaur project	the life and times of a dinosaur
le project Maiasaura	vie et moeurs d'une dinosaure[16]

Though gender was not grammatically inscribed in the word "Maiasaur," it was semantically included in its translation: "the good mother lizard." This was explicitly noted in texts on the pedestal base for the pewter Maiasaur. However, gender was also signalled in the name's Linnaean generic form, *Maiasaura*; the "-a" suffix signals feminine gender, while the more familiar "-us" (as in *Tyrannosaurus*) signals the masculine.

Just as the pewter Maiasaur acted as a polysemic three-dimensional touchstone to the exhibition – preparing visitors simultaneously for a tactile experience, an approachable familiar creature, as well as something rare, not to mention dynamic, in the curator's sense – so the accompanying exhibit text, printed horizontally on the mounting table base, provided additional indexing keys to the time-space and eco-behavioural particularities of this dinosaur presentation – all the while focusing on one individual Maiasaur (the typographic emphasis approximates the original):

> **Maiasaurs** (my-ah-sores) **were large plant-eating dinosaurs** that lived about **80 million** years ago. By comparison, humans evolved less than 5 million years ago.
>
> We are showing a model of a single **maiasaur** here, but maiasaurs were **rarely alone**: they lived in enormous **herds** of up to 10,000 animals.
>
> **Maiasaurs** are noted for the care they gave to their young. Their scientific name means "**good mother lizard**."

In these limited but potentially dense cues, the exhibition plays the "individual" in its "social world," its uniqueness among the masses – a virtual prototype of liberal philosophy. Here, almost from the outset, are the suggestions of the Maiasaur's "life and times," its nurturing capacities aligning in an uninterrogated way with its gendered nomination as the "good mother." There is no mention of nurturing "fathers" or males. There is instead a matter-of-fact point about who does the caregiving in the life and times of this dinosaur – it is the good mother. Indeed, that impression of maternal benevolence reached visitors, as in the case of a woman I spoke with who was visiting the museum from the USA:

> BN: What would you say the story is of that exhibit, what were they trying to convey to you?
>
> VISITOR: I would guess, I would have to say that when you think of dinosaurs there were some very good ones and that that particular one was very good with her young.[17]

Was this Henson's dynamic? Was it Ross's familial relevance? The clear sense of this dinosaur is that it is "her" nurturance that makes her "good." The configuring of family, feminine gendering, mothering, nurturance, a "gentle" and "adorable" dinosaur, articulating with "families" as "natural" audiences for this exhibit, was becoming increasingly obvious as a working phantasmatic order of the display. Some weeks later, while looking at the ROM's webpages on the Maiasaur Project, I came across the page presenting this very translation of the scientific name once again. I noted the effect of this semantic work coming through the URL itself:

http://www.rom.on.ca/palaeo/Maiasaur/maiamom.html

There *she* was, interpolated into the web address: "maiamom."

As discussed previously, Andreas Henson was playing up the contrast of meat-eating and plant-eating dinosaurs – but there was also a gendered spin in his equation. In another conversation, he told me that he was utterly conscious of the potential for gender identification in all of this:

BN: I want to know ... how conscious were you? ... Were you thinking in detail about the political dimensions of choosing the "friendly" good mother lizard as the opposition to *Tyrannosaurus rex*, being a contrast between a feminized association and a masculinist power logic?

AH: Yeah, absolutely.

BN: You were?! ... So who were you discussing this with?

AH: ... Actually no one. [*laughing*]

BN: No one? You mean, you weren't talking with the exhibit planners or any of the designers?

AH: No, no ... I came to them with this logic, this vision ... whatever you want to call it ...

BN: And you knew ... you were playing all this out?

AH: And I think that was part of the initial resistance to it ... because they weren't used to thinking about it that way ... in a very typical mainstream way, dinosaurs were thought about in this rather nasty manner, as these robust, vicious, aggressive creatures ... and to have this other thing, that dinosaurs may have been caring parents, who didn't just drop their eggs on the landscape and wander off ... that there might be some kind of community structure, family if you want, parental care, and so on ... that's a totally different image of dinosaurs ... So basically – and this is a very important metamorphosis – because you basically get away from the notion of dinosaur as monster to dinosaur as animal, capable of complex behaviour ... So, from primitive brute, you get to this sort of civilized, developed organism ...[18]

Here Andreas Henson, using a logic of developmental progress in his contrast between "brute" and "civilized" dinosaur, had also taken credit for introducing the greater complexity of the saurian world into the ROM's display economy. It is also clear that he did so with full awareness of how that would be underwritten by cultured, binary gender associations. With a consummate understanding of the deployment of tropes, ironies, and hyperboles, Henson claimed to consciously extend what he thought would play against his sense of a dominant public cultural logic that imagined aggressive-to-passive, masculinist-to-feminist relationalities. His idea was that if these logics were deployed smartly, the translations would help bring a sensibility of the greater complexity of the saurian world. His calculations, however, had not been able to predict how interpretive planners would fill out his account, underwritten as it already was with a specific normative, heterosexual binary.

"Family" Audience/Dinosaur "Family"

Andreas Henson was also careful to point out to me how the nuance in the "family values" story was an interpretive turn from the more elemental dichotomy he was trying to make:

> [I]n the script, however, they ran with that notion and you get family values – "the perfect time to raise a family" ... That, I no longer had any input in ... that just sort of happened ...

So, having brought these "logics" forward – from gendering that signals shifts in intellectual modes to audience and market potential – Henson now appeared to be distancing himself from the narrative consequences that developed once the planners began to activate the logics in their own fashion. Henson continued, offering his thoughts on what may have taken place:

> Observing in our current social-political context, that's the emphasis on "family values," the notion that because family values as traditionally defined – i.e., "father knows best" – are allegedly deteriorating, society at large is falling apart, and you know we soon have [*shifts to ironic tone*] roaming bands of god-knows-what [*chuckling*] on the street ... So, I think there's a tendency on the part of many educators to try and preserve the status quo, as in "oh look at this miracle of family"... and I think, even this poor dinosaur was co-opted into this political agenda ...[19]

Henson – reciting, with his regular searing irony, what sounds rather like Arthur Conan Doyle's fear of declining male centrality in the family

– had missed how both he and the "educators" were actually working with similar strategies of imagining. From all indications, the planners' motivations appear to be just as much located in their particular phantasmatics of "family" audiences and what would be relevant for them as had Henson's when he imagined the exhibit's child audiences dividing up the dinosaur universe into vicious killers and friendly creatures. Scientists and planners were using the same strategies – and for both, these were readily collapsing science/society, nature/politics, materiality/phantasy – and this by way of the syntaxes of human kinship reckoning. For the planners, the translations moved to the inscribing of common "family structure" across the human-saurian divide – something they knew that the history of scientific accounts from Horner, and further cited by Henson, would uphold at least sufficiently for most visiting publics. Such implosions, fusions, and articulations by scientist and planner alike produced *relevance* that required them to imagine the audience of the communications. As Sharon Macdonald remarked in her studies of exhibition development at London's Science Museum, "the very fact that a communication is 'for the public', and that it embodies a specific vision of that public, shapes the kind of representation made."[20]

The gendered positioning of the key translators, the curator Henson and the planner Ross, is also instructive. Both had imagined how audiences would think about the exhibit, but along slightly different axes. Andreas Henson, the remarkably self-aware scientist-curator, is a leading expert on dinosaurs, and is among the estimated one hundred dinosaur palaeontologists with academic or curatorial appointments in the world – almost all of them men.[21] The notable androcentry of this field of disciplinary interest is counterposed by a similar historically contingent situation in the gendered make-up of the ROM staff. There, Henson was among a predominantly male curatorial staff who had a significant degree of authority in scientific/public knowledge translation. Set against this group was a predominantly female museum programming staff charged with translating the curatorial content for the public. This gendered economy of knowledge production and translation was something of a joke among staff in exhibit programming. One of them stated it with an anthropological and Lévi-Straussian twist:

> My joke was, "It's still a hunter-gather thing! … The curators go out and bring it, and then we have to cook it and prepare it. They bring in the raw stuff and we have to cook it … Actually, I think some of the reason … is

that museums tend to be lower paying than a lot of other employers. It is a funny quirky place. There are a lot of men. These communications departments are almost all women. Media relations is all women … All the press people who do the press releases. There's a *lot* of women in the museum.

A gendered knowledge economy – expressed directly through wage differentials and professional training disparities – had permeated the very making of this exhibition, and with it the naturalistic ordering of dinosaur life that would articulate most with the imaginary of its anticipated audiences.[22]

Of course, the stewarding of every detail in the making of an exhibition is a practical impossibility, while translation is a practical necessity. As a consequence, Henson's gendered binarisms passed into the exhibit, subjected to a series of further translations en route. Here was a very literal train of translations in materially based phantasmatics, traceable back through the planning process to the logics of the palaeontologist, through to Horner and Makela's publication on "evidence of family structure among dinosaurs," through the revolution in thinking about dinosaur bioenergetics, and beyond.[23] At every step, the phantasmatic orders were also underwritten, however selectively, by the materiality of fossils, by applied scientific techniques in field locales, by graphic verification in technical publications, and so on. Such trajectories of gendering would in all likelihood have been redirected away from such possibilities in the times of Osborn and Doyle and their networks of natural ordering.

On the whole, the women who made up the exhibit programming staff confirmed what Henson had said about their initial resistance to the nurturing image of the Maiasaur, though they also recognized that they were up against bigger, meaner dinosaurs in the consumerist world of public culture, museums, and leisure entertainment. The manager of exhibit programming, Wendy Madsen, had recalled conversations among her staff:

We would say, like, "This is great, Hollywood's doing the 'man-eater,' and we're doing 'the mother lizard'" [*laughing*] … "Can't we change the head on this thing that looks like a camel …" We really were joking about it like that, just to say, "Oh, my god!" – I mean the whole thing is that there's this *T. rex*, and it's this … *monster, it's this monster* [*growling in imitation*] that will eat you alive, and that's what it's doing in Jurassic Park … It's the ultimate man-eater, the ultimate predator, the ultimate threat … So, what do

> we have? We've got *this dopey mother*! ... So certainly there's a great interest in dinosaurs, and certainly babies are always big sellers with families ... And we had thought ... we were promised that there was a baby to this thing ... not just a mother, there was baby ...[24]

But such consumer sentiments also make clear that human and nonhuman concerns for kinship and familial relations were deeply at play and readily transposed in this exhibit. For the exhibit planners and the webpage designers, just as it had been for the original scientific team who named *Maiasaura*, this dinosaur was thought of in terms of a mother, a "she," a "maiamom." Equally to that point, she would certainly not be this man-eating monster that over and over seemed to hover ominously in the near reaches of public/scientific imagining. Visitors were often very much aware that the museum was working against such images, as in these comments of a fourteen-year-old (Sarah) and a twelve-year-old (Andy):

> BN: Why do you suppose they have a Maiasaur exhibit here in this museum?
> SARAH: Maybe to go more one-on-one with a type of dinosaur ... People will learn more about a single thing, and learn more about what scientists know.
> BN: Is this the kind of dinosaur you would choose to do this with? To do an in-depth story?
> SARAH: I don't really know. Maybe something like the *Tyrannosaurus* or Raptors would have more public appeal.
> ANDY: But the Maiasaur would break the stereotype, true?
> BN: It would break the stereotype of?
> ANDY: Of the big, long, green things ...[25]

Here was a simple but stunningly acute analysis by two young visitors about how the exhibit was designed for its anticipated audiences. They understood the curator's and the planner's subterfuge, that *T. rex* – and other dinosaurs glossed as "big, long, green things" – were iconic forms needing to be displaced with alternative ideas and forms of life. The Maiamom was doing her work, even though in the course of the project's development the organizers would find out that there was no baby dinosaur, that actually there had only been a handful of juvenile dinosaur bones found in association with the skeleton of the adult. Nonetheless, this rather storybook-ish approach – with good mothers and their babies – had sedimented itself into the process, aimed at an audience of children with adult companions. It appeared that such children were assumed to be familiar with the "day in the life" genre of storybook writing (e.g., a day in the life of "a baby killer whale," "a tiger," "a firefighter," or "a puppy").[26] Among the most popular of North American feature films on dinosaurs for young children throughout the 1990s and later have been the animated film *The Land before Time* and its several sequels (the first of which are, notably, productions of Steven Spielberg and George Lucas, who brought us the film *Jurassic Park*). The films base their action in relation to a dinosaurian time-space known as "the great valley," a sort of Eden-like verdant utopia, impervious to marauding "sharp teeth" – the latter being the filmmaker's child-code for carnivorous dinosaurs, which nonetheless do make their way into the storylines, generating fear, tension, and drama. Children visiting the ROM and its Maiasaur exhibit helped me understand the connection between those films and this exhibition, as in this exchange with three school-aged girls (ages five to seven years) who spoke of the film's nearest Maiasaur equivalent:

> KYLA: And my favourite one on *Land before Time* is Duckie ...
> BN: Duckie ... ?
> ALLISON: Duckie is a water one.
> EMMA: Duckie is a little water one and he is green.
> KYLA: One of those duck ones ... with the face.
> BN: Oh.
> ALLISON: And it's really small, like that. And the mother is huge.[27]

The Horner and Makela accounts had travelled widely into the terrain of popular culture and were here carried back into the legitimation space of the ROM – strong articulations to particular evidence-related phantasmatic-materializations.

The approach to storyline was becoming ever clearer: this exhibit was more or less a domestic life biography of "a dinosaur" in its life-world in other bygone times, the life and times of this particular dinosaur, this particular Maiasaur. Indeed, this seemed quite reasonable; Jennifer Ross, the storyline developer for the exhibition, had told me that her first guiding text for the exhibition had been a children's book.

Elementary Structures of Canadian/Maiasauran Kinship

> Kinship is not only about relations between and among humans, and not even about humans in a sense that is based upon, or drawn by analogy from, animals, as was Darwin's quintessentially modern claim. Perhaps

in speaking of kinship in the postmodern era, or of postmodern kinship *tout court* it becomes necessary to think about the constitution of what can stand as a kind or type at all.

Anthropologist Sarah Franklin on the "Idea of Relation"[28]

Sarah Franklin's remark about kinship – not in terms of "kinds" or fixed analogues but rather as always emergent possibilities for enacting relations – becomes all the more potent in the face of *Maiasaura*'s familial relations as enacted by curator and planner alike. All of these curatorial, planning, and translation moves around the syntax of family would similarly become apparent in continuing further through the exhibition space. From my experience during the gallery walk-throughs, the technicalities of science as hard-won demonstrations of unadulterated fact seemed relatively absent thus far. Ahead, however, lay labels, specimens, and cabinets, reminding me that this was a museum in a more traditional sense, a place of authority and seriousness, of things normally beyond the everyday reach of most who would visit here.

The area was subdued but theatrical, with spotlighting to highlight the cases and the texts. The pewter Maiasaur, in a similar fashion, had been dramatically lit, casting sharp-edged shadows, increasing its visual allure and its value (as precious metal if not as an official specimen). Akin to a department store show window with an array of merchandise, this space of fossil showcases also spoke to public sensibilities of the "displayed worlds" which so regularly inform consumer life.[29] Though referring to a science exhibition which directly borrowed on the "supermarket" metaphor, Sharon Macdonald noted the practised connection of shopping and display viewing in which consumerist logics are "there already in assumptions about consumption as a key means of expressing individuality, activity as choice, objects as commodities … and museums as part of the marketplace."[30] In this instance we were dealing with an exhibit not on shopping and food consumption by people, but rather one in this component resolving around the varying food choice of different dinosaurs!

Performative frames of experience are recognizable as everyday dimensions of contemporary social experience when such linkages of the theatrical with museum display and with consumer shopping experience are considered. A messy spectrum of conjunctions including <object + display>, <commodity + commerce>, <specimen + spectacle>, <performer + stage> are hooked together in the embodied sense of performative practice in this particular nexus. Without that practised sensibility – and facing the messiness and exigency of the museum's planning and development techniques – how well could the *Maiasaura* exhibition extend itself in relation to its varied, if wishfully predicted, publics? I had to wonder whether the interpolation of consumerist praxis within the instruments of exhibitionary mediation was not already so highly entrenched as to be quite intractable. The distinctions between market, audience, consumer, and visitor blur all the more in such a place of familiar encounters.[31]

So, in the same instant that the exhibit further fetishized the entry Maiasaur, lending it an aura of preciousness by the choice of material, scale, and theatrical lighting, it also continued to press forward the proposition that the Maiasaur was an object which could give entry to the familiar – an everyday experience – and simultaneously entry into a differently configured and animated nature that the exhibit-makers had sought to share. The exhibit would continue to articulate this stand-in dinosaur to the human world by increasingly deploying familiar tropes from human domestic experience – tropes which reassured us that this dinosaur had a domestic life as well. As the psychic distance between viewer and object was reduced, affinity and connection were increased. Scale and tactile contact, along with such domestic sensibilities and everyday practised experience of displayed-world engagement, combined to provide the visitor with a very non-threatening introduction to this dinosaur. One might even develop a sense of communion with these creatures – a far cry from Osborn's striding menace of maximal evolutionary energetics, from whom we rather would be compelled to feel alien, vulnerable, and quite far apart.

Inside the collections display area of the exhibition (figure 13.4) (the section which planning documents labelled "The Cretaceous Period") were the following: two ranks of specimen-bearing glass cases (known by designers as "freestanding showcases"); and three central pedestals (known as "table cases"), two of which also contained specimens, with the third bearing only a graphic element. Inside each freestanding showcase was a printed text signalling what was being looked at. One label indicated "Maiasaur Relatives"; in French, "des Parents de Maiasaura." The terms offered a literal impression of kinship, of saurian affinities. In the showcase on the opposite side, the contrasting texts read "Maiasaur's Neighbours"; in French, "des Voisins de *Maiasaura*." Interestingly, the cabinets were mostly full of skulls and jaws, as noted in this schematic of this first segment of the exhibition (figure 13.5).

The showcase pedestals had what the design diagrams called "ribs" – more or less scroll-cut wooden forms – at the base of each showcase.

Figure 13.4. Fossil specimen gallery. Source: ROM Photo, reproduced with permission of the Royal Ontario Museum, © ROM.

I mused over the inference of being not just in Walter Tomasenko's "living, breathing environment" but indeed in the belly of some beast, knowing full well that I was in air-conditioned rooms in a prominent museum at the heart of Canada's largest urban centre. Given its position in the space, the first table case along the central axis seemed a likely place to find the next node along the ideal knowledge-guiding path of the exhibit – though I later noted that visitors often only randomly looked at this case among the several available, if at all. I glanced at its contents and sure enough found some opening statements from the guiding narrative. The labels declared what was here:

> Original bones from Maiasaur (juvenile tibia and humerus)
>
> World Premier: You are looking at recently cleaned fossil bones of *Maiasaura peeblesorum*. The ROM's Maiasaur is the best specimen of its kind found to date. It can take more than two years to separate an entire skeleton from its rock matrix and many surprises can occur in that time. You can watch this process at the other end of this exhibit.
>
> [Blackfoot Nation Land, Northwestern Montana USA, about 80 million years old.]

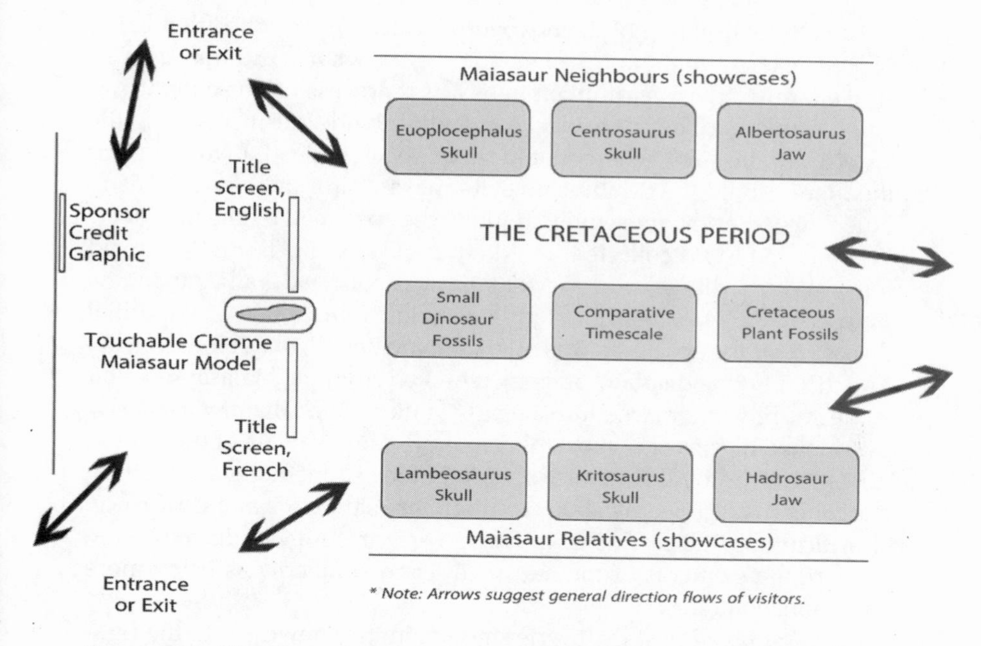

Figure 13.5. "The Cretaceous Period" or "Maiasaur Relatives and Neighbours," exhibition entryway and showcase gallery, schematic.

Though I was uncertain of who would read these texts, this narrative anticipated the totality of the exhibit, noting the "process at the other end of this exhibit," which was actually no longer there. At the time of this walk-through, the preparation lab had been removed from "the other end of this exhibit," the fossil "cleaning" having been completed some eighteen months earlier. It appeared that budgetary concerns trumped care for the specimen and the "process" of its preparation and study.

The contents of this case were changed from time to time over the course of the two-year specimen preparation "phase," displaying the elements that had most recently been prepared. The first to have made their way into this cabinet were the foot bones of the specimen – as previously mentioned, the skull of the beast never did make it.

The off-hand noting of the discovery locale, "Blackfoot Nation Land, Northwestern Montana USA," is notable for what it did not say. Unspoken, unwritten, yet implicit here was a process of museum acquisition – this may be from Blackfeet Indian land, but it was now the "ROM's Maiasaur." Where would this unpacking lead, I wondered at the time? Would the exhibit convey to me, a visitor in kind with many others, whether the museum was given the specimen by the Blackfeet? Whether the ROM collected it? Whether or how much the ROM paid for it? Why a dinosaur from Montana was collected, rather than one from Canada? As already noted, a very incomplete tale of the initial discovery of the specimen was offered in a video display. However, before this day that display had been removed from the Maiasaur exhibit space entirely and placed into a nearby gallery, such that it was mostly overlooked by visitors – most told me they were unaware of the video's existence. This fossil showcase section of the exhibit attended to different concerns – concerns about comparison, sameness and difference, classification, and family living. These were, presumably, deemed more appropriate elements of the narrative – and somehow as being more about palaeontology.

Having moved past the pewter model, in the showcases to the right were the "Maiasaur relatives" – three other sorts of duck-billed dinosaurs, highlighting what for the ROM was an internationally significant collection of this particular dinosaur group. Here again were the translation terms from Jennifer Ross's circle diagram. It was clear from these creatures that the Maiasaur was not alone in its life and times. Like the human audience for this exhibition, this dinosaur had "relatives" – part of the security of life in Maiasaur's world of familial social relationships.

Curiously, even as the "traditional" family is considered to be eroding in everyday Euro-American practice, it was somewhat more deliberately conserved in *Maiasaura*'s analogical nature. "Analogies," as Marilyn Strathern remarks, "are pressed into the service of … keeping values intact," this taking place when such analogies, including those of kinship, are "carried by people across domains often because there is some argument to be pursued."[32] The argument here, of course, is the doubled one: show dinosaurs to be related, to have families, thus appealing to presumed audience sensibilities and experience (the interpretive and marketing argument); and show them to have more complex sets of relations so as to trigger more of "a sense of a dynamic process" (the curator's argument). All of this was on offer, playing off the recirculated, now commonplace phantasmatics of older analogies and arguments of the sort Osborn and his twentieth-century inheritors advanced: Dinosaurs were alien, like us only in that among them were those that hunted other dinosaurs and killed and so were superior, and those that foraged plants and fled the killers and so were rank and file, fodder. It was the dynamic properties of kinship that now showed how *Maiasaura* and "her" ilk could both equal the great carnosaurs and, one might think, even begin to undo the Osbornian analogy.

To extend this human social-worlding effect in the Maiasaur exhibit, just as any person/dinosaur should have *relatives*, so in living in a given locale they should also have *neighbours*. Opposite the "Maiasaur relatives" cases were the "Maiasaur neighbours" cases. As with human experience, domestic life for this Maiasaur was not only about having relations and sharing a social world with those displaying recognizable physical affinities – it was also about sharing it with those who exhibited physical difference. I found myself musing on whether the human relations syntax might extend further: could this be an analogy to everyday human life in Toronto, a city touted as one of the most "multicultural" in the world, a place of care and concern for difference?

Those who are different from the family-forming ego of this story become "neighbours," including herbivores like *Euoplocephalus tutus* (an armoured dinosaur) and *Centrosaurus apertus* (a horned dinosaur), and even carnivores like the Tyrannosaurid *Albertosaurus sarcophagus*, the latter noted in the accompanying texts as "the terror of the many plant-eaters of the Late Cretaceous." Here was a reminder that some neighbours may be less friendly than others – the very sort of note struck in the "Family"-rated *Land before Time* films. Haunted by the older analogy, this was both a partial recollection and a partial turn from the fiercely competitive envisioning of former dinosaur world-makers, like Osborn and Knight, who inscribed the natural life of Spencerian/Darwinian struggle in their exhibitionary and artistic projects, pitting carnivore against herbivore in repetitive "battle" scenes. Here instead was something reminiscent, for me at least, of the story of the Canadian multicultural experience, combining a value for genealogical commonality (relatives) with a tolerance for racial and ethnic difference (neighbours). Nonetheless, there was an ego-identity in this tale – the "good mother lizard." In this space of varied sensory experiences, the predominant thematic of cozy domesticity and the "fact" of mothering kept returning.

I recalled how the interpretive planner, Ross, had stressed that the key to an exhibition was in making it "relevant," to animate it somehow for a visitor. On first consideration, bones and skulls and skeletons on their own lacked sufficient animation in the strict sense. Yet there was relevance here: the relevance in the "life and times" story, and specifically in the building story of affinity and relation.

Still, it has to be said that bones and skulls do signify on their own in the museum setting: they offer a sense of relative scale, they contrast, they are real or copies, they are mounted as technical objects, they stand as exemplars, they are parts for larger wholes.[33] These particular items are presented in their plexiglass cases as informational black boxes – where knowledge of them is assumed as already known to those who present it but requires elucidation for those who look upon them. Apart from the relations story, then, a secondary, undergirding logic operating in this display was that the didactic information included with these little elements of the authentic was true scientific knowledge – knowledge *in* to science through the specimen, knowledge *out* to the public through the spectacle. In relation to the entire exhibit, this space operated to convey a partial sense of the material basis for knowing about the life, times, and relations of this kind of dinosaur, while suggesting that knowing would gradually be imparted. Sharon Macdonald summed up this sort of operation succinctly:

> One effect of science museums is to pronounce certain practices and artifacts as belonging to the proper realm of "science," and as being science that an educated public ought to know about.[34]

In a very practical sense, the untouchable specimens presented here attested to a proper ordering of life known through science: *Maiasaura*, like the humans visiting this exhibit, occupied a mostly friendly world of neighbours and relatives – with the occasional lurking terror, shades of the tyrant king.

To some extent the gallery of dinosaur skulls and jaws had also capitalized on scale – but only suggestively, in the sense that the entire animal forms were not placed on display, only their skull elements. The greater promise of these showcases, then, remained in offering the authenticity of the fossils as the ground on which true relationships were to be known. Of note is that only some of the bones were "real," "original" fossils, while some were casts. Yet visitors repeatedly reported to me on the requisite authenticity of this material, as in the case of fourteen-year-old Tracy:

> It looked real … it should be real … it's supposed to be history.

If it was not real, or simply a real-enough facsimile, would it then not count as history? With a similar faith in the museum as site of the authentic, Marisa, an eight-year-old, pointed out:

> They sort of looked old and raw, and also because, like, why would it be in the museum if it's not real? Where else would they put it?

Unable to imagine another suitable locale for real things, for these respondents museums were a last remaining repository for the authentic, something that could evoke the most reassuring sense of reliable knowing, enough so as to count as real.[35] Yet another eight-year-old, Jenna, remarked:

> Well it's probably real because, um … I think they had a silver band and if you passed it the alarm would go on so it would be very, very precious …

While there was no silver band, realness nonetheless equated to extreme value, preciousness, the need for security. Continuing our conversation, Jenna also told me why it was ultimately important to have these precious things in the museum:

> You show them, so other people can see them … After that, they can learn about them, just to see how big they are and um, see what they look like …

Here was the same motion, performed by curator, planner, designer, and eight-year-old visitor alike: the move from the intrinsically valuable specimen as scientific thing you see to the article as spectacle from which others can learn; from visual testimony to educated knowing.

The fossils did not act alone, however. In this rather traditional display area, the showcases themselves were doing a particular sort of knowledge-training work, acting upon the responses of the visitors. The plexiglass enclosure allowed viewing, but not touching. A separation between the precious thing and the viewer had been achieved. This was all in keeping with the other galleries in this institution, which coursed through natural histories of the world and the region, through civilizational histories of the "West" (e.g., ancient Egypt, the "Near East," classical Greece and Rome, Europe) and the "East" (e.g., dynastic China, classical India). In every instance, precious glass-encased objects, artefacts, and specimens were displayed to the passing viewer. None of this is that surprising in relation to the Western history of museums, which have long served to authenticate, present curiosities, and produce awe and inspiration.[36]

In this setting, however, the semiotics of the ultra-modern pewter dinosaur custom made for this exhibit took on a redoubled potency. Its gleaming silver quality ensured that it obtained the aura of a collectible piece, and yet this object was not contained in a glass box. Rather, it was quite accessible, fully touchable. This gallery imposed a distributed sense of what could be touched, how, by whom, and to what end. A doubled possibility of embodied engagement was offered to the visitor. First of all, in this exhibit of the rare and the familiar (or the familial), direct contact with the simulated and representational thing through the touch offered the chance to commune with this created world. But second, separation from the original behind glass or through a video screen, accessible only by looking, rebuilt distance from and respect for the original, the specimen, the thing made exceptionally real. To be sure, specimens and artefacts are meant in most circumstances to be untouchable. That is, in part, how they are invested with power.

One strand of the tale I was beginning to piece together from the juxtaposed elements thus far in the Maiasaur exhibition went as follows: *once you have been attracted by the wonder of the total animal in touchable, precious miniature form, the first thing to be known is fossils of skulls, the real, all the more precious things; looking face to face at them reveals similarities and differences in the animals, and these in turn reveal relationships and segregation of kinds, relationships, and difference that will be at once familiar and surprising*.

At every step, at every stopping point, a variable mediated connection was forged between thing and visitor. Through the medium of a shareable imaginary, the full set of input knowledges and practices provided in the translations of scientists, media producers, technicians, writers, managers, and promotions specialists came to be processed in such a manner that it would connect with the knowledges of the visitor, allowing modest room for what Ernst Gombrich called "the viewer's share."[37] This was a partially trained dialogic procedure, the trained intra-action, put to work in the co-fabricating of Maiasauran familiality.

A crucial extension of this emerging imaginary, therefore, was genealogy and kinship as means for defining affinity and difference.[38]

In this instance, the promise of the showcases was their suitability for comparative typological arrangement. Comparing heads of "relatives" allowed for the ordering of affinity and similarity. Comparing heads of "neighbours" allowed the ordering of proximate alterity and difference. While a far cry from phylogenetic systematics – the now-standard scientific-technical practice for ordering organismal relatedness and evolutionary diversity – an elementary form of comparative anatomy and classification was being performed here. While the exhibit stopped at evolutionary, physical linkages between the animals, visitors could quickly read social meaningfulness into this biology. One twelve-year-old visitor responded to my question:

BN: What is the name of some of the Maiasaur's relatives? Do you know the name?

KATIE: I can't remember what the names were.

BN: Do you know what general kind of dinosaur Maiasaur is?

KATIE: Well, I could sort of describe it, but I am not sure if there is a technical name for it.

BN: Or a common name?

KATIE: Well, it is leaf-eaters who take care of their young. I'm not sure if there is a common name for it.[39]

Placid, leaf-eating, maternal domesiticity was the relational key. The project of technical comparative anatomy and phylogenetics was only a side story of this segment of the Maiasaur exhibition. This was an exhibit intended to be about one kind of dinosaur, or more emphatically, one individual dinosaur in a complex bygone world of storybook "neighbours" and "relatives," scenery, babies, and mothers.

3: Into the Time-Space of a Maiasaur Neighbourhood

> What I take away is the way that they showed the earth, the evolution, eighty million years ago. You go back and see how everything fit into place and how it was, and where the dinosaurs were.
>
> Martin, college geology student, age twenty-two[40]

Up to this point in walking through the exhibition, the computer and audio-visual interactive components of the display were yet to be encountered. There were sounds emanating from rooms beyond, but so far the display continued as an array of static elements. After the pewter

Maiasaur, the next most highly used tactile element – though not conceived as an "interactive" component – was the graphic timeline situated at the centre point of the showcase gallery, between the Maiasaur's neighbours and relatives. The display now began to orient visitors to the story of interpolating the Maiasaur and these other dinosaurs in time and space.

To reinforce the earlier, possibly overlooked message that the Maiasaur had lived eighty million years ago, this pedestal graphic presented a "chronology" with key times noted – from "the origins of dinosaurs" leading ultimately to "today" (see figure 13.6). As with the model Maiasaur, signs of tactile interaction were more than evident: the most touched parts had been worn down by hands and fingers pointing, tapping, circling – gestures of indication which orient us in our time/place with them in theirs.

While many of the exhibit's communication goals shifted, Jennifer Ross stressed how temporal orientation was essential to "cohesion":

> I remember the task of bringing it into some kind of cohesion ... We had no problem attracting people here, but what would I like someone to come away with? I said to myself, "I would like them to know what a Maiasaur was ... what the Cretaceous is." Just that. They don't need to know about the Jurassic and all that stuff ... So I used the word "Cretaceous" a lot. I hit that "eighty million years" when this one was supposed to be alive, and Maiasaurs in general. Just try to ... drill home a few simple things. I thought you can't possibly come away with everything.

Chronotopic gestures would locate *Maiasaura* – that was the minimal expectation, and, based on my interviews with visitors, it was one that the exhibit did indeed meet.[41]

Interestingly, the dominant flow of intelligibility was indicated by tactile wear patterns on the timeline itself. The most worn "moment" was that labelled as the appearance of humans. I imagined a visitor thinking, declaring, "Here, this is the beginning of us. We are human!" The next most worn was the appearance of dinosaurs – the imagined thought now being, "And that point was the beginning of them ... when the dinosaurs' world began!" The third most worn area was the extinction of the dinosaurs – "There, their world came to an end ... But we live on and we know them from their remains." The fourth most worn area was the Maiasaur time of eighty million years ago – finally, the creature of this exhibit has been located, the wear pattern

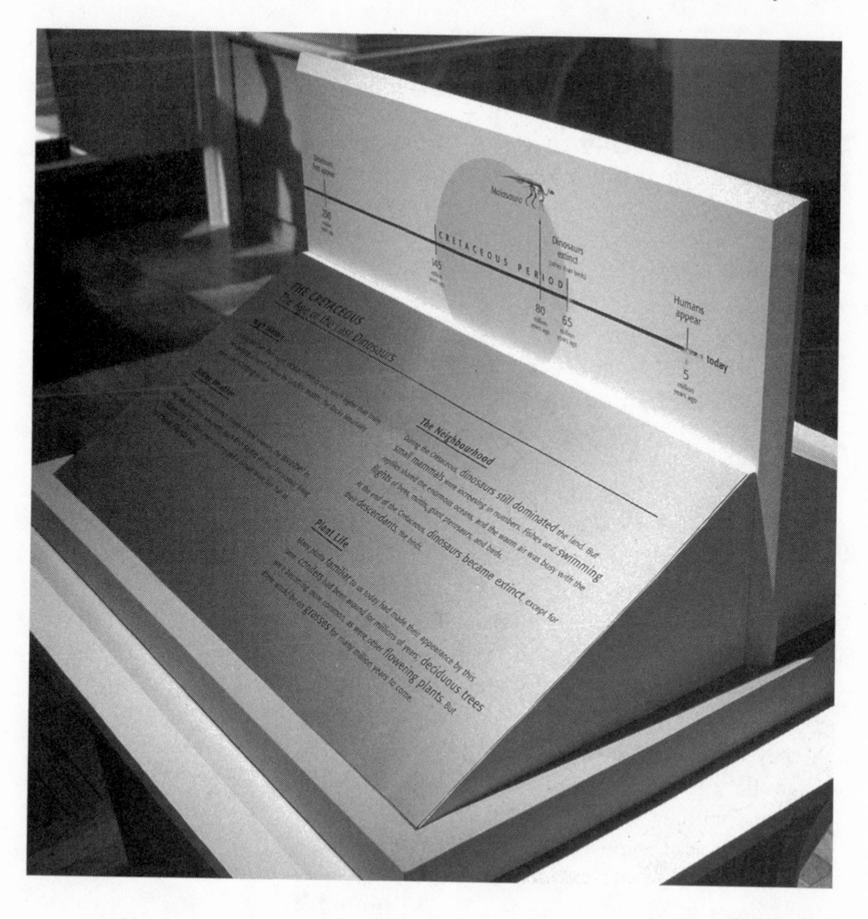

Figure 13.6. Pedestal display locating Maiasaur in Cretaceous time-space. Source: ROM Photo, reproduced with permission of the Royal Ontario Museum, © ROM.

suggesting the gradual zeroing in. A graphic circle demarcates the Late Cretaceous, literally circumscribing a space, a *space* which signifies the *time* of Maiasaurs.

In effect, this highly engaging interactive element had worked as a navigational device, the means of orienting visitors to the chronotope of the Maiasaur. It had permitted an imaginary "leap" back to the time-space of a dinosaur, its relatives, its neighbours. This, then, was also necessarily a tale about us *and* them, but also us *as* them, about our similarities with them as well as our differences from them. It was an orientalist tale in the general sense suggested by Edward Said, but set out in a workaday format for its contemporary situation, its current politics/natures of Mesozoic time-space.[42] Museum curators might very well deny that such comparative work occurs in their research or exhibitionary practices, but here it was operating freely in what is considered by many to be Canada's leading museum of nature and culture.

The display did not need to use the terms "us" versus "them" directly. Rather, they were embedded workings of the display. They were elements of its phantasmatic tissue. It was the visitors who made me realize these very anthropic characteristics of *Maiasaura*, as in the case of fourteen-year-old Katie, whose knowledge of dinosaurs had encyclopaedic qualities and who especially identified with duck-billed dinosaurs:

> They were more family-type dinosaurs, and I am really connected to my family and relations and so were they. I think they said by the model, there, that they travel in packs. With people, we live in cities. So they're a sort of sociable type animal, just like us …

The Domestic Nature of the Maiasaur Family Life

The bodily agency of tactile contact was brought fully into play at the first of the two multimedia interactive consoles: the *Maiasaura* life and times videos, known to the exhibit developers as "A Visit to the Cretaceous" (see figure 13.7). The second and larger console, which featured the loud running images I have already referenced, is discussed in the next chapter.

This was a rare encounter at the ROM in the 1990s: an interactive console. The console and video unit were constructed as a single integrated cabinet. The front console extended forward, angling towards the visitor, almost like the instrument panel of an aircraft or a video game or – from 1990s popular television culture – like the helm of the Starship *Enterprise*.

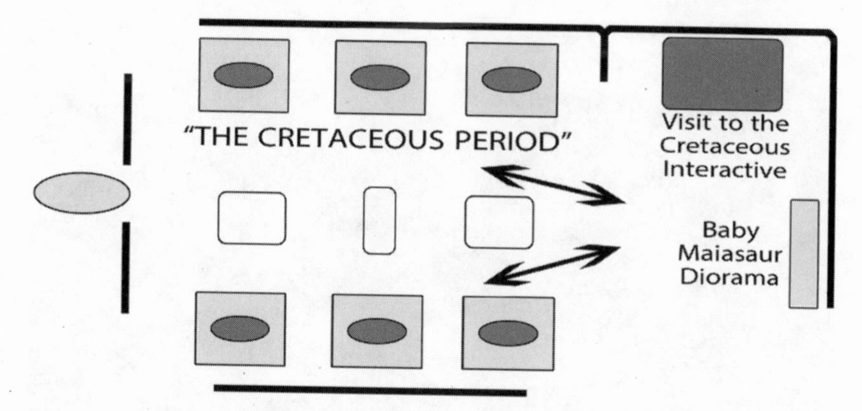

Figure 13.7. "The Cretaceous Period" – Location of "A Visit to the Cretaceous" interactive and baby Maiasaur diorama, schematic.

The default screen image which appeared when the console had not been activated by a visitor was a slowly transforming title image – referred to by some of the exhibit developers as a "screensaver."[43] First, a rock face with a slab apparently missing, the title "ghosting" into view, word by word – *the – Maiasaur – project*. Then, in the rock face's opening, the subtitle materialized in similar fashion: *the life and times of a dinosaur*. The gentle, digital rise and fall glissando played softly through its cadences during this almost hypnotic sequence. The subtitle then faded back and the form of a skull materialized, largely filling the slab outline – it was a skull reminiscent of the "Maiasaur relatives" already met. Finally, suggestive of some ultimate act of magical resurrection, the skull form cross-faded and transformed into a fully fleshed out form – also a face already met, the same face as the pewter Maiasaur model. These apparitional materializing acts were about coming face to face *with a face of the past*, a "looking glass world" (to borrow Stefan Helmreich's terms)[44] where a saurian individual existed in a lineage of relatives, a neighbourhood – much like the visitor looking on (see figure 13.8).

Above the screen, if you looked up to notice, was a sleekly stylized moth-like form in chrome mesh. This was one of the design elements meant to cue the meeting of the organic and the technological,

extending the effect of the chrome-mesh title screens and the silver Maiasaur model. I had not noticed this rather subtle feature until one of the children I interviewed told me that his favourite element of the exhibit was "the ceiling" –alerting me to the limits of my own attention, the reach of children, and how their attention works differently as a consequence of their stature, witnessing the exhibit from a lower vantage point.

The interactive console was designed to be low enough for a small child to reach up to, and it presented an array of eight round, palm-sized buttons. The illuminated buttons were enticing, with backlit images to draw attention to the choices offered. There were three different legend texts: "A Cretaceous Neighbourhood," "Cretaceous Continents," and "A Maiasaur Family." In the centre was a computer "joystick," familiar from home computer games. The repeated actions I witnessed here were touch, pause, watch – view – select, touch, manipulate, watch – view again, select again, and so on.

I watched one child make the first selection. "A Cretaceous Neighbourhood" was activated on the monitor – cutting off the transforming Maiasaur rock-face/skull-face title sequence – and animated dinosaurs,

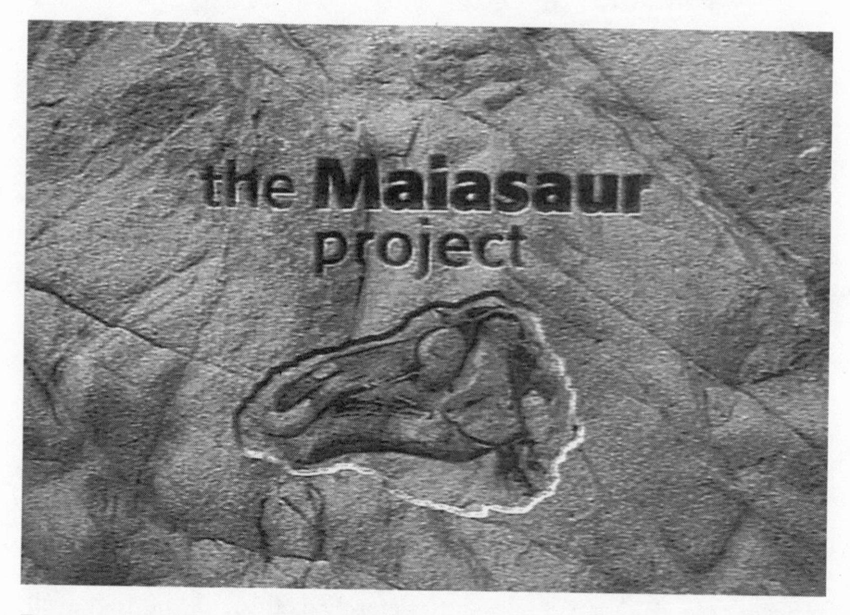

Figure 13.8. Dissolving frames of illustrated Maiasaur "face" in rock face. Source: ROM Photos, reproduced with permission of the Royal Ontario Museum, © ROM.

plant fossils, lists of scientific names of new plants appeared amidst an image montage. Then the gentle tones of a woman's benevolent, warm voice, which I had been hearing at a distance, began to rise and fall. I picked out fragments of the narration, ending on the chorus with which I had become so acutely familiar – "a perfect time for raising a family":

> The Cretaceous Neighbourhood, eighty million years ago … The Rockies are just rising … A shallow sea covers much of central North America … The weather is warm year-round … Plant life is teeming … Old kinds of plants live alongside new kinds … Broad leaf trees crowd out the evergreens … And for the first time flowering plants abound … the Late Cretaceous teemed with life … It was a perfect time for raising a family.

The showcase with fossil plants began to make narrative sense in relation to this "Cretaceous Neighbourhood" video story on the "explosion" in flowering plants. The agency of the story here was in its ability to create a seamless bridging between showcase and video cabinet, the former containing fossils, the latter containing scenes. That agency conjoined objects of the present with reconstructed worlds of the past – and back again – enticing exhibit visitors to apprehend all the more how such objects and reconstructions operate in this museum exhibit's chronotopic arrangements. Henson's technical point about the co-evolution of diverse ornithopod dinosaurs and flowering plants had precipitated into this domesticated outcome from planner Ross – "a perfect time for raising a family." More to the point, this was fully in accord with all the logics of palaeontological gendering, family structuring, historical oppositions of carnivorous (i.e., mean) and herbivorous (i.e., friendly) saurians, audience imagining, popular culture, and co-evolutionary hypotheses – all that the socio-technical, public/science network had invested. This was an exquisite moment – not of "co-optation" as Henson had suggested, but of implosion, of precipitations at the nexus.

As to the exhibition overall, its carefully crafted narrative apparatus appeared to be remarkably cohesive – even if, as I would later find out, most people visiting did not read as closely as this, nor did they walk the idealized path or get close enough to properly sample all the components aligned according to this highly articulate narrative.

I also learned that many people fill in what they see and engage with their own knowledges, cobbling together rather personalized narratives and scenarios. Nine-year-old Alex described to me what he knew:

> [U]sually dinosaurs when they were living, they weren't alone. Like, they had other dinosaurs close to them. Nearby. Same as people ... actually, the Maiasaurs were in big groups. There would be a few families in each group. I'm not sure but I think the *Maiasaura* – some animals today, they kept the other families' babies. Maybe the *Maiasaura* did that.[45]

This was typical of many children visiting, who could recount to me surprising, odd things about dinosaur behaviour, elaborating what they witnessed in the exhibit – and often reaching well beyond the messages intended. In Alex's mind, these nurturing dinosaurs might even care for dinosaurs not of their own family!

Evan, a five-year-old, told me,

> [W]hen I was there and ... that thing which is always running ... I peeked around the corner and saw it digging its nest with its head! I didn't know it did it with its head. I thought it did it with its nose![46]

Such a behavioural detail considerably exceeded strict technical work in palaeontology. Nonetheless, because of this video presentation, Evan now "knew" his error.[47]

Notwithstanding his extensions on the story, the bodily actions of moving, peeking, and comparing that Evan shared, along with touching, selecting, and looking, were anticipated as part of the experience. The entire exhibit, along with its interactive console, was meant to subtly direct interaction and connection-making. While the left-to-right options on the interactive console (following literate European reading conventions) were typically chosen in that order, the central buttons with their associated joystick captured greater interactive attention. The central buttons were titled "Cretaceous Continents," and they presented visuals of the globe with continental configurations which were continuously reconfigurable by manipulating the computer joystick. Choosing an option for a global or North American view, the visitor could manipulate the handle, moving a cursor along a time scale ranging from eighty million years ago to "0" million years ago. Terrestrial shapes of "America," "Africa," "Asia," and the "Pacific" would travel across the screen and transform over the implied passage of time.

Much as the timeline had temporally zeroed in, the North America selection would zero in spatially, aiding in locating the area where the Maiasaur lived. There on the screen was the "shallow sea" gradually covering "much of Central North America," as mentioned in the video. By following the transformations from the familiar present to the alien past, the visitor established the total world moment of *Maiasaura*. The imagining of a visit, of travel across time and space to a Cretaceous Neighbourhood, had been achieved through this device. As two university students (twenty and twenty-one years old), who were enrolled in a geology program, told me,

> MARTIN: The way that showed the earth, the evolution, eighty million years ago. *You go back* and see how everything fit into place and how it was, and how the dinosaurs kind of where they were.
>
> AMY: *To see time*, the changes …
>
> MARTIN: You could actually see it ... I have seen pictures of stages, but never step-by-step how it came together.[48]

The third selection from the console was named "A Maiasaur Family." The animated imagery from this video was familiar to me from a 1980s CBS series called *Dinosaurs!* hosted by America's voice of the NASA Space Program, Walter Cronkite. The 1980s had seen a plethora of documentary films on dinosaurs produced for broadcast on Canadian, American, and British television networks. These films had picked up on the "revolution" in scientific thinking about dinosaurs, incorporating all manner of propositions: "hot-blooded" dinosaurs, herding and migrating dinosaurs, intelligent dinosaurs, biochemical analyses of dinosaur bones, dinosaur-bird evolution, pack-hunting dinosaurs, and indeed, from Jack Horner, parental care, eggs and babies, and nesting behaviour in dinosaurs.[49]

The obviously modelled dinosaur "face" that appeared in this movie segment[50] was somewhat different from that of the metallic entryway Maiasaur – no explanation for this was given. In what appeared to be morning light filtering through the canopy of a deciduous forest, the animated sequences included an adult Maiasaur pushing earth around what appeared to be a "nest"; a "baby" dinosaur stumbling to stand on wobbly legs; and the adult dinosaur touching the baby gently, nuzzling. I followed the narration – the soft female voice speaking soothingly again, with a storytelling tone of reassurance (square brackets cue what I read as an immanent meta-narrative):

"Daybreak, northwestern Montana, eighty million years ago …"
[This is a "day in the life" moment in ancient "Montana," Blackfoot lands no longer noted …]

"It's the start of the Maiasaur nesting season …"
[This is iconic Maiasaur time, a time of reproduction, nesting season …]

"By the thousands, the adults begin preparing … Eggs need constant guarding …"
[Parenting in this densely populated neighbourhood has its stresses – unseen predators, the "terror of many plant-eaters," may lurk here, looking for eggs …]

"Newly hatched babies, helpless at birth, will triple in size in two months …"
[Babies are vulnerable too, but (scientific knowledge shows) they grow quickly …]

"Long after the Maiasaur's extinction, their stories have been recorded in fossils …"
[Museums collect fossils, which are kept so "stories" like this may be properly known.]

"Scientists will name them the *Maiasaura*, 'the good mother lizard'…"
[Respected authorities name the dinosaur in accord with the properly known stories.]

Though no direct statement as to gender had been given, and recalling the nickname "Henrietta," I was increasingly persuaded that the adult presented was significantly a "she," a good, caring mother – though of course no one actually knew if ROM specimen #44770 was female, a mother, caring, or otherwise. The signs collide: scientific name, kindly female narration, gentle sounds, the single adult in the video image, domesticity. Child-rearing appears as normal, female work in this bygone natural world.

Visitors to the exhibit commonly read this sort of normative gender assignment, but often reworked the emphasis in surprising ways. Three very enthusiastic young girls[51] – who had particularly cued into the touchable pewter Maiasaur – revealed how readily they could interpret and refashion the model and the exhibit narrative into complex scenarios, elaborating the gendered logics of the display in novel ways:

BN: What did you think of the model when you touched it?
KYLA: It had spikes on the back, right? Little spikes.
ALLISON: I think so.
BN: Now, what kind of dinosaur was it? Do you know?
KYLA: It wasn't a Maiasaur.
BN: You don't think so?
[all say "no"]
BN: And how big was it, approximately? Can you show with your hands?
KYLA: It was about … it was this long. This tall.
ALLISON: It wasn't very big.
BN: A little smaller than each of you … then, right? Smaller than you guys. Did you think it was a model of a grown-up dinosaur, or a …
KYLA: [somewhat inaudibly] I think it was a dad dinosaur.
BN: You think it was a dead dinosaur?
KYLA: No, a dad! Because it had little spikes on the back. And I didn't see spikes on the back of the mom.
BN: Okay. And if you see spikes, why does that mean it might be a dad?
KYLA: Because, for two reasons. First, I think, 'cause I didn't see any on the mom one. And second, because I think the dad has to protect the mom and the babies.

ALLISON: But what if the dad is not there?
KYLA: I don't know.
ALLISON: I think that it is for both of them.
BN: ... oh, so you think that spikes would be on both of them?
ALLISON: Yeah.
BN: What do they have to protect the mom and the babies from?
KYLA: Enemies.
EMMA: The enemies.
BN: [turning to address Emma] What were you going to say before you heard "enemies"?
EMMA: *Tyrannosaurus rex*.
BN: *Tyrannosaurus rex* is the enemy?
KYLA: Yeah, meat-eaters.[52]

I thought to myself, "All of this, from the simple touch of a metallic model dinosaur?" Of course, much more was at play – the entire exhibit, its narrative, the dialogic engagement of the children. Still, the materialized phantasmatics of the nexus loudly came through: families, domesticity, gendering, good plant-eaters, enemy meat-eaters, complex ideas about dinosaurs as opposed to a more exclusive sense of "vicious killers" alone. These girls refigured and recombined the display information with their own imagining. Indeed, they contrasted the "goodness" of the Maiasaur with the character of the *T. rex*, even remarking that "*Tyrannosaurus rex* eats his babies." All the same, the gendered meanings of the Maiasaur displays appeared to align generally with their senses of heterosexual gender role assignments. For example, there apparently had to be dads and moms. The dads had to protect moms with babies, and spiky backs somehow helped to do that. The enemy – predictably enough – was Osborn's "king of the tyrants."

A second account came to me from a fellow anthropologist in the Toronto area, Ken Little. Little's son Will, at age eight, had long been a dinosaur fanatic and had visited the ROM's Maiasaur exhibit in recent months. Over breakfast one morning, Will began to expound on the behavioural complexities of *Maiasaura*:

WILL: With the *Maiasaura* dinosaurs, the mommy dinosaur stays at the nest and takes care of the babies. The daddy dinosaurs go out and kill!
KEN: But Will, you know that the Maiasaurs were all plant-eaters ... they didn't kill other animals. And didn't the boy dinosaurs take care of the babies too?

WILL: Yeah, okay, the daddy took care of the babies too, but the daddy still had to go out and get the food.

KEN: Wait a minute, Will, think about your mom and me – that isn't the way it works for us – we both go out to work so we can get food.

WILL: Yeah! ... that's because you're people. Maiasaurs are dinosaurs![53]

Will had clearly outwitted his father, the learned cultural anthropologist, who all too quickly presumed that Will would respond by mirroring the daily life of their domestic human world in his view of that of *Maiasaura* dinosaurs. Will, however, knew how to collapse biology *and* social anthropology, recognizing that while dinosaurs and humans may have some similarities, there is something in the respective and particular merging of natures and cultures that makes them distinct. Will's view describes a human/animalian order, and one that ascribes sex roles in slightly differential ways to people versus dinosaurs. What circulates here, all the same, is a particular sense of gendered family relations, modelled upon multiple more or less uncertain sources.

However much adult companions might have attempted to intervene and set things straight (as it were), the regularity and autonomy of these sorts of children's accounts still largely reified mainstream gender logics. The shift to "sophisticated ways of thinking about dinosaurs" in the final analysis had been produced as Henson had predicted, though not necessarily in the manner or direction he had imagined. Nonetheless, his imagining of a hegemonic sensibility among museum visitors corresponded quite well with the responses I encountered. Here indeed was Henson's sense of the gendered opposition between masculinized vicious killer dinosaurs and feminized friendly maternal dinosaurs. But did this meet the fullness he had been seeking in his aim to generate "a sense of a dynamic process"?

Henson knew that the question of interpreting maternal care – or even parental care in the ungendered sense – was contested among vertebrate palaeontologists in the case of *Maiasaura*. For curator Henson, that uncertainty should not be allowed to undermine the utility of this dinosaur, whose representative was ROM specimen #44770, as an ally for shifting public senses of dinosaurian biological complexity. He told me:

> The key reason behind its exhibit potential was that Maiasaur was, in a way, a *catalyst* in our new thinking about dinosaurs ... *Maiasaura* was a dinosaur where the majority of researchers believe you actually have the first actual

> tangible evidence of paternal or maternal or whatever ... parental care in dinosaurs. Sort of taking them away from our standard view that reptilian dinosaurs wouldn't do this, whereas a more bird-like or mammal-like dinosaur would. Admittedly, "A," a naive dichotomy – some reptiles actually do look after their young – and "B," an intangible one, because to this day it's not really clear how good the evidence of parental care in Maiasaurs actually is ... some people have actually expressed serious doubts about the classical scenario as proposed by Jack Horner.

Yet Horner's classical scenario would carry this exhibition. One can only ponder what might have happened if such a contestation had been displayed. Could these children have still transformed the creatures into the domestic figures of feminized, maternalized herbivorous dinosaurs bringing up babies, of boy dinosaurs always wary of the marauding meat-eater, or if not that, then out on the prowl for meat themselves?

Again, the enrolment – and reification – of gendered sensibilities that was articulated with and through the work of Horner's scientific practices was what would prevail. Predicting audience responsiveness yet again, curator Henson insisted that presenting scientific debates and uncertainties simply would not work as good public communication:

> We've tried it, time and again, in lectures, in exhibits ... People are not satisfied with it ... People come because people have been conditioned over time to think that science works like religion ... and really conveys truths. And, when science fails to deliver ... since most people don't understand science in the first place ... it's like "what's the whole point – we don't understand these people anyway, so now they don't even tell us truth, why don't we just go and burn them all?" [*chuckling*]

Notwithstanding Henson's somewhat anxious admission here that the Horner scenario was contingent, this exhibit would, in his terms, present it as "truth" – so he would not fall victim to some form of public immolation, perhaps? The conveying of a "truth" was fulfilled, as a college geology student indicated:

> ANDREA: We watched that little film ... and it talked about the habits of the dinosaur, and I thought, "I didn't even know we knew that stuff." The last time I was here I was a kid, and I think we only knew fifteen dinosaurs.
>
> BN: So it was quite foreign to you to find out about a lifestyle ...
>
> ANDREA: Uh-huh! I didn't even know that they knew the living habits![54]

Result: the nexus of Mesozoic nature would be selectively configured in this legitimate public materialization, aiming to align the way expected audiences would typecast dinosaurs with the project of advancing Henson's "more sophisticated way of thinking about dinosaurs" – elements of his dynamic process.

Horner's "Classic Scenario" as Articulator

Horner's "classic scenario" was a remarkably potent articulating device, in the sense that it was the right scientific story to enrol public sensibilities and phantasmatics of binary gender and binary dinosaur kinds. Though only in an oblique and partial way, this was what Henson had intended. Such a scenario aligns with genres of children's storytelling and conventions of popular fiction and cinema, easily conjuring the Hollywood-promulgated accounts of babies separated from mothers (as in *The Land before Time*) and absent, yet somehow heroic fathers. I recalled again how the missing animated herds of Maiasaurs, which Jennifer Ross had so lamented, might at least have given a sense of multitudinous sociality to counteract the replaying of such reified stories of dyadic domestic life. The animated life-world left underplayed in this 1998 exhibit would be completely overshadowed by 2010, as new CGI-animated nature films such as PBS's *March of the Dinosaurs* would build on fast-advancing palaeontological research on mass herding and mass migration behaviour in dinosaurs to generate what is now a commonplace scientific/public understanding of Hadrosaurian (duck-billed dinosaur) behaviour and sociality.[55]

The conviction that such familial storytelling could perform effective < science + audience > articulations was equally held by the exhibit programming staff of the ROM. One of the interpretive planners had co-authored a visitor research study entitled "The Effects of Technology-Based Devices at the Royal Ontario Museum: Observations and Visitors' Perceptions."[56] The study resulted in a conditional endorsement of the use of interactive media. The "Maiasaur Family" video was classified as a prime case of a "Didactic Presentation," which they defined as:

> A technology-based application (audio, visual, audio-visual), which presents conceptual information that is relevant to the exhibit topic, in a factual, straightforward manner.

The "Maiasaur Family" video was therefore seen to be "factual, straightforward." In the context of "technology-based application," the video content was factualized. The visitor research study went on in very precise terms to present evidence on how the factual information could be made to be "relevant":

> Nearly half of the visitors who used the "Maiasaur Family" A.V. reported that they greatly enjoyed its story line, which dealt with the raising of the young. In particular, this seemed to have tremendous appeal to family audiences, who often commented that they were able to relate this theme to their own lived lives ... Therefore, in planning the design of such devices, practitioners should place as much emphasis on the story line or the narrative content as is given to its technologically dazzling elements.

The author's prescription for interactive displays was simple enough: use them not so much to dazzle, but rather to cause relations, to articulate. The way to do this was by emphasizing "story line or the narrative content." The author's proof: "the 'Maiasaur Family' A.V." Interpolating the study's empirical results with all that I have been presenting, a long sequence of oddly connected articulations can be reconstructed: the fossil eggs and bones of baby dinosaurs in Montana; Horner's nesting scenario; the CBS *Dinosaurs!* documentary; the find of what would later be ROM #44770 by Sherry Flamand; the collection of this specimen, "Henrietta," by Bearpaw Palaeontological, Inc., and its subsequent sale to the ROM; the enlisting of matter from this history by Henson; the imagining of "family" audiences; the translations into an exhibit; the selective foregrounding and embedding of Maiasaur as the family dinosaur; the engagement by visitors with these elements as an interactive story of mothering, babies, touchability – the particularly familiar, familial life and times of a dinosaur.

Through the planned and contingent articulations of curator and planner, ROM #44770 had brought this history of "good mother" inscriptions and storylines along with it. Through all of this, it was the malleable phantasy of "family" which had been circulating, coming to be reperformed in the reconfigured Mesozoic nexus. The visitor moved through the exhibit as a "family," interacted as a "family," and encountered a dinosaur "family." The audience targeted was the audience produced. In many regards, that audience lived literally *in Maiasaura's world as Maiasaura now lived in theirs.*

The ROM's acquisition of a *Maiasaura* specimen, "the best specimen of its kind," had enabled all of these unruly translations. Moreover, a highly regarded research collection had been expanded. The ROM's outstanding collection of Hadrosaurian dinosaurs now included the

"Good Mother Lizard," ROM #44770. The $250,000 paid was more than an investment in a fine, unstudied specimen. It was also an investment in the articulation of cultured science with the ROM's most coveted audience: middle- and upper-income Canadian and tourist consumer "families" who have choices of how to spend their leisure time and money. The ROM had advanced its position, both as a major research museum and as a viable player in the highly competitive leisure industry. The lab was gone now, diminishing the potential of articulating ROM #44770 to sophisticated palaeobiological accounts. The nexus of articulation had shifted away from the specimen in its lab, now moving to the exhibitionary story of family-living dinosaurs. Henson's dynamic process had barely survived the ravages, mutated in unexpected ways by the exigency of readings by planners, designers, marketers, and visitors themselves.

To close this chapter, I turn to the final component of the "Cretaceous Period" section. To the right of the "Visit to the Cretaceous" console was a small glass-fronted scale-model diorama or "habitat group" – of the conventional sort, in contrast to the Working Lab as an effective diorama inhabited by a live technician, as discussed in the previous chapter.[57] Depicted here on the flat backdrop illustration was a clearing with tall deciduous trees in the background. A standing mount of a baby Maiasaur skeleton was frozen in the foreground space of the inset display case. It was about the same size as the pewter Maiasaur at the entrance. Two sections of text at the foot of the skeleton read:

> **Hatchling Maiasaur.**
> The legs of baby Maiasaurs were not strong enough for them to run around and find their own food. However the wear on their teeth shows that they were eating plants which must have been fed to them by adults.
>
> **Maiasaur Nests.**
> Vast Dinosaur nesting grounds were found in the early 1980's in northwestern Montana. Each nest held as many as 25 eggs. The hatchlings evidently stayed at the nest for several months, like baby birds.

The scientific authorization of a syntax connecting fossils, nurturance, babies, and parenting arrests us yet again, a conflation of the politics/natures of scientists and planners allied to the museological material they are charged with collecting and animating for visiting publics. In this display, the backdrop image is in proper perspective and scale to give the illusion of scenic continuity with the skeletal form in front. It shows an adult with a rather spindly baby hanging gently from her beak – in the manner of a mother cat with a kitten. She lifts the baby over a nest with what I can't help now but recognize as "her litter" of babies. I interviewed visitor Andrea, age twenty-one, as she was leaving the exhibit. She told me she didn't know the meaning of the name "Maiasaura." I told her it meant "good mother lizard," to which she responded:

> Ahhhh! So, is that why it said "raising a family … it's a good place to raise a family"… or something like that?

Andrea understood.

14 Technotheatrical Natures: Maiasaur's World, by Default?

Spectacle by itself is Disneyfication ... If you can put it into context, so that there's meaningful ideas and authority behind it, then you take it to the next level ... To me that's the heart of it ... if you can do that, that's admirable in a museum ...

Walter Tomasenko, former manager, Digital Media Services, and creative director for the Maiasaur Project

The ROM staff is well aware that, for all the science supporting the computer graphics, "no one really knows what the Maiasaur looks like or how it moves," admits Walter Tomasenko, creative director of the ROM's Digital Media Services. Consequently, notices to that effect will accompany exhibits that rely on "reasonable inferences." Henson, however, believes that any issue of interpretive liberties is overshadowed by the life multimedia injects into the fossils. "The animations give people a feel of what the living breathing animal would have looked like, which is a very difficult conceptual transition for the lay person to make," he says. Tomasenko reaches for a more theatrical comparison. "It's basically like Jurassic Park," he says, "except we have to be accurate first and entertaining second."

Joanna Pachner, reporter, *Electronic Link*[1]

So what of the multimedia components of the Maiasaur Project? The latter of the two quotes above speaks of the relations between multimedia production and the generation of scientific knowledge in the Maiasaur Project – the pivotal tension explored in this chapter. Then, in the next chapter, this is put into higher relief by considering that which resolves the tension: the *factish* of the pewter model Maiasaur.

From the second quote above – drawn from an article in a specialty magazine designed for the North American computer graphics industry – media producer Tomasenko points out how "no one really knows what the Maiasaur looks like or how it moves." Curator Henson does not directly refute this point, but instead implies that while the "lay person" may have difficulty in making the visionary transition from the fossils to the living Maiasaur, there are, by default, those who are properly equipped to minimize this difficulty – scientists like him, experts with their techniques. What lies between the finished vision and the specimen, then, is this rather large collection of suggestible and authorizable imaginings, and of multiple techniques for making this "difficult conceptual transition" – a transition to something which, as Tomasenko says, "no one really knows"; an enduring *jamais vu* that is characteristically answered in the palaeontological enterprise with a reconstructed *maintenant voir* which in turn extends into museological, literary, and filmic milieus. As discussed at length in previous chapters, here the malleable space of phantasmatic/systematic otherworld-making between the specimen and the spectacle is made apparent and put to work yet again. In this chapter, however, it is possible to consider more emphatically the conjoining of specific actors in the common work of fabulating dinosaurs anew – especially in the context of tensions over what vision and techniques of materializing are put to work. This is achieved through measures of deliberate choice, combining with measures of serendipity and the subsequent cobbling together of imaginative possibilities.

Here I take up Isabelle Stengers's use of fabulation, after Gilles Deleuze, to think about what emerges. Stengers posits fabulation as a sort of crossover act, where figures and possibilities shift in unexpected, usually dramatic ways, emphasizing lines of flight – that is, unexpected possibilities in reimagining, rematerializing. The question here is whether, in the course of the technotheatric engagement of multiple actors in the constituting of Mesozoic natures, we find that the nexus of specimen and spectacle closes down or opens up to the possibility of fomenting at least partially novel fabulations – that is to say, some modestly unexpected *lines of flight*, to again use the propositions of Deleuze.[2] While it might seem that *Maiasaura* is conventional in its familial kinship configurations, when set against the *T. rex* genealogy it proves to be an alternative nature to think with, offering at the very least an enticement to the possibility of refabulating at the interstices of the specimen-spectacle apparatus. However, as will be discussed here, the fabulation of the Maiasaur Project, having turned as much on the tensions encountered, was arguably more *fabulation by default*, rather than fabulation as full-blown speculative, collective engagement.

The quoted statements of the Maiasaur Project's media producer Tomasenko and curator Henson are remarkably conflicting and yet extremely complementary positions on the status of knowing, and of the relation between the materiality of fossils and the virtuality of mediated visions. This begs the question: which comes first? Do you start with the spectacle and add the context of the fossils, as Tomasenko suggests in the first quote above? Or do you start with the fossils and build the context around them? Developing an answer to this conundrum has vexed museum specialists just as it has palaeontologists, becoming a source of tremendous tension and anxiety. In many ways it is the modern matter of concern that constitutes and rationalizes the very relevance of museums, not to mention the propulsive workings of modern knowledge production. It is what precipitates in this tension that is of interest, for such conditions both sustain what has come before and open up, modestly or dramatically, what new and unexpected creatures may come next – a multiplicity of other dinosaurian possibilities, including the curious example of *Maiasaura*.

This chapter discusses the most explicit example of spectacle and virtual creation work from the Maiasaur Project, the "Meet a Maiasaur" interactive computer animation theatre, exploring some of the vexing questions behind its political and "market" effectiveness as public display, as well as its effect on the shifting status of the museum itself. I also consider how it worked in relation to the other components of the exhibit, in particular the complexities faced in building a relation between the theatre's "living, breathing environment" and the fossils and evidence from which those visions were meant to be fashioned. This discussion also completes my ethnographic walk-through of the exhibition, arriving at the concerns of the next chapter and the puzzle of the "factish," and all that came together for it to exist. Decisively avoiding the choice between fetishism and facticity, the factish is something that is at once of the specimen and of the spectacle, a fusion that refuses the divide between specimen and spectacle and so gives resolution, articulation – albeit a resolution that bears careful scrutiny if we are to identify the concerns it concentrates and mobilizes.

The chapter is organized into three parts: first, a continuation of the ethnographic walk-through started in the preceding chapter; second, an examination of the way the "Meet a Maiasaur" theatre was developed and how it came more into alignment with marketing and corporate concerns in contrast to the concerns evident in other components of the exhibit; and third, a discussion that returns to the shifting politicality of museum natures.

The ethnographic description resumes where I left off in the previous chapter – that is, moving from "The Cretaceous Period" gallery into the highlighted interactive segment of the exhibition, the "Meet a Maiasaur" theatre. Contrasting sketches of the space as it was laid out over two different time periods are shown in figure 14.1. The first is from the time the lab was in place, the second from a time after it had been replaced by a mounted Maiasaur skeleton.

"A Living, Breathing Environment"

> To a generation weaned on interactivity and rapid-fire visual images, specimens in glass cases can be downright boring. So the museum's digital media services are using the tools of multimedia and interactivity to bring an 80-million-year-old Maiasaur to life.
>
> Sheila Cameron, CityTV reporter [3]

Peering around the corner now, and hearing occasional deep, drumming sounds, animal-like breathing, and the shrill cadences and exclamations of excited children, I saw ahead of me a mounted skeleton of what must have been the *Maiasaura* specimen (see figure 14.1, noting the closing of the former lab entry by way of the mounted skeleton).[4] This, I surmised, was what vertebrate palaeontology collections manager Simon Kilgour had told me was ROM #44770 – what this exhibit was supposed to be all about. To the left, a huge animated "family" of Maiasaurs was projected onscreen – two adults with a baby trailing behind, the group rushing headlong through a woodland.

The connection between the parenting stories from the previous space and the imagery in this space was fortuitous in this instance. If I had rounded the corner at another moment, a very different set of actions might have been playing. The sixteen selections of animations that were offered showed *Maiasaura* in eight walking or running sequences, four sequences of what were called "behaviours" (e.g., drinking, making sound, and appearing alert), and four others of *Maiasaura* in a sequoia environment (see figure 14.2). The majority of these animations showed a solitary Maiasaur. The button array was the same as that

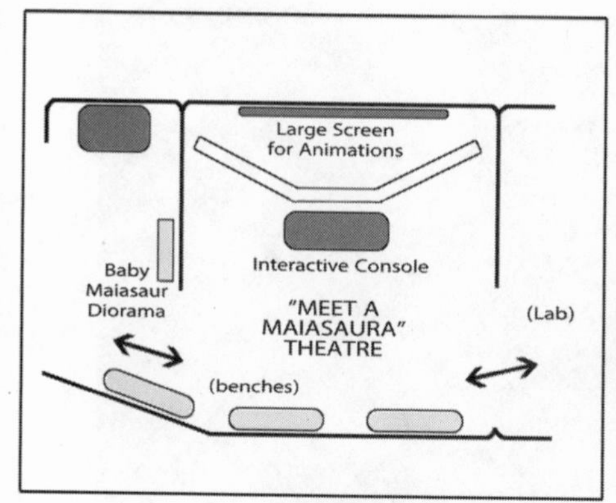

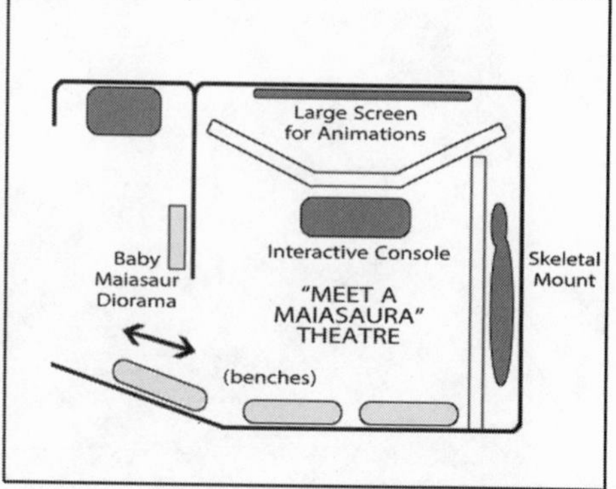

Figure 14.1. Shifting layouts of the "Meet a Maiasaur" theatre space, schematics. Left: May 1995 to May 1998. Right: May 1998 to September 1998.

in the "Visit to the Cretaceous" interactive – a series of backlit plastic buttons with still-photo transparencies showing a single frame of what was to be seen.

Knowing that these images were a technically and imaginatively reconstructed form of the animal, I was intrigued by how they were made to appear, move, and sound. The colouring was greenish with gradating bands of yellow – a sort of camouflage, I wondered? When walking, the Maiasaur moved on all four legs; when running, it lifted up onto its hind legs, tucking the front limbs in close to its body. The drum of footfalls was used in all the sequences of the animal moving. Though large and swift, this creature did not seem at all threatening. One child remarked on the Maiasaur's temperament on the basis of these animations:

> It was gentle, it wouldn't, like, fight back, it would run instead, because in one of the videos you see it running really fast.

I scanned the images for signs that were supposed to be of scientific interest, or which had some connection to Henson's original curatorial proposal.[5] Interestingly, one of the few special features of the new specimen ROM #44770 was a horny, beak-like structure, which had been

Figure 14.2. Computer graphics view of the Maiasaur from the "Meet a Maiasaur" theatre. Compare with stance of pewter Maiasaur, figure 13.3a. Source: ROM Photo, reproduced with permission of the Royal Ontario Museum, © ROM.

highlighted in the original curatorial proposal, yet the images placed no special emphasis on this structure. They did, however, include the feature which several children who had petted the pewter Maiasaur had mentioned to me based on their touching of the model – a fairly heavy-looking flap of skin under the neck, the dewlap. It had not struck me until hearing the children mention this anatomical feature that the form of the model and those in the animations shared much in common: the posture, the sense of weight of the animal, the skin textures.

Then I was reminded of Andreas Henson's points on this: the exhibit was meant "just to give a little bit of a feel of how this animal looked." But Henson also admitted, "We don't know the colour, but we know a lot of other things." Chief among these, he noted, "every example of dinosaur skin impression ever found … is always made up of these polygonous scales." Here indeed were indications of these scales. But again, I only knew to look for them first, because of my conversations with Henson or from reading his curatorial proposal, and second – and this was very intriguing – because children had told me of the skin texture several times in their interviews! There was no textual or verbal cue drawing my attention to these things – but children combining touch and sight had indeed found some such connections, making articulations through their pragmatic engagement with the physical model and with the animated images. The physical fossil of skin impressions was not presented, but some knowledge of them was relayed through these physical, media elements.

In some instances the computer-animated animals moved across what effectively looked like spotlit spaces – practically vaudevillian. One sequence showed the head leaning to the whitened "pool" of light, apparently drinking – I heard one child exclaim, "It's drinking milk!" The Maiasaur could also be seen snapping its head up quickly, looking about pensively, its nostrils then distending in an unusual manner. These were all only limited ranges of motion or "behaviour." I had been told that the visual animation of every little gesture had involved a complex of decisions and a great deal of technological manipulation and labour. For most of the visitors I watched here, such technical matters were to be left unknown. This was a show of finished visions, along with all the other technical, showcase, and specimen trappings one would expect in a museum.

Jennifer Ross had explained to me just how much labour there was behind all this digitally created animation and modelling, suggesting as well the resources that had to be pulled together:

> We needed something called a "rendering farm"... It's not like regular film animation where you can finish your animation two months ahead of time and do all your editing and sound mix, and then run it. We had to finish this thing *eight* months ahead of time. Because it then all had to get rendered off of the hard drive in the computer onto some digital video. There may be another step in there. This thing, "rendering," is transferring digital information. It can take months. It became clear as panic built and the deadline got closer – this has nothing to do with your public perception of dinosaurs, but it is interesting – they were going to need a lot of help. Walter [Tomasenko] had good contacts at Alias and got them to donate software and time – this all would have cost us a fortune. We had computers and people at the Science Centre, at the U of T, at Sheridan College. Everybody was rendering for us. There was *a lot* of nail-biting going on.

Ross's comment, "this has nothing to do with your public perception of dinosaurs," is correct insofar as the public would never see the "panic" and "nail-biting," or the huge "rendering farm." But, as I had learned already, in terms of resources committed, schedules and budgets being pushed to their limits, visual elements being dropped, and communication objectives being compromised, the creation of these images did indeed have much to do with "the public perception of dinosaurs." Farmed-out rendering achieves certain technical imaging possibilities while closing down others. Such rendering turns under tight time constraints or aligns with Annemarie Mol's idea of "enactment," as remarked upon by John Law, who noted how "to talk of enactment is to attend to the continuing practice of crafting. Enactment and practice never stop, and realities depend upon their continued crafting – perhaps by people, but more often … in a combination of people, techniques, texts, architectural arrangements and natural phenomena (which are themselves being enacted and re-enacted)."[6]

For example, the budget and time constraints even affected the animations that would be chosen for presentation, some of which were considered unfinished. In these, a "wire-frame" computer graphic form of the dinosaur moves through the landscape or across the black screen-space, giving a sense of technical wizardry, a sense of what lies behind this illusion of animation.[7] The intention had been to replace these unfinished renderings, but some clearly remained, without any explanation offered in the exhibition. The brilliant blue lines making up the dinosaur form glowed with computer fluorescence on the screen in this very dark theatre. Strange swirling cross-marks, presumably guide points associated with technical aspects of graphics production, hovered in proximity to the wire-frame animal as it moved. The wire-frame figure was set against a photographically exact forest background, and as it moved, just as with the finished and fleshed-out animations, each footfall was audible as a deep drumbeat. There might as readily have been a wire-frame or a fully fleshed-out Maiasaur before me and the other viewers. Sounds of breathing could be heard. The wire-frame head moved to the ground, there was the sound of shuffling leaves and shrubs, a snapping noise; then the head lifted and a low grumble and a guttural noise could be heard as the ghostly, graphical creature swallowed the vegetation (see figure 14.3).

Referring to the wire-frame animations, a fourteen-year-old visitor, Sarah, told me how easy it was for her to incorporate these into a set of everyday logics:

Figure 14.3. The "Meet a Maiasaur" theatre with blurred in-production wire-frame image and cross-marks on screen. Source: ROM Photo, reproduced with permission of the Royal Ontario Museum, © ROM.

> There was one exhibit where they had a sort of a Claymation.[8] That was okay, but I liked [in the other one] the way they showed the running, and using the structural, skeleton process to figure out how they would have run … they were showing the structure inside it at the same time. You could just tell it was much more professionally done. It looked much more realistic … [it was like] *Jurassic Park* …

Interestingly, and whether or not she understood what they were precisely, the wire frames caused no confusion for Sarah. She took them as intentional structures, skeletons which somehow aided in the determination of running motions. Or perhaps she also had understood the wire-frame lines as true elements of the organic structure of the animal, as though it were some sort of cyborg. Sarah's imaginative engagement ensured that nothing was missing and allowed her to create an individual, situated coherence out of the display she experienced then and there.

In what was by now becoming quite routine in my conversations, the reference standard (in her case, the "realism" standard) of *Jurassic Park* was summoned again. Even though *Jurassic Park* is never referenced directly in the exhibit, it was present in Sarah's complicated conjuring of this reality. Precisely how the animation was like *Jurassic Park* was unstated, but the parallel she drew suggests it is that of revealing the process of production – or animation, of film, of organic dinosaurian life itself. Sarah displayed a quick ability to sort through and remix visual indicators of the digital and the organic to generate an enacted coherence.

Returning to the imagery once more, the giant animation screen blacked out after each twenty- or thirty-second sequence. The baseline background music for the title-rock-skull transformation seen in the "Visit to the Cretaceous" interactive returned, with an electric piano or other computer-generated music playing sonorously – again, as if to perhaps suggest an unfolding or a revelation. The tones rose and fell, ending with a gentle but deliberate high note, sustained as a kind of pinnacle point, an exclamation. The imagery of transformation rolled visually on the screen, then the text emerging onto the face of the rock, then inside the outlined rock face, the ghosting sequence, the skull, the "face" of *Maiasaura*. As I paid closer attention, I found that the beak-like structure appeared to be depicted on the transforming skull, and then it was fleshed over. And now, as well, an extra textual element was ghosted onto this giant screen – the sponsorship credits, here in the boldest place of any sponsorship credit in the entire museum:

Sponsored by Padullo Integrated Inc.

Animation software donated by Alias

Assistance with animation rendering: Centennial College, the Bell Centre for Creative Communications, the Ontario Science Centre, Sheridan College Faculty of Arts

A graphic panel on the wall near the entryway to the exhibit had presented sponsor information as well. I did wonder whether some visitors might read these texts and think about this. My attention typically numbs to such sponsor credits when viewing museum exhibits and I suspected this would be the case with most people visiting the display. Indeed, in spite of the presence of these elements, even Phil Thomm – the preparator who had spent most of two years working in the exhibition laboratory, and who passed by these elements many times – told me he had no idea who the sponsors had been. I assumed that this textual information was being presented not so much for the public as for the sponsors themselves.

As the sponsor list faded, the default soundscape once more looped back seamlessly on itself, a mostly imperceptible repetition – until someone touched the button and another animation appeared: a spotlit Maiasaur walking slowly across a darkened stage-like space (figure 14.4). The scale of things on this screen was even greater than it was in the "Visit to the Cretaceous" videos, more suggestive now of something magical, transformational, the power of the gesture to conjure giant, hidden mysteries – this was about bringing the dead to life. I thought back to some comments which Andreas Henson had made about his feeling that many people expected science and scientists to perform acts tantamount to spiritual revelation:

> It's almost a priestly function … this person who has revealed knowledge and communicates truth to the general public. Well, frankly, that's not how science works.

To this point, the exhibit had not remedied the sort of potential misreading of which Henson spoke, but seemed as if it could only contribute more to it.

For the moment, at least, the space of museum-located palaeontological imagining was yielding to a sense of mystical revelation and resurrection. The rib-like structures which had been placed at the base of the showcases in the Cretaceous gallery were repeated here, now elevated to the walls – vague significations that the visitor was supposed to be, somehow, within the beast itself. Tomasenko's terms of producing a "living, breathing environment," which literally conflated the animalian with the spatial locale, took on greater meaning. With the harmonics playing, we, the visitors, standing in the darkened area, waited for something momentous to be imparted.

Yet at other moments, the space also operated like the dark space of a video game parlour, though on a much larger scale and with a more refined and spartan aesthetic. Here, however, the animated beings on

Figure 14.4. "Spotlit" computer-animated Maiasaur. Source: ROM Photo, reproduced with permission of the Royal Ontario Museum, © ROM.

the game screen dwarfed the player at the console. Two eleven-year-old boys, Simon and Ben, on exiting the gallery, told me how things worked in there:

> [E]verybody's always racing to get to the button first … I got there first and then he got a bunch then I got a bunch … and then, like, six other people.[9]

Still different sorts of interactions also took place. In another moment, a man sitting on a bench at the back of this room called out to a girl, about seven years old, who was standing at the front, telling her to compare the skeleton beside her to the graphic animations being projected. She pressed the button and a running Maiasaur appeared on the screen. She turned around to the man, whom I now guessed to be her father, and said, "I picked it!" Other children began to crowd around the console. After looking up at the creature running before his eyes in the dark space, a boy with the girl who had pressed the button turned back to the man: "Dad! It's like a *T. rex*!"[10]

Though absent in the exhibit, *T. rex* re-entered the scenario again and again through so many of the visitors' imaginative encounters. Several visitors told me they thought that, somewhere in the exhibit, a *T. rex* had been presented – some thought it was the model, some said the big-screen animations, and several saw it in the mounted skeleton of what I knew to be *Maiasaura*. Ramona Asuncion, a visitor from the US who was probably in her forties, declared how utterly flummoxed she was by the visions around her:

> It's really kind of weird, it looks like a *Tyrannosaurus rex*, it has more that build when you look at it ... but they were meat-eating ... It's the body ... it's the head ... the body is ... I'm saying that only because, when you get to the head, the head looks different, the head looks more like ah ... the ah ... Brontosaurus ... like that with the eye socket, I mean with the eyes, the kindlier eyes, you know, whereas ... the remainder of the body is like that of a *T. rex*, I mean that's what it strikes me as being ... I thought it was surprising when it said it was a plant-eater ... I'm looking at those little tiny forepaws, I mean they weren't tiny but they're smaller than the hindquarters and that's typical of what you would see on some of the meat-eating dinosaurs ... at least that's how it strikes me, from what I've seen ...[11]

Here was Ramona, utterly mired between her expectations of seeing a *T. rex* and the differing saurian form before her! Was she experiencing the early stages of the conceptual shift which the curator had anticipated – the stereotype of the vicious killer dinosaur being displaced by the friendly plant-eater? Was she beginning to think in "a more sophisticated way about dinosaurs"? To me, she seemed confused, incredulous, and nervous about yielding to the possibility of participating in the fabulation of a creature that just might *not* be *T. rex*, the great icon of the twentieth-century epoch of dinosaurs among us.

Perhaps Henson's gamble to fabulate with Horner's *Maiasaura* was working after all in this instance with Ramona. Isabelle Stengers suggests that fabulations tell us about what it is to be part of an epoch while also how not to be determined by that epoch, as when alternative phantasmatic *in-ventions* overflow the epoch's own *con-ventions*.[12] Sympathizing with Ramona's confusion, I was persuaded that she was struggling hard to address what she encountered as overflowing the convention of "meat-eating" dinosaurs, to discover how she might not need to be determined by what, in effect, was an inheritance of the Doyle-Osborn epoch. Rather than a moment of revelation for Ramona, this loomed as a moment of joining another adventure that challenged the conventional dinosaur that had inhabited her expectations, her imagination.

There were many instances of this mixing of recirculated and emergent characteristics in the ways visitors responded to the "kindlier" plant-eating Maiasaur envisioned here. This was not restricted, however, to expectations of anatomical composition, as had been the case with Ramona. Fourteen-year-old Katie also used *Tyrannosaurus* as a comparative for discussing dinosaur behaviour, as I discovered when asking her if she thought *Maiasaura*, "good mother lizard," was an appropriate name for the dinosaur in this display:

KATIE: Yeah, it is – but I don't think for just that dinosaur. Because, even the *T. rex*, they do take care of their young to some extent. So it is not something which is unique to that dinosaur.

BN: Yeah – do most people know that the *T. rex* take care of their young?

KATIE: No … because it just kind of looks like a savage monster, that will eat anything. So it is hard to picture it having a nest, and nuzzling babies, and stuff.

BN: Where did you first find out about Maiasaur?

KATIE: Here.

Here with Katie was an interesting flexibility to exceed and bend categories and expectations in ways that were a challenge to the much older Ramona – that is, to allow not merely for a different dinosaur in *Maiasaura* compared to *T. rex*, but for a different *T. rex*! As with so many children and younger people visiting the exhibit, here I was encountering a flexibility of imagining, a willingness to adventure with ideas, as Stengers might suggest, that dissolved the captivating powers of a prior Mesozoic nexus. As with Ernst Gombrich's and Ludmilla Jordanova's understanding, the "beholder's share" – of the intra-action now with the exhibit – was activated all the more readily by children, the "emancipated spectators" in Jacques Rancière's terms.[13]

But returning to the theatre, the boy who had exclaimed "It's like a *T. rex*!" had swung back to the console, vying to hit the next button before other children could. Listening to the lumbering rhythm of the footfalls, he swayed along with the dinosaur, lifting and dropping his feet on the spot in mimetic sympathy – a mimicry I witnessed children performing many times in the theatre. The other children began to press in even more. The man stood quickly, possibly dismayed by the arriving crush of bodies. He called and clapped his hands together, "Okay, you guys … let's go." Interrupted in their mimicry, the children turned and walked off quickly out of the theatre space, past the showcases of skulls and the entry screens without so much as a glance, scurrying after their father towards other awaiting exhibit experiences.

These interactions of a "family" group engaging the exhibit highlight some of the sorts of actions that were commonly repeated in this "Meet a Maiasaur" theatre space. The most recurrent pattern was that children would rush to the console and begin almost automatically to press buttons – the way that small children rush to press elevator buttons, or older children aggressively bang at video game consoles – while the accompanying adults stood or sat at the back of the generally open space, watching or conversing with the children interacting at the console and screen. A complex of social interactions was activated, with the interactions, and intra-actions, of one person with the display prompting novel and imaginative interactions with others in the group, drawing in conventions, disturbing expectations, allowing for or interrupting sustained engagement.

There were several actors (and non-human "actants") at play here: the computer graphics animation sequence on the screen; the people at the console viewing the sequence; the console with its selection array; and the people standing back and watching both the images and those interacting with the console, who also looked back to the folks (often adults) behind them for assurance, conversation, dialogue, and verification. Those were among the regular and regulating features of the "Meet a Maiasaur" theatre. The exhibit apparatus itself, while conveying a particular configuration of one dinosaur, was also effecting complicated socio-natural play. In that play, visitors would draw upon and bring to bear knowledges from elsewhere – the theatre could as readily conjure the content of a video game, a "beam me up" scenario from *Star Trek*, the behavioural logics from Hollywood films, or previous notions about dinosaur icons like *Tyrannosaurus*. In moments of adult-child communication, the adult would sometimes insert moral points about human life – Maiasaur typically becoming a natural example of "good mothering."

Another recurrent, embodied mode of engagement I witnessed in the galleries was in how some visitors would look back and forth between the skeleton and the animated imagery. Amy, a twenty-one-year-old, remarked:

> It was really neat, the sound effects and stuff. Because I was standing there looking at the skull, and hearing this stuff from the movie, that was fun. Seeing the similarities. I really liked that.

In this comparative play between skeleton and animation, visitors would also imagine and visualize. Renée Akhtar, twenty-four, explained to me that she had been captivated by the skeletal mount with its skull displayed directly adjacent to the giant screen:

> [I]t just helps you visualize like how big dinosaurs are or were, or what they really did look like, otherwise ... you can't tell something like that from pictures or the movies or whatever ... it just helps you visualize!

In her use of the skeleton for visualization, Renée was behaving very much like curator Henson when he imagined the living dinosaur while handling a fossil he had found in some eroded badlands. At the same time, the visions on the screen beside the skeleton provided a techno-scientifically authorized visualization, something that would serve to constrain the imagining one way or another. In his youth, Henson told me, he had visualizing aids in the works of widely published palaeo-artists like the Czech Zdenek Burian and the American Charles Knight.[14] Visitors to this exhibition had the walking, running, friendly good mother lizard in addition to the litany of popular productions of television and feature films. The complex, personalized admixing of dinosaur possibilities taking place here was occurring through the juxtapositions of skeleton, animation, display instruments, recollections of visualizations from other venues, and living moments of fabulation as everyday practice.

The visitor articulation possibilities in engaging with this exhibit seemed extremely complex, and often messy and unpredictable. This was the sort of imagining going on with the mounted cast of ROM #44770. What sort of visualizing and consequent articulations would have taken place for these visitors had they encountered a working laboratory instead? Perhaps it would become clear that many technical steps were at play in the practices of moving from skeletal form to animated on-screen creature? On the other hand, I recalled the silhouette diagram of the skeleton on the wall outside the lab which had precisely depicted the pose of the skeleton in advance – that is, as it would be mounted in the display *two years later* – even as the bones were only just being removed from the matrix. These bones, at the very least, were not the precise fossils and knowledge workings that had given shape to the posed skeleton. Rather, the fossils of ROM #44770, real and telling as they might be, were stand-ins, instruments for showing how laboratory preparation of specimens took place, allowing for the imagining of what kind of work preceded the reconstructions on display.

The exhibit did have knowledge-constraining effects on the direction of phantasies, however, and again these developed through the juxtaposition of various display elements to produce the narrative effects which interpretive planner Jennifer Ross had conceptualized. Simon, one of the two boys who had been operating the console as though it was a video game, recounted to me what he believed the exhibit had been about:

> Mostly the exhibit was about the *Maiasaura* … and it was about the evolution of the world … there was a lot of stuff in there about, um, evolution of everything … like how the world changed so much … like, there was less water and then there was more water, less, more … And it was eighty million years ago … and it was almost completely covered in water, big huge sections … and the *Maiasaura* lived near the water in North America … the world for the *Maiasaura* was like a jungle … because in the other videos, some of the buttons were marked "environment" and you push and it showed it as jungly with water …

Notably here, Simon's narrative was largely that available from the "Visit to the Cretaceous" interactives into which he had freely and readily aligned the "Meet a Maiasaur" "environment" selection depicting the creatures in a "jungly" place. He went so far as to look for "water" in the "Meet a Maiasaur" animations after hearing the "Visit to the Cretaceous" interactives instruct that the Maiasaur "lived near the water." Both boys also went on to tell me, first, that dinosaurs weren't that interesting to them, in spite of the excitement over what they had just experienced; and second, that it was the exhibit media that forced them to be interested:

> I think the interactive stuff was great … I didn't want to read like all the labels, but when you push buttons and it tells you … it's easier … Yeah, and some dinosaur books get pretty boring, I think, but if it's like on a giant TV, it's hard *not* to watch … and it shortens it so you don't have to listen for too long …

The interactives had appealed greatly to these boys – as they had done to the majority of both the female and male children I had witnessed in this theatre. What appeared to affect children most was speed, thrill, stopping power, and immersiveness – the very characteristics of computer gaming experience.

These two children, however, had almost entirely ignored the "front-end" entryway and hall of skulls and showcases into which the designer had put so much energy. As designer Sam Enright noted in one comment, referring to the "front-end" component or "Cretaceous Period" displays:

> [I]t would have been much better with more money ... but a very large portion of the budget was spent on the multimedia ...

Here, however, was a case – as the planning documents had suggested – where the visitor should be able to "get in and get out" quickly. Perhaps they knew that children would be setting the pace. Action would beget action. The exhibit relied on the hope that kids (like these) would be able to rapidly assemble the elements into a totality – and in this instance just such a rearticulation was indeed achieved.

The multimedia producer, Walter Tomasenko, had attempted to intervene here, predicting that children would treat this as a video game. But, interestingly, it was this likeness to video game competition which appeared to best create the conditions for articulation! Tomasenko's intervention was to cause the button lights to go out once a selection was made and the animation was playing. This ethno-methodological design feature may have slowed down the engagement, but only enough to momentarily interrupt the usual competitive free-for-alls of children pounding the lighted buttons. Once again, it had been the interpretive planners who had predicted the fuller intricacies of visitor behaviour in this space. Responding to the written plans, Jennifer Ross had described this:

> We expected visitors to be in small groups ... We just know ... that most visitors do not come alone ... We anticipated that this exhibit would bring family groups, mostly. Or at least it would be an important part of the audience, therefore you had to plan stuff so they could cluster around it ... or point things out to each other. When they are playing with a video ... three or four of them ... they can have room for each other. Oh yeah [*now reading the Interpretation Statement*], "constant interruptions will be normal, as will random sampling of different exhibit effects"... They will point out points of interest to each other. Ask each other questions [*reading again*], "Exhibit elements must be dramatic, streamlined and easy to comprehend, and stay close to the central theme"... and then, "People should be able to get into an experience without preamble, but should feel satisfied getting out quickly." They couldn't get too in-depth, and so on.

In a designed, fly-through, socially active, embodied encounter like this, the planners made equal effort – and with great acuity at that – to ensure there were just enough cues to keep such playful knowledge engagements from straying too far, to keep the visitor leashed down to the central story of this dinosaur and specimen.

The ROM also offered guided walks through its galleries with a variety of programs, providing another means of directing visitor knowing. Some walks were led by volunteer docents, some by teachers on the museum's education staff, still others by paid interpreters. One docent by the name of Virginia, about sixty years old, was a long-time member of the museum. She told me how she would proudly bring visitors here to the "Meet a Maiasaur" exhibit as part of her "great women at the ROM" interpretive tour. *Maiasaura* would join Egyptian mummies and Victorian "ladies" as exemplars of feminine achievement on display at the museum. Yet again, the narratological flexibility of dinosaurs was being put to work, in spite of the explanations provided.

As discussed in relation to the back-and-forth comparison of animation and skeleton, visitors – and now guides, too – would continually fill in their own ideas, narratives, or visualizations when translations and explanations were not provided explicitly by the display. In a different instance, another guide led a group of seven middle-aged and older adult visitors into the theatre. The mounted skeleton was in place off to the side. The guide responded to a question, directing the imagining to unseen, serious actions elsewhere:

> GUIDE [gesturing to large screen with moving Maiasaur animations]: What you see here are computer-animated reconstructions of the Maiasaur. This is not Disneyland here. This is all based on sound and thorough research. The only thing we don't know is the colour of the dinosaur as the skin is not preserved …
>
> WOMAN VISITOR [pointing to mounted skeleton in the same room]: Is this the original that it's based on?
>
> GUIDE: Well, yes, but this is a cast, a copy. The actual fossil cannot be displayed as it is needed for research.[15]

If anything, the animations juxtaposed against the mounted skeleton in the museum setting had acted upon the situation, drawing out the question and requiring the guide to put the elements into reasoned context. The central imperative was to suggest a sequence of connections between authentic fossil and authentic representations – the original specimen for research, the cast for display, and the animation based on the original specimen. We now know that these correspondences were imagined ones, but very effective ones all the same. They worked as a chain of articulations that at once allowed for a sense of the accomplishment of palaeontological reconstruction and the securing of museological authority in generating these outcomes, as well as the authority from which the guide could now borrow in relaying this account.

In this instance of simple conversation, the guide and the visitor had understood each other consummately, knowing what the appropriate questions and the appropriate answers should be. It was important to the guide to note what these animations were not – "This is not Disneyland here" – just as it was important for the visiting woman to interrogate the extent to which the "sound and thorough research" was "based on" an "original." The use of a copy for display was then justified with confidence by the guide in noting that the "actual fossil … is needed for research." By making this simple series of connections, the guide had also distinguished the authenticated animations from those of "Disneyland" (or, say, of *Jurassic Park*), understanding that the "quality" of those outside animations might appear, for all intents and purposes, to be just as expert and finished as these.

Curiously, and amplifying the common imagining of the museum function taking place here, the guide was in part mistaken in her comments. No research on the material was taking place, though the skull was being readied for further preparation and planned future study. In fact, nothing on this specimen had been published to date, and very little in this display was actually based on the specimen in the first place.[16] Rather, this was a moment of uttering and producing the expectation to fulfil the action of being a museum – a performative moment to be precise. To make that most intelligible, it was equally important to use Disney as a foil. That foil, as I discovered, was crucial not just in such public outcomes, but in the very making of this exhibit, which, for this institution, was "probably the first project where media was incorporated as a key part of the design."[17]

~

It was becoming increasingly apparent how multiple actors and elements had affected the natures emerging in the "Meet a Maiasaur" theatre. There was no "given" knowledge, but rather an amalgam of many knowledges at play – some expert, some not. A key actor, of course, was the exhibit's multimedia producer. The following discussion takes its cues from two lengthy interviews I conducted with that producer – and former Digital Media Services manager – Walter Tomasenko. His accounts, while lauding many of its achievements, revealed much about frustrations and contestations in the making of the exhibition, something that was consistently hinted at, talked around, and sometimes blurted out in many of the other interviews informing this research project. Most importantly, his commentaries underscore the complexities in fashioning such a project, while highlighting many of the key informing logics operating in this especially challenging project of leading-edge spectacle and leading-edge specimen display.

"An Exhibit Like No Exhibit You've Seen Before"[18]

> If the goal is to get people to go, "Oh! Cool," and discover it, that component works just fine ... If the goal is to instruct and to inform, then it didn't reach that goal ... But the purpose of exhibits is such a spectrum ... there are some who believe they should be only entertainment, others, that they should be only living textbooks ... I think it should have the full range in the exhibit, so it provides a rich didactic and an entertaining experience ...
>
> Walter Tomasenko, former manager, Digital Media Services, and creative director for the Maiasaur Project

Walter Tomasenko was the director in charge of developing the interactive multimedia displays for the Maiasaur Exhibit, of which the large-screen animations were preeminent. He explained to me how, by the mid-1990s, "Canada [had] about 60 per cent of the computer graphics industry [with companies like] Side Effects, Prism, SoftImage in Montreal, Alias in Toronto." With an understanding of this industrial backdrop, when asked how the Maiasaur Project exhibit was first conceived, Tomasenko pointed out that he had been working on computer graphics and interactivity for some time, even before the new curator of dinosaurs for the ROM was hired in 1992:

> I think the idea is, in some ways, self-evident and attractive. Andreas understands media more than many palaeontologists ... when he sees video projection, computer animation, interactive technology, he thinks, "Ah! let's do this." ... I had worked at the McLaughlin [Planetarium] before it closed ... and developed quite a number of prototypes of various types of media ... showing prototypes of dinosaur animation ... ways to

> manipulate aspects of a dinosaur, whether its sound, its gait, its physical structure ... I had then shown a whole grab-bag of this to ROM's senior management and beyond ... So the idea came about [in] a variety of different ways ... basically [the museum] just "came" to it ...[19]

The belief in an inevitability here, of a technology that just needed to be exposed to the right people, is clear enough as a taken-for-granted position of Tomasenko's. However, this is somewhat different from the sense given to me by the curator that the project happened principally because there was an outstanding specimen available. Two strong egos were are work here, both positioned close to those with the power to decide on allocating resources to this exhibit. Tomasenko eventually acknowledged this:

> [D]epending on your point of view, the confidence to put the money in and to produce the Maiasaur Project internally emerged ... because of the success of our other media projects ... or others would say it came from the curatorial side ... now they had the specimen, so some would say it started there ... So really the idea came from a number of different sources rather than just one ...

It was indeed the case that the exhibit conception and materialization did not, and could not, have had any singular visionary or authoritative source, scientific or otherwise. Rather, it came out of a series of semi-fortuitous performative collisions: of high-tech media-maker with media-literate dinosaur researcher, with the appearance on the institutional fossil market of a very complete dinosaur skeleton, with the recognition by managers in Exhibit Programming that it could be interesting to combine new media with the display of specimens, with the lab-mediated potential to illustrate or suggest how technical work is mechanically linked to physical reconstruction, with the story-weaving work of the interpretive planner, and with still other agents and knowledge-driving and recirculating forces in the history of this exhibit, this specimen, and the history of dinosaur cinematic productions.

Senior managers in the museum had taken this exhibit to the Board of Directors as well to explore private-sector funding opportunities. Andreas Henson recognized that the inclusion of interactive media would make the entire project more "sellable," facilitating what he saw as the critically important working laboratory to be brought into the display. The interpretive planner had noted to me how the fundraising "pitch" was being discussed around the museum:

> [P]art of the ability to do the fund-raising was that "we are going to do an exhibit like no exhibit you have ever seen before. This is going to be a really super high-tech exhibit, really different ..."

So, the very point that the museum could go interactive with advanced multimedia was crucial here. The linkages were quite prolific, and not atypical of the sorts of "unholy alliances" – in the words of Canadian museum theorist Robert Janes[20] – that are increasingly required of museums and other cultural institutions as governments continue to withdraw from their previously mandated funding roles in the museum and heritage sector. Some extensive business alliances were put to work in the exhibit's development: the Board's connection with the principal sponsor, Padullo Integrated; the sponsor's connections with broadcast television as media partners and with the computer graphics company Alias Communications, which donated close to $200,000 in animation software; and Alias's longstanding connection with the ROM's digital media production staff, most notably Tomasenko. In the background, of course, fuelling such media networks, were the corporate consulting and production linkages between the digital animation industry of the Toronto area and the major US players in animated film-making: Disney, Industrial Light and Magic, Dreamworks, and others.

The sponsor credits I had seen on the "Meet a Maiasaur" screen could now be read as significant acknowledgments, not simply those of some disinterested funding arrangements. They also signified a large network of social, technical, economic, and advertising alliances mobilized when one key sponsor had been located for this exhibition. I eventually learned what today is virtually a given in the life of public institutions: that most sponsors at the ROM finance projects as a result of and in order to produce networking benefits, both social and entrepreneurial. Those close to the fundraising for this project, while professionally discreet in their discussions with me, suggested in oblique yet unambiguous terms that "it was probably the sexiest project on the go at the time," speculating as well with regard to the principal sponsor that "advertising firms are savvy to what's hot, what's sellable, what people are interested in." Indeed, one sponsor remarked that the operating principle was that corporate sponsors often see "sponsorship [as] a marketing vehicle ... your marketing department can be driven by sponsorship, and sponsorship can be driven by marketing opportunities."

The Marketing and Advertising Department of the museum worked quite closely with the ROM Foundation, the museum's fundraising arm. In managing the private funding campaign for the Maiasaur Project, the Foundation had the most intimate sense of what the sponsor's alliances and interests were. It becomes all the more understandable that the erstwhile *pro bono* advertising productions contributed by the media partners disproportionately emphasized the interactive and multimedia elements of the exhibit. Similarly, the simple little shift for promotional purposes in the subtitle of the exhibit, noted earlier – from "The Life and Times of a Dinosaur" to "An Interactive Display" – is now recognizable as a signal of the reorientation of the museum, in this project at least, to the interests of private-sector marketing alliances.

The significance and complexity of the dichotomy encountered in media reporting (see chapter 10, "A Really Big Jurassic Place") – which had opposed the scientist and the lab against the digital media producer and the interactive displays – were becoming all the more pronounced. That dichotomy had also expressed a somewhat messy, but practical and functional division in the making of the exhibition – and was its most repeated source of tension, in fact. In one direction, the interactive multimedia components of the Maiasaur Project had made the strongest social and technical connections with corporate, fundraising, and marketing concerns and their related networks. In the opposite direction, the Working Lab and the conventional "life and times" story with the showcase elements had made the strongest social and technical connections to the palaeontological and museological enterprise and its conventional networks. In other words, the "public as market" and marketing and sponsorship concerns aligned readily with "sexy" high-tech interactive multimedia. The "public as audience" aligned more with the Maiasaur family story and the Working Lab. After all the complex work of articulating these dichotomous elements, the functional divide within the museum persisted – if, however, in new, dense configurations that would be more or less imperceptible to the museum visitor.

My point in reviewing all of this is to illustrate how these two exhibitionary impulses became a source of considerable tension, which would reconstitute the possible "natures" of *Maiasaura* that this exhibition would offer to its public – the natures in which these publics were participating in continually shifting ways. The anxiety and challenge faced by all was how to manage the division and sustain the concerns of all involved. One heterogeneous network of connections reached to corporate marketing interests and "sexy" spectacle. Another net of connections – or "vascularizations"[21] – was reaching to fossils and the socio-technical actions of palaeontology. Both networks *plus* the tensions at their intersection were reshaping a relevant, authorized presentation of the Mesozoic and were impinging on how and what would be configured. As I suggested in the first part of this book, such vascularizations necessarily have force on the production of what may come to count as "nature," altering what is supposed to be based solely upon the erstwhile *tabula rasa* of palaeontological evidence – the fossil found in the ground. It is the interaction and intra-action of networks – rather than some general play of "context," "text," or "sub-text" – that literally materializes and brings about the performative Mesozoic worlds at their nexus, at the zone of implosion (as Donna Haraway put it).

In addition, this movement towards new media technology had been envisioned by people like Tomasenko as something which could produce other sorts of organizational effects. In keeping with wider moves in the advancement of information technology in industry, Tomasenko saw digital media technology as something which could be used in every aspect of museum operation – foreshadowing what in the 2010s has in fact become commonplace in administrative, financial, and communicative management in museological and, indeed, almost all institutional/corporate practices. He gave examples of how such technologies could eventually permeate the entire work environment of the ROM:

> … in the galleries … as an interpretive tool … travelling exhibitions, in which you can multiply the artefact and make many exhibitions out of it … a visitor information system … where you could use the technology to help orient the visitor and help sell things … a data management system … integrating collections management with data management … a tool for research, where the technology could be used to run simulations and do science … a revenue-generating source … we would sell our expertise, consult, produce animations … for example, I went to Saudi Arabia with three partners to design and build a museum and multimedia complex in Riyadh … and then also to stretch aspects of the technology … small screens, large screens, domed projections … to actually use that as an attraction in itself as a domain of "location-based entertainment" which, for better or worse, is a trend we're seeing …

I should note that "location-based entertainment" – Tomasenko's example of one "for better or worse" trend in museums – is an industry term for "ride films," such as those associated with IMAX large-screen theatres – one such film being the computer-animated film *T. Rex! Back to the Cretaceous* released in 1998, not long after the Maiasaur Project's life began at the ROM. Tomasenko's main point, however, is that if you come up with an idea which distributes and integrates the technology throughout the institution (and indeed transnationally, even as far as Saudi Arabia), then "the idea becomes self-evident." People would, with what then becomes almost "second nature," buy into it, see its potential, and begin to use it.[22]

However, recalling my earlier discussions on competing concerns in the development process, even in Tomasenko's account the inevitability of the adoption of this technology was not so easily won:

> My experience of this at the beginning was it was a real fight, and people claimed ownership [of the project] from the beginning ... I think it was frustrating for a lot of people ... there was the 3D design component of it ... those people had really tight constraints to work under ... there was the interpretive planners who are charged with getting information about it ... there's the computer animation guys, like myself ... there's actually several media components ... there's the curator, who in his very difficult schedule was working on a couple of digs at the time ... there's the administration ... there's the actual building and fabrication of it ... People were wrestling over really silly things, myself included ... I had a certain vision about how I thought the animation could work, but had certain constraints in terms of rendering time ... I believe there should be someone charged with the vision of it ... and the responsibility for the creative vision ... But you see there were maybe five different people who felt they could have done that ...

This comment confirmed what others involved in the development of the Maiasaur Project had told me previously: that there was as much wrestling and cobbling together going on in the making of this exhibition as there was a fluent, tightly managed, harmonious collaboration. One has to wonder, however, whether Tomasenko was not merely stating what is always the situation: that even if only one person is "charged" with the vision, the outcome will be that of a collective made up not merely of people, but of received and unexpected non-human media and collection components available in the craftwork of making an exhibit.

At the same time, the collision of visions, concerns, and elements invites a winnowing through, a falling out of some possibilities and the firming up of others. Here, I found new significance in a remark on the instability and contingent shifting of goals over the course of the development process that had been made by the exhibit programming manager, Wendy Madsen:

> You see, that's where there can be tremendous conflict and tremendous problems … that stated or unstated, there can be seriously different goals … And it's not even that they're conflicting, it's just that you can't accomplish them all … So you have to make a choice …

Worthy of note in this somewhat fractured collectivity of very sincere workers is how, in fact, this process showed a marked absence of the sort of top-down vision management that had been characteristic of Henry Fairfield Osborn in his rule over the public exhibit development at the AMNH three-quarters of a century prior. Much as some may have wished for such a singular visionary (however unachievable in any case), that approach may well have been consigned to the past – notwithstanding the cult of the individual that is sometimes assigned to corporate enterprise, as often noted in the case of Steve Jobs and Apple. One argument, then, is that a positive fallout in democratic terms may occur in collaborative exhibits, where a more widely accessible, open-ended outcome is produced, allowing more open valences for differing visitor engagements and concerns to be brought into the mix. That said, the presumption of visitor concerns and particular sorts of audiences in this project had still led to quite a constraining, conventional result: the good mother, friendly dinosaur, family-life scenario. Another argument is that the equitable merging and blurring of public and private, specimen and spectacle had been achieved. Facts, however one might construe them, did not really enter this picture in pure form, but were always and already augmented – and would, in the situation of this exhibit, be further reconstituted.

The more collective, team-based approach without an apparent helmsman meant that the resultant struggles over resource allocations and stories to be told would have far more force on the exhibitionary outcome than the single-vision approach that Tomasenko sought. The challenge to ensure strong communicative articulation of components would become more of a challenge as control over its execution was distributed. In the final analysis, the "Meet a Maiasaur" theatre appears to have been the key visitor centrepiece of the exhibit, as opposed to the exhibit team's expectation and rhetoric that the lab would be the focal point. Moreover, the explicit communicative articulations between the two were fairly weak, even though there were implicit ones.

Perhaps most telling in Tomasenko's commentaries was his description of the moment that had provided him the greatest sense of accomplishment in relation to the large-screen "Meet a Maiasaur" theatre:

> [T]he best experience I had was at the end, eight hours before opening, sounds in place, speakers in place ... morning of the press release ... and there was a kid that happened to walk in and tipped up on her toes and pressed one of the buttons and then went ..."*Wowww!!*"... then she pressed each one of the buttons, one at a time, and just was taken by it, because it was so big, and so loud, and ... it really did it ...

The initial effect had been achieved exquisitely for Tomasenko – the very sort of total engulfing in the scale and the *sheer spectacle* of the mediated experience, the potentiality signalled in the exhibit production team's name for this segment of the exhibit, "Meet a Maiasaur" – though he acknowledged that the task of aiding visitors in "getting to know" this specimen had been given over to other segments of the exhibit, most notably the lab and the "Maiasaur neighbours and relatives" segments. Satisfied as he was with this moment of the awestruck child communing with the exhibit, Tomasenko was also dismayed that the experience did not, for him, go beyond that point. Responding to my queries as to what happened, he made it clear, "the animation was successful, but the interpretation got voted down."

That which "got voted down" would have made a significant difference to the articulations between the theatre and the other elements, and, in Tomasenko's opinion, with the visitors. He had proposed a far more intricate array of media to aid the "interpretation." Additional media elements would show an "anatomical Maiasaur" in the round with the flesh added to its bones (very much like that seen in human medical anatomy books), along with different skin colourings for the dinosaur – rather than a prescriptive singular one which might suggest an unreasonable security in what was known. Amplifying the importance of enrolling the visitors' vision, he went on, "I also wanted close-ups of eyes, big as basketballs, to encourage people to look." Many of Tomasenko's suggestions appeared to aim at breaking down any sense of separation between the humans looking on and the virtual dinosaur – this dinosaur could look back, and like us, it had an anatomy which could be presented in a fashion comparable to human anatomy. Tomasenko wanted to push virtuality to the limits of the technological moment, to bring this dinosaur to life, to make the theatre into his "living, breathing environment." Though operating in different conceptual registers – Henson conceived of sophisticating the visitors, Ross of creating relevance for them, Tomasenko of immersing them – Tomasenko's vision, like that of Henson, like that of Ross, was comprehensive or, perhaps more accurately, sought comprehensiveness.

The most important technical elements in Tomasenko's view, however, were to be an additional pair of television or "slave" monitors which would create even greater densities of visual information and interconnection:

> I designed [the theatre] to have the large monitor, and smaller "slave" monitors, in French and English, with interpretation that would direct you to look at certain things … Those got cancelled because of interpretation … to me that's one of the goals of effective media in cultural institutions, to encourage a person to pay closer attention and to discover some aspect of the real thing … I believed and hoped that a few things would happen … there would be a dynamic between the computer-generated dinosaur and the lab, so people would go back and forth between the two, so people would make comparisons and contrasts … I had hoped to create an immersive experience … not a television experience, but an environmental one … I had hoped the slave monitors would give directions and pointers, so that people would pay attention and notice details about the dinosaur, notice the movement, the texture of skin, the movement of the neck, and some of its other kinds of behaviours … the slave monitors, unfortunately, didn't happen … I think in many museum exhibits, one needs some direction on what to do, what to pay attention to … Without it here, you'd go, "Well … there's a picture of a dinosaur … okay, so what?" … I really think the loss of the slave monitors was a serious blow … a profound mistake in the interpretation …

Clearly, the slave monitors were potentially crucial devices for producing spectacle-to-specimen continuity, for Jennifer Ross's "relevance," and for articulating the lab to the theatre.

When asked what the outcome actually was in the end (as opposed to what he knew would work), Tomasenko responded dryly, emphasizing the impoverishment of the result:

> What we actually got was some behaviours – so running, walking, drinking, making sound … And then we got a bunch of semi-realistic renders through a sequoia forest …

My initial thought was that an opportunity to link the "dynamic process" of the curator, to build relevance with the computer-generated dinosaur, had evaded the production team. To use Bruno Latour's terms, the team had failed to enrol a possible ally – notably the ally of Tomasenko's little slave monitors – in the project of creating a stronger, more integrated exhibition – indeed, a more integrated vision of Maiasaur. An emerging anxiety was that dazzling spectacle rather than the specimen (i.e., science) was now driving the museum-located exhibition. Missing the chance to enrol the slave monitors so as to articulate specimen and spectacle was considered, in retrospect, a crucial error.

On the other hand, if these connecting apparatuses and visualizations had been incorporated, it remains unknown what specific effects they would have had on the performative outcome. Would the human-like family lifestyle have been that much more material and forceful? Would the illusion of a one-to-one linkage between the specimen in the lab, or later the mounted skeleton, and the animated forms have appeared that much more seamless? Or would the complex socio-technical and historical actions of "the making of a Maiasaur" have been brought into play? Whatever the specific outcomes, there is little doubt that the inclusion of these tools would have had significant effect in the generation of stories, relations, so shifting the ways in which Maiasaur natures could come into public being.

A retrospective comment from Jennifer Ross acknowledged with consistent, admirable humility and sincerity how the decision to leave out the small monitors had ultimately weakened the exhibit:

> [T]he head that ought to roll in an immediate sense if it doesn't work is [mine]. People are very nice here, and if something doesn't work, they are really nice about it. But I can tell you – on that giant video, for instance, with the giant dinosaur, I said, "Let's keep it really clean and not put up text and not explain why we are doing this shot as opposed to that shot, and not tell people what to look for." I now think that was a mistake. I don't think that people really – you know, did that kind of focused looking, where you are really saying, "Woah, look at what happens when it turns a corner and puts out such-and-such a foot, and the tail whips around behind it." You know, or you just simply say, "Oh, it just turned a corner." That kind of thing. I think it was mistake – I didn't want to interfere with it or anything. I just wanted it to be this experience. I now think, for instance, that I could have done more there. *C'est la vie.*

Given that an entire collectivity was at work here, along with other anxieties and contingencies, Ross's assumption of sole responsibility for this outcome seems harsh and unwarranted. There were others at the ROM who sensed that a much more serious and vexing issue was being contested – one in which the very status of what counts as a museum was at stake.

"Disneyfication" – The Entertainment-Spectacle Complex Absorbing the Specimen-Spectacle Complex

What other possible significance could the loss of these monitors have? Tomasenko viewed the situation as one of institutional conservatism rooted in an abiding Luddism:

> … first, the resistance to change … At the first staff opening … people were all lined up against the back wall with their hands behind their backs, or just leaning against the wall … seeming to evaluate it, some with their arms crossed … one person would saunter up slowly, press a button, and walk back … they didn't even mingle … I was amazed … I mentioned it to a friend … who said to me, "It should be obvious to you … for them, this represents the future and the lack of security, because technology is stepping in, so the skill set, is totally different for many, really an antithesis"…

Rationalizing his project even further, what Tomasenko desired instead was that the managers and staff in the museum would understand that integration of media systems could produce a "return on investment":

> … very few museums have a "return on investment" model … so, few measurable goals – for tracking effectiveness, to increase attendance, to leverage resources, improve education, to get folks to the website, to manage better production pipelines, to educate staff, and so on. That's the return-on-investment view, but I think you get several kinds of reactions from people working in museums, which are far more front-end: "Kids love it, it's a draw, kids show other kids … it brings the perception of the museum into the '90s and the twentieth century" … at exactly the same moment [*speaking ironically now*], "it moves the museum toward the Disneyfication of the collections" … you know ... "theme-parkitude"… "you can even pick up your Mickey ears on the way out …"

Tomasenko was presenting what was, in his estimation, a means of avoiding theme park identification through better "pipelines": integrate, build a media infrastructure, put digital vision at the core of the institutional process, make it the core of the ROM's business model. Yet it was the exhibit manager, Wendy Madsen, who located the issues in much more worldly trends as a festering source of anxiety among many who were perhaps even more committed to principles of maintaining what for them was the public distinctiveness of a "museum":

> [I]t's fascinating with what's happening … because the museum is being included in the "entertainment field"… it's "leisure dollars," and decisions about what you will do with your leisure time … "Will I go to Canada's Wonderland, or will I go to a movie, or will I go to the museum"… It's a fascinating area to try and sort out … You know, there's all that marketing jargon about your "market share," your "brand," what makes you "unique," and Universal Studios and Disney are creating museum-like experiences, and they keep saying, "The one thing we don't have is the scholars and the collections." You know what? They'll have them … they're going to have them so soon. Because it's cheap relative to the kinds of costs they encounter – There are a lot of unemployed experts out there, and you can buy collections … [*laughing nervously now*] … I think it's a very, very interesting moment …

Indeed, several visitors had cited theme parks (including Canada's Wonderland) as an appealing alternative to visiting the ROM. The critical point which Madsen was presenting here rather complicates Tomasenko's comment and his general proposition that "spectacle by itself is Disneyfication."[23] Though implying that there was some great technological revolution underway, Tomasenko rightly suggested that the historical work that museums do is to put spectacle "into context, so that there's meaningful ideas and authority behind it," arguing that, by doing so, "you take it to the next level." The historical situation now, however, is that Disney itself is in a position to "take it to the next level" – and with that to take over many of the special roles that Tomasenko considers the sole reserve of the museum. The question, then, is how will public cultures of nature fare in the face of such continuing transitions to consumerist, bottom-line, profit-margin thinking?

Madsen is correct in gesturing to how very interesting this moment is, and dinosaur fossils may be a key litmus test for many reasons: they are collectable, they are rare, they are available on the market, and the entertainment free-for-all associated with dinosaurs and palaeontology has been intensifying over the last several decades – with the Spielberg/Crichton *Jurassic Park* productions being exceedingly inflated examples. The concurrent "landmark" case directly implicating museums and the scientific community, however, was that of "Sue," the notably "female" *Tyrannosaurus rex* specimen sold through Sotheby's in 1997 to a consortium including, among others, Chicago's Field Museum and Disney Resorts, Inc. That corporate alliance produced an investment portfolio that could pay out $8.3 million US for this single specimen – in other words, "Sue" cost approximately thirty times what the ROM paid for "Henrietta." Museums are now being put in the position of developing liberalized fiscal alliances that increasingly look like the corporate mergers so familiar in the process which capitalist economic impulses install under the banner of "globalization." Under the acquisition agreement, Chicago's Field Museum would split up the whole skeleton, sending half of it to Disney World to be prepared – Maiasaur Project style – before the paying public, while the other half would be retained by the museum in Chicago. Indeed, Disney actually "headhunted" the Maiasaur preparator Phil Thomm to prepare Sue in their reversioning of the ROM's Working Lab!

On one axis, then, the overriding and utterly vexing issue is the ongoing erosion of the museum as publicly funded, publicly effective institution – and the issue of the museum being drawn ever more into the circle of market competition with the entertainment and leisure industry. Tomasenko and Madsen share a liberal consensus in resisting "Disneyfication," but with differently nuanced concerns. Tomasenko's utopianism sees market participation as laudable so long as the dedication to the object and the collection is maintained. Madsen's point considers that as the highly capitalized theme park, film, and entertainment industry moves into the realm of museological practice, publicly funded museums will be increasingly marginalized.

With that as a backdrop, the unspoken and apparently intuitive position of interpretive planners to soft-pedal multimedia was possibly less of a Luddite response than an implicit uneasiness about the risked *capture* of public museum functions by the marketplace. Such a capturing, Isabelle Stengers notes, is achieved almost magically, a kind of sorcery with no sorceror.[24] Madsen's suggestion was that museums were at risk of being beaten by the media industry at their own game – the very project of articulating the specimen and the spectacle. The complementary point is that this move is enhanced by and is accelerating the gradual displacement of state interests in the activities of otherwise "public" museums.

As if to salvage as much as he could from the project as he saw it, Tomasenko spoke at last of the several gains that he felt were made with what was established:

> Real benefits?... it created a relationship with a big software company, Alias ... got a bunch of free software, and access to high-end hardware ... visualizations of chariots in the Saudi Arabian project ... working on plate tectonics ... I don't think that would have happened without that infrastructure in place ... that requires high-end graphics hardware ... Software requires people to be thinking about incorporating media from the very beginning, rather than as an afterthought ...

Arguably, from the first public mediations of dinosaurs in 1854 in Crystal Palace Park, dinosaurs have continually required people to think about incorporating media into the making of saurian natures. Software and dinosaur reanimation could become handmaidens in the new milieu, as had concrete scale models in the mid-nineteenth century. As highly capitalized Disney- and Hollywood-driven animation, ride films, and theme park attractions have become commonplace for consumers, the institution has been pulled again, further and further into ever more sophisticated and costly modes of visionary and experiential production. The pressure with such moves is to incorporate the technology not just into institutional practice, but into the way people actually think, as Tomasenko noted. The distance between the technologies of spectacle and the techniques of palaeontology has become ever greater. It is little wonder that the media producer saw the loss of his proposed slave monitors as a "serious blow," a "profound mistake." It signified for him not just a loss of communicative power but a loss of articulation power, the undermining of all the modern, utopian potentialities which new technology, in his partial perspective, could offer to society. The consequences, of course, are multi-edged.

Fallacy of Misplaced Dualism: Situating the Science-Spectacle Divide

Most of the preceding discussion has addressed the increasing slippage between public museum spectacle and industrial spectacle. Other matters become apparent when multimedia production is considered along a second axis: that of its relation to scientific knowledge. Tomasenko also pointed out how fickle he felt the entire process of reconstruction and animation had been when it came to matters of grounding the result in any kind of coherent body of technical palaeontological and geological knowledge. Here, he speaks of the "interactive continents" display:

> What was interesting is that we produced the animations in-house direct from U. of Texas and U. of Chicago data, and had maps over ten-million-year intervals, showing palaeogeographic information of the time … I realized how much hokum was in that … because of the way that these maps were generated … there's just so much conjecture … the difference between one researcher's map and another's … one has North America split into two parts, the other into three parts … How come? … Well, because someone found one aquatic fossil in the middle of this continent, so there should have been this massive one-thousand-mile lake in the middle! …

As encountered in the interpretive planner's activities, in the expressions of the palaeontologist, in the designer's wish for elemental and modern continuity, and now here once again in the media producer's frustrated efforts to tie the exhibit down to reliable "facts," the returning and resonating issue was the common urge to articulate, to create a seamlessness that was total, flawless, as real as the real world of things it was all supposed to be based upon. That realness was, however, astonishingly elusive:

> I really enjoyed working on the evolution [display], which showed difference in skulls … but it's misleading as hell … there's no continuous steps, there's explosions … and it's not more advanced or more evolved in the sense that you're leading to more complexity … so the morphing and evolution of one skull to another is really misleading … *but*, it's really pretty … makes for a great interactive …

Museum-based communicators were caught in this tug-of-war over "getting it right" and achieving some kind of formal satisfaction. Over and over, Tomasenko's lament was that he was let down on *both* sides – by the exhibit planning and budgeting decisions that prevented him from adding didactic technologies to articulate with the audience, and by incoherence in the "science" that prevented him from anchoring his visions to the fossils. Specimens, the very repertoire of things in an object world which were supposed to animate his animations, were becoming increasingly disconnected. Tomasenko's concern can be summed up thus: *how could one go about creating these lives without the material to do so, and without the means then of saying how you knew what to do?* Put another way, the risk for Tomasenko was literally an inarticulate dinosaurian nature.

On one hand, Tomasenko's position betrayed this general anxiety and tension about articulation, but in a more specific way, it also pointed out the way that limits on the scientific technical knowledge may be filled in by the media specialist – much as the children engaging the fragmentary manifestations of knowledge in the exhibit had done. This was entirely consistent with what has gone on throughout the history of palaeontologist-illustrator working relationships: where one of the collaborators is unable to provide the visionary input, whatever it may be sourced in, the other will fill in the blanks, as Rancière would argue. In the game of palaeontological reconstruction – for scientist, artist, or both together – the eradication of the phantasmatic is impossible and even antithetical.[25] Tomasenko explained the constraints he faced and the solution he applied:

> [O]n the science, I talked to a few other dinosaur palaeontologists ... And their Maiasaur would have been totally different in terms of its movement, its gait, its colour, its behaviour ... and yet they each had equally valid cases ... I needed input on things, like "how does the dinosaur walk ... which leg moves first? ... does it walk like a mammal?" ... So, [the curator and I] went to the zoo and had a look at some animals ... Andreas may have consulted some people on biolocomotion ... no one could give answers or reasons ... I was looking for information from trackways ... apparently it can't be done ... So, to come up with something, we did some experiments with the inverse chromatic solutions with Alias software ...

His last comment here signals the partial "solution": fill in the blanks with the technologies of spectacle, of industrial imagining, quite literally of "Industrial Light and Magic" – for the software technologies used here were practically the same as those used to animate Lucas's and Spielberg's creature- and monster-populated films. The technical apparatus available to the animator, and the speedy immediacy with which it could be applied, far outstripped that which was (to some extent) available to vertebrate palaeontologists – though here we know that the media producers collaborated with their zoo visit to come to an agreement on the appropriate animalian model to start from. From there, the animator would make the choices with the aid and guidance of technology where the scientist had provided the initial answers.

At this point, then, I have come full circle, returning to points made in the first part of this volume about the historical trading and collapsing of fiction and science, of phantasmatic knowing and material knowing. Moreover, it is at this point that the slippage between palaeontological dinosaur reconstruction, museum dinosaur display, and dinosaur animation in the film industry once more becomes unavoidably evident. Borrowing now from the pragmatist propositions of philosopher Alfred North Whitehead, the natural history museum milieu, modernity's knowledge arcade par excellence, points to what might be called the *fallacy of misplaced dualism* (an instance of Whitehead's "fallacy of misplaced concreteness").[26] Specimen complexes and spectacle complexes are brought into interchange as a matter of everyday action – this is techno-theatre. Where scientific specimens and science practitioners of all sorts fall short in providing the full materialization sought by sound museological exhibition-making, then media possibilities, animation techniques, and the apparatus and personnel of spectacle-production are brought to bear – and vice versa. Mesozoic politics/natures precipitate as much in the collapsing together of specimen and spectacle as they do in the co-emergence of phantasy and materiality. But rather than live by the fallacy of misplaced dualism and despair at the error it holds out, what dinosaurs in the specimen-spectacle complex (of a piece) offer, much more promisingly, is the chance to fabulate anew.

This slippage, wrought through the tension between the work of palaeontological reconstruction and the work of market-oriented media tooling – what we might call *the specimen-spectacle tension* – turns out to be a mode of the classic tension of modernity, most exampled in modern capitalism. Deleuze and Guattari spoke of this as a dimension of the "apparatus of capture," and of how the tension is manifest as "undecidable propositions," as was the case of the relation between media producer Tomasenko and curator Henson, placing limits on a more dramatized possibility.[27] They are together and apart at the same time, and at the interstices of their productive work there appear those material figures, those touchstones, that resolve for a time and by default the tensions wrought. The crucial touchstone for that slippage and tension from "The Maiasaur Project, Life and Times of a Dinosaur" – interestingly enough, and to which I have been referring repeatedly – was our precious, diminutive, dewlap-bestowed, pewter Maiasaur.

15 *Mirabile dictu!* Factishes All the Way Down

If you ever watch one of those 1950s science fiction movies … the scientist provides "facts," like, "This is a giant ant. This giant ant will take on LA." There's no equivocating there. It's not like, this giant ant will go south and destroy San Diego, no, this scientist has said it would destroy LA and, *mirabile dictu*! … it goes and takes out LA. That's science for most people – it's a nice, kindly old man who goes and tells the truth. It's almost a priestly function, this person who has revealed knowledge and communicates a truth to the general public. Well frankly, that's not how science works.

Interview comment, Andreas Henson, January 8, 1998

"Give us the facts," "reveal the truth," *mirabile dictu*, "wonderful to relate," "marvels be told": these are the catchphrases for the pressures and responsibilities placed upon scientific experts. As Andreas Henson remarks above, the burden to tell truths is a heavy and disconcerting one, especially when he knows all too well that "that's not how science works." That germinal understanding leads me back to thinking on my original problem in addressing the Maiasaur Project. If "science" does not work to reveal truth, then might these many accounts of the Maiasaur Project, when gathered together with the preceding accounts of *Tyrannosaurus*, Mesozoic performativity, recirculating scenarios, and much else, suggest an alternative and more pragmatic account of "how science works" – one founded on practices of articulating dinosaurs, in this instance of articulating the *Good Mother Lizard*?

I have spoken of the scientist's wish for conveying the dynamic process of palaeontological reconstruction, but that was set in motion in relation to so much more: the marketer's competitive goals to succeed in museums in the slipstream of *Jurassic Park*'s successes at the movies; the hopes invested in the meticulous action of a laboratory; the diplomatic efforts of the interpretive planner to translate across many knowledge experiences with special attention to those aligned with middling Canadian family living; the participation of audiences in constituting a fuller, if also personal, vision of the living creature and its relations; the tensions of media and curatorial producers and their complex of tools and procedures. Through this I have tried to bear witness to the precipitation of an altered Mesozoic, one inhabited by dinosaurs whose forms and ways of living themselves bear witness to all the efforts of all those agents, including the fossils that have come together to allow that very precipitation to take place. This is the work of articulating dinosaurs, and of the anthropology that I have invited readers to join me in exploring.

A consistent feature of much that I have drawn upon has been museum-situated actions of palaeontologists and their museological and extra-museological cohorts engaged in the animation of dinosaurs as palpable once-living life forms. The efforts of those in the civil institution of the museum, and most importantly of the scientists, remain a vital concern, not just for me but for the thriving of institutions of civil expertise, for they are all implicated in the recirculating, interrupting, and articulating of dinosaurs in always-entwined specimen/spectacle histories. A significant element of my coda – more than a conclusion – will be to recover the mutual indebtedness and the intricate reciprocities of cinematic and palaeontological practices of animation. It is here that the pewter Maiasaur becomes a most helpful example of what Bruno Latour has called a "factish," a crucial emergent, articulated actor with special powers to effect further articulation in both familiar and potentially unexpected ways.

So, in these two final chapters I will make an attempt to bring together and offer provisional resolution to the multiple movements laid out in this book. I will do so in four moves, with special attention to two actors: *Maiasaura* as it came into being in the "Life and Times" exhibit; and Henson the scientist, charged as spokesperson for *Maiasaura*'s adequate coming into being at the ROM. The first move is to consider the modest factishes that emerged: the exhibit-readied pewter Maiasaur, a precious implosion of multiple histories in modelled, reanimated animalian form; and to a lesser extent its counterpart, the specimen ROM #44770. The next move will be to consider how the material *enunciation* of *Maiasaura* in the ROM exhibit, as a distinct creature, turned on more than a renunciating of *Tyrannosaurus* and its situated being in order to make evident a more dynamic expression of dinosaurian life. As importantly, this enunciation also turned on the playing out of an array of ramifying articulations and factishes, some extending quite beyond the active reach of the curator or any other individual involved in realizing the exhibit. Most notably, this extension drew science and popular cinema idioms and techniques into museological practices and outcomes. The factishes considered, therefore, act doubly to extend curatorial knowing as well as cinematic knowing, creating new tensions and possibilities for exhibited *Maiasaura* natures.

In the final chapter, "Just Trying to Be a Scientist," I make the third move of revisiting the particularities of how natures, saurians, scientists come to be "civilized" – that is, brought into civil existence in the milieu of public museums. Lastly, and to bring attention back to the effectiveness of curatorial scientists in museums and the articulating of dinosaur life, I will consider the consequences and challenges facing curator Henson in fulfilling his self-understood role of scientist in the face of emerging factishes, signalling in turn the wider pragmatic and speculative potency available to publicly engaged practitioners of palaeontological reanimation and articulation. Through this I signal a hope for scientist-curators to elude capture by the market-oriented impulses of the specimen-spectacle complex by renewing their commitments to the meticulous work of articulating specimens in a more fully situated manner – one that is responsive to the multiple milieus in which curators undertake their museological and disciplinary, field and laboratory practices.

More Than a Rex Object? The Specimen/Spectacle Factish

In the opening quote of this chapter, curator Henson expressed his anxiety over how scientists are compelled to respond to public expectations for enunciating unequivocal truth: "It's almost a priestly function, this person who has revealed knowledge and communicates a truth to the general public." This sentiment has long been played upon in ROM marketing campaigns. Similarly, Henson's concern over the fallibility in efforts to give pure truth has haunted humanists as much as it has scientists. Noted feminist theorist and activist Adrienne Rich famously remarked:

> In speaking of lies, we come inevitably to the subject of the truth. There is nothing simple or easy about this idea. There is no "the truth," "a truth"

> – truth is not one thing, or even a system. It is an increasing complexity. The pattern of the carpet is a surface. When we look closely, or when we become weavers, we learn of the tiny multiple threads unseen in the overall pattern, the knots on the underside of the carpet.
>
> That is why the effort to speak honestly is so important.[1]

There was no discernible attempt at lying in the efforts of the ROM staff to bring about an adequate formulation of *Maiasaura*. Nonetheless, it is far from an easy task as a scientist to speak honestly, as Rich suggests, when caught in the tensions arising where the complexity of forces articulating specimens intersect with those articulating spectacles.

The tension raises a further question: what is it that lies between Andreas Henson's facts that can never quite be provided by the scientist who is asked to pronounce marvellous truths – *mirabile dictu!* – and Adrienne Rich's truths-in-the-making, her analogic carpet-forming practices with their hidden, laboured-over knots, what she calls an increasing complexity? Certainly, it is not simply lies, untruths. But neither is it precise truths. Rather, truth has to be some "thing," so to speak, that can be offered more or less as a faithfully spoken proposition rendered by the scientist; something which simultaneously confers the robustness of that scientific honesty by dint of the meticulous labour and increasing complexity involving many "things" and agents that have brought that proposition, sometimes spoken of as "fact," into existence. Nothing happens in a flash – nor do dinosaurs. To be effective, such meticulousness demands a hesitation, a slowing down, as Isabelle Stengers put it, to allow the extensive complexity to be apprehended.[2]

This is, of course, what I have been referring to, on one hand, when I use the terms Mesozoic performativity and performative dinosaurs. Witness the two protagonists – perhaps we should say antagonists – of the extended tale of this book: the phantastically articulate and much-laboured-over creature *Tyrannosaurus rex* that emerged in the early twentieth century and the more recently articulated *Maisaura peeblesorum*, emerging in the late twentieth century. Both far exceed simple facts just as they exceed simple fetishes – this being the very point of attraction that the ROM sought to exploit in one of its poster campaigns for a temporary exhibit of a large *T. rex* skeletal mount. The campaign was cleverly entitled "Discover the Joy of Rex" (figure 15.1), with the equally clever characterization in one poster, "More than a rex object," and in another, "Your kids shouldn't have to learn about rex on the streets."[3]

Latour's idea of factishes – however clumsy the neologism might feel – seeks to capture what can be reckoned of dinosaurs as material

Figure 15.1. "Joy of Rex" campaign pamphlet. Source: ROM Image, reproduced with permission of the Royal Ontario Museum, © ROM.

realities: they are creatures that confound the choice between "fact" and "belief" for the simple reason they are neither simple facts nor simply imagined beings. Latour notes how we "can retrieve the factish from the massacre of facts and fetishes when we explicitly recover the actions of the makers of both."[4] To be sure, dinosaurs are exemplary of Latour's proposition of factishes. Writing with reference to experimental factishes, such as Pasteur's lactic ferments composed in the lab, or to aspects of the Freudian unconscious, or to mathematical theorems (and, I argue, this works as well for dinosaurs qua dinosaurs), Stengers expresses succinctly how factishes arise as beings "endowed with an autonomous existence."[5] They are positive expressions of factualized phantasy, phantasmatic facts, and provide an affirmative resolution to queries of philosophers, like Paul Boghossian, who asks:

> How could science have made it true – made it a fact – that there was a big bang, or that Jupiter has sixteen moons, or that dinosaurs roamed the earth? Was not the solar system waiting for Kepler, were there not dinosaurs before there were scientists?[6]

The answer to Boghossian is an inescapably qualified "yes" – this affirmation is because there also eventually existed astonomers Copernicus, Galileo, and Kepler for the solar system; cosmologists Alexander Friedmann and Georges Lemaître plus cosmographers such as Humboldt; and then natural historians Cuvier, Owen, Marsh, and Cope for dinosaurs. All spoke, in some measure, *with* the cosmos, proper to their disciplinary engagement.

But, as I have noted at many junctures, this is well-worn terrain in Science Studies. On point in this volume, it has been quite well established by now that fact/belief always collapse in the scientific or public composition and articulation of dinosaurs. So it has been since Gideon Mantell's 1825 description of an iguana-like *Iguanodon*, Richard Owen and Benjamin Waterhouse Hawkins's 1854 Rhinocerine *Megalosaurus*, Cope's energetic kinetic Dryptosaurs, Osborn's supremely battling *T. rexes*, Horner's nurturant good mother saurians – and so it continues with clever, speedy pack-hunting *Velociraptors* ("swift plunderers"), elaborately ornamental *Kosmoceratops* ("ornately horned face"),[7] the whimsically named *Mojoceratops*,[8] the fabulous *Raptorex* ("king of plunderers"), or the "dawn of time" dinosaurs *Eoraptor* ("dawn plunderer") and *Eodromeus* ("dawn swift runner"). While their names are richly indexical, signalling the intention or whimsy of those who named them, each of these dinosaurs will have its own particular history and practice of material/phantasmatic constitution (its own performative efficacy), each will have its scientist-allies, each belies an enrolment to some manner of more private or more public concern, each has its assemblage of specimens, and the animating concerns of each could be brought to the fore through an articulation-forensics, such as I have attempted here with *T. rex* and *Maisasaura*. Name and articulate a dinosaur, and you also begin the work of considering factishes and all that they assemble together, all that they make possible, all that they close down. Select a factish, and you avail yourself of the opportunity to think with and trace the messy forces and complex of articulations that brought that autonomous thing into existence. Those forces in turn allow the dinosaur factish to survive – and recirculate – until conditions cause the disintegration of the very articulations that gave the being its autonomy.

By the time Henson acquired the specimen for the ROM, *Maiasaura* had already achieved a significant measure of autonomy, accruing from the prior palaeontological work. ROM #44770 was already in itself a factish, an assemblage from revised fabulations of Henson and his technical team, engaging with the palaeontological fabulations of Jack Horner and Bob Makela, who offered the first technical description of *Maiasaura peeblesorum*. Factishes, as in the case of this or any palaeontological specimen, allow us to explore the work of articulation and disarticulation *tout court* where the politics lie not simply in the public natures positively composed and brought forth but also in the obscuring of the articulating forces and elements that allow the specimen, and its translated display counterparts, to emerge intact. The power of the specimen to speak was at first amplified, then muted with the presence, followed then by the loss, of the Maiasaur Project's Working Lab.

Without the lab and its directly allied factish ROM #44770 – and even to some extent with the lab in place – it was the pewter Maiasaur that offered a most positive, persistent materialization in the Maiasaur Project exhibit, and which offered so much in connecting things and people and ideas together. It was, from the moment of its conception through to its physical production, both fetish-like and fact-like in its constitution. I turn attention more closely to this more intimate, touchable figure in order to recover the actions making it into such a poignant factish, one that would in turn work upon visitors in their exhibit encounters, and one indebted to complex genealogies of popular and technical remediation action.

Chapter 13 contoured how the idea of featuring a pewter Maiasaur in the official entry to the exhibition had issued not from the curator or from interpretive planner Jennifer Ross, but rather from 2D designer Sam Enright, who had been inspired by the audience-capturing effect of a metallic model of the Statue of Liberty at the site of the statue in New York. Enright had noted further to me the milieu in which a Maiasaur model found its relevance:

> [T]he idea came up through an opportunity … an artist was modelling the dinosaur up with Andreas to have it digitized for the computer animation … He only took it to a certain state because all he needed was the basic structural, muscular framework with which to translate to computers and they would flesh out the creature, with Andreas's input … So we got this artist, with Andreas obviously influencing the final details, to make this thing look as realistic as you could speculate it looking, *without* skin colour … which I think they know least about …

This highly technical work of producing the computer graphic animations in the "Meet a Maiasaur" theatre had a crucial requisite: the making of an accurate scale model. To meet that need, a very accomplished animal sculptor, Manfred Tolman of Toronto, was commissioned by Walter Tomasenko to sculpt a pair of three-foot-long model Maiasaurs. One of these became the original from which the pewter touchable Maiasaur was cast, and the other, as Enright noted, became the digitizable source model for the making of the computer graphics animations.

Factishes to Articulate Curatorial Knowing

As readers will recall, in more than one sense the action here resonates with the moment of interchange eighty years earlier, when Osborn, Christman, Ditmars, Brown, and Matthew worked together on the posable, battling *T. rex* maquettes, seeking to reanimate miniatures with an evolutionary telos and with concerns about vitality that seemed potently relevant to them, and especially to Osborn with his imperious animal/human propositions, inspired as they were by bizarre, elitist fears of the degeneracy of his own racinated clique of powerful, wealthy New Yorkers. Although now the model-making work would be for digital animation, it quite echoed the prior moment of Osborn and his cohort, where the model-making techniques of public museum science display would eventually migrate into the content and techniques of analogue, stop-frame film animation through the later association of Ditmars and Willis O'Brien when they collaborated on films, including *Evolution* and *The Lost World*.

Curator Henson explained that he collaborated with artist Tolman, and according to both, Tolman had closely followed Henson's instructions in the envisioning of the Maiasaur. In that process, Henson had to envision the mounted skeletal form and stance of the Maiasaur and the fleshed-out creature. The gestural form imagined in this collaboration would eventually be repeated: in the finished mounted skeleton, in the rather stiff postures seen in the "Meet a Maiasaur" animations, and in the comparably smooth form of the metallic, pettable Maiasaur. That in turn provided the same form as the official illustration on the advertising material. Indeed, the way the animal model was configured physically is the form which appeared again and again through the exhibition.

The pewter Maiasaur acted tremendously upon the children who encountered it. It was mentioned repeatedly by visitors as something they remembered most from the exhibit. In short, the visual continuity between display elements in this reconstruction was quite remarkable. As designer Enright had told me, "It builds an integrity, a consistency, a repetition, a certain positive redundancy in imagery." Here was visual design coherence at its simplest.

Yet, the irony in the development of the exhibit, in spite of the importance of the imaginative materialization of the Maiasaur figure, was that only suggestive fragments of the detailed technoscientific and display work of articulating these forms, from specimen to finished creature, were elaborated in the exhibit. Of course, these ethnographic accounts are an attempt to augment this. Yes, preparation work upon the specimen did take place, but specifics of that work had little effect on the conformation of the displays in the remainder of the exhibit. The exception, of course, is the interesting dewlap underneath the reconstructed head and neck of the creature, which was only apprehended as the specimen was being physically prepared. Notwithstanding this particular point of articulation, Henson's intended dynamic process was subject to rerouting, and often obscured as a result. The pewter Maiasaur, therefore, would have to operate and was operating all the more fetishistically, in the sense noted by Donna Haraway: "fetishes ... produce a characteristic 'mistake,' fetishes obscure the constitutive tropic nature of themselves and of worlds."[9]

While fashioning the form of this figure required significant input from the curator to provide technical instructions on dinosaur anatomy and physiology, and while this instanced an act towards articulation, it was *articulation to what Henson knew of* Maiasaura, *not articulation to the specimen itself*. An article on the computer graphics production for the Maiasaur animation illustrates how many hidden connections or articulation points to the curator's knowledge were enabled through the making of this model. It also reveals the tremendous expenditure in labour and technologies that informed those animations, suggesting as well why the animation component drew such a large portion of the exhibit's budget. I have inserted numbers at points in the text where curator Henson's guidance played intimately into the craftwork, via his expert advisory role:

> A model builder, working closely with the curator, [1] produced a scale model of the animal in resting position ... The geometry went through several refinements to ensure its scientific accuracy [2] and to minimize control vertices for ease of animation.
>
> Next, artists produced several skin textures ... The skin and beak textures and color were derived from fossil evidence and discussions with curators. [3] An illustrator relying on fossil samples created the backgrounds, such as a Sequoia forest and the young Rocky Mountains ...
>
> With the animations blocked out in detail on storyboards, the animators worked the geometry into position and rendered a series of low-resolution motion tests. Curator Henson [4] helped the animators direct the animal's movement based on the analysis of dinosaur joints and bones. (As a zoologist, he found even this state-of-the-art technology fell short of creating full fluidity and realism of movement. He also foresees possible controversy over his interpretation of the creature's speed, which he based on stride measurements from footprint and track way evidence.) [5][10]

At each noted juncture, the curator engages the sites of tension, moments of intervention, directly and modestly shifting the kind of outputs possible in the technical work of digital animation. The parenthetical point, regarding the creature's speed and stride, anticipated a controversy that never arose so far as the public would be concerned, even if it was noted by Henson as he advised the animation work. Nonetheless, the gait of the creature witnessed by visitors would be significantly sourced in Henson's studies. While the Working Lab had been intended to reveal the dynamics of palaeontological work and only managed to do so in a partial and haphazard manner, this rich and complex translation work of arriving at the pivotal animated creature-outputs – and notably the visualizing input of the palaeontologist – remained hidden from the visiting public in the display. Of course, some visitors did remark that they had a *sense* that there was a lot of technical work behind the animations, but this was in part because some of the animations were incomplete, not because positive accounts were offered. The digital animation wire frames, or splines that would usually only be used as reference points for animators in the process of visual production, were still visible in the "Meet a Maiasaur" projections. They remained unexplained, unsituated, simply there for whatever interpretive consideration a visitor might bring to them – among these were thoughts that they were crucial techniques of palaeontological reconstruction.

The paradox here is that the model, in many regards, prefigured the mounted skeleton. It became clear that before the lab was finished, before extensive preparation had taken place, the scientist, aided by the model builder, *already knew* what the dinosaur should and would look like. I recall the visitor in the gallery asking the docent if the animations were based on the skeleton: in fact, both the animations *and* the skeleton were based on this complex set of intra-active imaginings of the palaeontologist with the sculptor, the animators, artists, media producers, and designers, drawing not so much on the specimen ROM #44770 as on existing imagery of *Maiasaura*, and on a number of technical articles on related hadrosaurian and ornithopod dinosaurs – some of these dating to the late nineteenth century, including, no less, the key article of Osborn's Lamarckian teacher, Edward Drinker Cope, though of course it was used to different ends.

Factishes Articulating Cinematic Genealogies

The constellations of phantasmatic-material forces resulting in the pewter Maiasaur and its multiple graphical reiterations in the display are denser even than this. The entire procedure of model-building, scene-making, and live-action animation also signals how the Maiasaur Project is an event extending the twentieth-century genealogy of literary, popular cinema, palaeontological, and museum exhibit trading which I reviewed in the first part of this volume. The Maiasaur Project is indebted to that history for its own generative possibilities, a conceit cleverly played upon in another of the ROM's poster slogans, "We discovered them long before Hollywood did," this one from the 1980s (figure 15.2). No doubt, the poster crafters were attempting to reclaim museological

Figure 15.2. Poster: "We discovered them long before Hollywood did." Source: ROM Photo, reproduced with permission of the Royal Ontario Museum, © ROM.

authority by this bit of fun, a knowing response to how cinematic and museological dinosaur spectacles emerged together, thriving upon each other, germinal actions in the specimen-spectacle complex.

In regard to the Maiasaur Project, the first and most transparent connection to Hollywood trajectories occurred as the ROM purchased the rights for sixty seconds of footage of animated nesting Maiasaurs, created by California animator Phil Tippett of Tippett studios and used in the "Maiasaur Family" video. That film, the first dramatized filmic animation of nesting Maiasaurs, had drawn upon scientific knowledge and visions from the early 1980s associated with the Maiasaur research conducted by Jack Horner, and those animations were in turn fed back into this mid-1990s exhibit.[11] It very much matters which scientists are in the privileged position in such dramatizations.

As I have already discussed, the trading between film animators and palaeobiologists also occurs through direct involvement in Hollywood film consulting and production. Jack Horner extended his influence in other cinematic dinosaur constructions, including while consulting on the *Jurassic Park* films. David Kirby noted how Horner's own preferred understandings of *T. rex* as a scavenger and of bird-dinosaur evolutionary relations settled into the *Jurassic Park* reanimation, much as Barnum Brown and Osborn's understandings had informed O'Brien's work seventy years earlier:

> Horner's position as consultant allowed him to help shape the visualization of a contentious scientific idea in a major Hollywood film – a film that Horner himself admits has clearly had a major impact on public perceptions of dinosaurs. As with *King Kong*, we can imagine what the film might have looked like had filmmakers chosen an opponent of the bird-dinosaur connection as consultant. Would we look at birds or dinosaurs the same way today if bird-from-dinosaur opponents – such as University of Kansas paleontologist Larry Martin or Wesleyan University developmental biologist Ann Burke – had served as *Jurassic Park*'s science consultant?[12]

Horner's easy movement between his palaeontological engagements and his filmic ones undresses the ROM poster's claim about whether Hollywood or museological science "discovered" or knew dinosaurs first. Yet again, we have to couple palaeontological phantasmatics with reanimation techniques, and retrace once more the complex genealogy of scientific/cinematic trade, backwards to the germinal horizons in New York in the early part of the twentieth century.

The animation technique used in the Tippett sequence for the Maiasaur Project was more or less the same as that used by cinematographic dinosaur animators starting in the 1910s: stop-motion animation, the technique that has largely been replaced in mainstream Hollywood film production by CGI (computer-generated imagery) animation or at a minimum merged with it.

An Australian animation studio's website describes the technique succinctly:

> Stop motion animation involves the manipulation of movement through time. Shooting each frame of film, the animator subtly alters the model's position with incremental changes constructing a movement over a sequence of frames. When the film is viewed at normal speed the character appears to move of its own volition. Of course the hard work done by the production crew is hidden between and beyond each frame of film to allow the illusion to be complete.[13]

The point about vanishing work in the service of complete illusion is not to be taken lightly. It mirrors Adrienne Rich's proposition on the hidden labour behind what counts as "true," finished, complete. In film animation, like exhibit development and museum science, the work of many individuals in the imagining process comes into play, work which in museums typically receives no visible credit. Film productions, at the least, credit many of those who have contributed in every showing. The unhooking of that labour – akin somewhat to Marx's and Benjamin's[14] now-conventional senses of the fetish character of the commodity – is what then gives the outcome its autonomy, its aura, its very reality. That point has equal purchase when aligned with a museum artefact, with a mounted specimen such as ROM #44770, or with a free-moving animation of a prehistoric creature. Matt Williams, another animation enthusiast, adds to this point:

> [T]he animator can, in effect, confer life on his subject and, by superimposing the model onto a full-scale, lifelike background … make us believe that dinosaurs do, in fact, exist.[15]

Put the autonomous animated character into a "lifelike" space and its autonomy from the hidden means and networks of its production is even further secured. A more enduring factish is generated in the setting of a natural history museum as the character is lent further validation by the faithful techniques and intensive analytic engagement of expert palaeontologists and preparators.

The move, from "make-believe" to "making us believe," is pivotal and common to palaeontological and popular cinematic prehistoric creature-generation. In the idioms of Bruno Latour and Isabelle Stengers, and whether we are addressing cinema or palaeontological reanimation, the matter is how "well-composed" the outcome is, how autonomous the reanimated creature becomes, and whether it is brought into public life convincingly and effectively.[16] When set into motion against the history of museums and palaeontology, it is clear that the work of stop-motion animation has contributed, in some measure, to the finished materialization of dinosaur palaeontology's object – the reconstructed dinosaur and its living world. Every dinosaur, whether palaeontological or cinematic, exists as a factish, as an exemplary manifestation of that which lies between fact and fabrication.

These respective métiers – the specialized crafts of film animation and palaeontological reanimation – have the general goal of composition and reconstitution in common, component actions in the work of *articulation*. Moreover, the two métiers are articulated to each other, as they clearly were in the Maiasaur Project.

The animation producer of the Maiasaur Project's "life and times" video, Phil Tippett, can be located as a key figure in the articulate genealogy of monster and dinosaur film animators (discussed in chapter 6), which began with Willis O'Brien in the early part of the twentieth century, saw professional engagements with Rarry Harryhausen in the 1960s and 1970s, and eventually led to associations with the animation production teams of George Lucas and Steven Spielberg. A recent web fan article noted Tippett's animation achievements:

> His creations had included the volcanic flying reptile of *Dragonslayer*, the elephantine Imperial Walkers in *The Empire Strikes Back*, and the gelatinous Jabba the Hutt in *Return of the Jedi*. These massive incubi had bone-rattling impact, menacing beauty, and the kind of horrific details that stick in a movie-lover's memory.[17]

Tippett's revolving involvement in monster, alien, and dinosaur animation films also marks in shorthand some important technological changes in animation which had a direct bearing upon the making of the Maiasaur Project. Tippett had worked for several years with Industrial Light and Magic, George Lucas's special effects company for the *Star Wars* films. Forming his own company in 1983, Tippett worked on the CBS-produced, Walter Cronkite–hosted series *Dinosaurs!* in which he used stop-motion techniques to produce the Maiasaur animations which would earn him an Emmy Award for special effects.

It was footage from this series that made its way into the Maiasaur Project's "life and times" presentation.[18] Notably, Tippett had produced these stop-frame model animations *without* computer imaging technology, though such technology would become available to him later when he collaborated with the producers of *Jurassic Park*, marking an important watershed in the transformation of film animation from analogue to digital techniques. Tippett's place in the genealogy of both cinematic and palaeontological animation work was reported to me in interviews with media producer Walter Tomasenko:

> So he went from stop-motion to CGI … I'm not sure which software he was using … And he went on to being instrumental in *Jurassic Park* … There was this really talented guy who had his own software … and used Alias stuff … did some incredible walk cycles … did some of the finest walk cycles, even before *Jurassic Park*, of the Tyrannosaur, and did full-motion simulation. Now he's working for Disney. So many pockets of activity. Every palaeontologist who had access to computers would be starting to get into it … Some are just creating surface and structure of it … then there is actually building the true structure of it … as the dinosaur moves, you get jiggling … Others use it as a research tool … like trackways … everything from basic research to "does it look good" …

Spielberg hired Tippett in 1991 as "Dinosaur Supervisor" to work with his special effects teams on the first *Jurassic Park* film, and with famed Maiasaur palaeontologist Jack Horner as principal palaeontological advisor. Tippett participated in producing some fifty different stop-motion sequences. The critical turn towards use of computer graphics also took place during that production, in which Tippett's studio took their models and attached computer-linked sensors to their joints to permit a model-to-computer digitizable interface. They later did the same with the actual animation technicians, who attempted to mimic how they felt dinosaurs moved, and most notably Velociraptors. Such techniques are still based on the relay of previously modelled images, animal analogues, and human-body actors wearing digitizable suits, speculatively mimicking the movements of dinosaurs.[19]

The *Jurassic Park* sequel, *The Lost World*, advanced computer-generated animation to its current technical norm, where the motion is almost entirely digitally created on-screen – aka CGI, computer-generated imagery.[20] Indeed, animation of dinosaur films tends now to default to total digital production, as in the case of the Disney family film *Dinosaur* (2001), the BBC series *Walking with the Dinosaurs* (2000), Discovery Channel's *Dinosaur Planet* (2004), and BBC's *March of the Dinosaurs* (2011). Video game animation technology has tracked the digital turn as well, and now there is an unimpeded flow in practices and experiences between television, video games, museum-based animation like that in the Maiasaur Project, location-based entertainment and ride films, and, finally, Hollywood SF film.

The "Meet a Maiasaur" theatre animations, similarly, superseded the Tippett visuals in this historical movement. Alongside this newer digital animation, these older-style images could easily be trumped, surpassed by a newer technological imaging process. The technical advancement of modernity, of the message of science as a progressive march forward, would be signalled as well by this contrast. But here in this exhibit, for now at least, the older Tippet animations – what some children I interviewed referred to as "claymation" – would serve the storytelling intent of the presentation, conveying an "older" frame of scientific knowing fashioned by late 1970s and early 1980s palaeontological concerns, along with the familiar, family lifestyle of Horner's *Maiasaura*.

Returning to key thematics of this book, the purpose in reviewing these connections to film animation histories is twofold. First, I am registering the genealogical relation of the Maiasaur Project exhibit to the phantasy/materiality trading histories discussed in the opening chapters of this volume. Second, I am highlighting the ongoing importance of some form of base-model as a key point of reanimation work in scientific-public transactions.

This also allows a continued tracing of some of the practices that are applied (and were introduced in the first part of this volume) when models are put to work doubly in the service of film animation and dinosaurian reanimation – to wit, the generation of the translational figure of the pewter Maiasaur scale model which channelled the informed, expert imagination of curator Henson, but always in conversation with the prior trade in image-making and development. In the Maiasaur Project, the omission of such a translational figure would likely have left an enormous chasm between the vision and the fossil material with which the palaeontologist worked – a cause for doubt. But it was not omitted, and its positive, material, animating presence translated and articulated much in the display as though it was the product of dynamic science, even though there was so much *other* dynamic science and public concern drawn from cinematic production that had co-shaped this outcome behind and beyond what this display could begin to present.

Factishes Disarticulate and Articulate "How Science Works"

While Walter Tomasenko had been seeking a technological point of media-to-audience articulation through the use of small "slave monitors" to connect the action of the animated beings in the "Meet a Maiasaur" theatre with the Working Lab, it was this other much more elaborate interactive translation device in the form of a miniature *Maiasaura* that was already having tremendous effects upon visitors to the exhibit. The pewter Maiasaur had been a remarkable nexus – to call that term forward again – in the Maiasaur Project exhibition. Clearly, it had acted in many different ways, and for children in particular it was one of the most memorable elements in the exhibit, as well as the element which connected their interactive, visual experience in the "Meet a Maiasaur" theatre with a tactile experience. Its stance and the stance of the mounted skeleton echoed each other. It was a physical manifestation of palaeontology meeting art, of amplified phantasmatics activating around the demands of computer animation production. Its sense of value as precious object was concentrated and exaggerated, so it behaved as a museum rarity, without being an actual museum specimen, yet one meant and thought to be based on the actual specimen from which it would undoubtedly borrow further authority. Its associated media-making history was hidden, that is black-boxed, within it. Its associated craft history was black-boxed. The inferences of the scientist and artist in making it were black-boxed. Its very shape and stance were the result of unseen arguments and agreements. Yet its taken-for-granted being and autonomy would enable weak but plausible suggestions of connection between the reconstructed skeleton, the life and times story, the "Meet a Maiasaur" animations, the family accounts, and the prompting of interactive engagement.

Behind its shining finish, its anatomical precision, its completeness, was contained much of the larger dynamic process activated by and through dinosaur palaeontology and its products, and specifically those circuited through the Maiasaur Project development. As key representative of an exhibit claiming to offer an experience about the scientific-technical process of taking a dinosaur from fossils in the ground to finished exhibit, this figure imposed a halting, gleaming precision of finished knowledge. The model served to affirm Henson's fears about expectations of ready-made truth:

> When I tell people in my sort of charming, post-modernist, neo-Popperian ways, that I test hypotheses, they go "What?! You're not telling us truth? What's the point in having you?" They don't understand how scientific stuff is developed ... [however], if you tell the public this is not an established truth, but just a hypothesis, they just won't buy it.

Henson suggests that the divide between the complexity of the work of science, including hypothesis building and testing, and the public demand for what he knows to be so elusive – "truth" – was significant enough that to make up that divide in a transparent manner would run up against serious obstacles. Nonetheless, he did make an effort, knowing that the demand to practise and sustain, let alone communicate, technical rigour is great on its own. Add to this the extra demand of offering workable translations for producing an extremely cost- and labour-intensive multimedia spectacle, a spectacle that developers hoped would be comparable to that of Disney and Universal. One consequence in the Maiasaur Project was the foregrounding of *Maiasaura* as a far more *approachable*, family-friendly dinosaur than its predecessor, *T. rex*, though yet a decidedly gendered creature.

This also points to how factishes may have both a positive and negative effect upon articulation, and on circulation by allowing certain networks to be brought into play, while "cutting" others, as Marilyn Strathern has described it. Strathern's example is new genetic diagnostic procedures that flowed from the "discovery" of hepatitis C, and how these displaced other diagnostic approaches – that is, other modes of articulating known phenomena through technical procedures. Citing Jacques Derrida, Strathern was noting "the way one phenomenon stops the flow of others,"[21] or, as she put it, how it *cuts one network while opening the flow of another*. Similarly, in speaking of the economic performativity of markets, Michel Callon uses the term *agencement*, after Deleuze and Guattari, to refer to this potential of one kind of enacting agency – the action of credit-granting, for example – to distribute widely, while potentially displacing other economistic practices as it stretches its influence.[22]

This was indeed the effect of the pewter Maiasaur, whose agency borrowed from that of ROM #44770, animated figures, and more. At the same time, while in its making it clearly articulated with the entwined histories of dinosaur animation from film and palaeontology, in its presentation its agency was to articulate with visitor experience, with the specimen, and with the finished images throughout the exhibit, where other knowledge potentials were activated. While clearly part of these cinematic-scientific media networks and with dinosaurian gendering networks drawn from the *T. rex* legacy, the pewter Maiasaur was, for better or worse, also able to truncate all those stories and actions that effectively made it so potent, in the instant that it launched others publicly. The effect was to articulate certain ramifications, while disarticulating certain others, including some that the curator had seen as crucial to his enterprise.

Enunciating Palaeontology's Kinder, Gentler Dinosaur

Objects on display are not for the most part presented as "objects of men's hands" but "appear as independent beings endowed with life."

Sharon Macdonald[23]

Conjoining the two tasks of unpacking and then mapping factishes has allowed a keener reckoning of how the flux of nature, society, publics, and institutional practices is of a piece, and always in motion – or, to use terms previously introduced, how these things are articulated as integral or confluent features of the specimen-spectacle complex in flux. This is flexible politics/natures all the way down, achieved by way of flexible factishes all the way down.

Factishes like ROM #44770 and the pewter Maiasaur precipitate as propositions of contingent "normative" dinosaurian being. Judith Butler's points on performativity bear repeating. The many "acts," reports, utterances instancing saurian natures are, as Butler says, "dissimulated" from the very condition of their own performativity. On the public front, however, the intrusion of individual visitor knowledges and phantasies – the intra-active "in-filling" of which I have spoken – into the dinosaur propositions suggests that even what we might announce as normative is inherently unstable, always open to interruption, rerouting away from any presumed norm.

Contingent norms and contingent materialization collapse into a transitional, but otherwise autonomous natural being in a given historical moment. A proposition that can be ventured in regard to the performativity of ROM's Maiasaur Project is that it instantiates the existence of familial, passive, nurturant dinosaurs, aligning such features with maternalist identifications of the feminine – a set of concerns that clearly articulate to the multiple milieus in which the exhibit came together. This exhibition articulated the systematically imagined social world of a particular dinosaur, and the institutionally imagined social world of the ROM's targeted audience-market: "families." Apart from the ROM team's pretence in not calling specific attention to *how* these intricate articulations were achieved, the exhibit nonetheless publicly offered the Working Lab in which the specimen would very technically and almost magically be "brought back to life." Then, by presenting the finished, "living, breathing" creature in an immersive, interactive animation theatre, it was as though the as yet unfinished work in the lab led to this explicit, authoritative resurrection of the "life and times of a dinosaur."

Much like Osborn's *Tyrannosaurus rex*, the ROM's *Maiasaura peeblesorum* and the events of its life and times rise up as autonomous figures of nature, with ROM #44770 and the pewter Maiasaur as their emissaries. The Horner-Makela scientific Maiasaur specimens discussed in *Nature* also worked as factishes of multiple unresolved technical, scientific debates and engagements with fossils. ROM specimen #44770 articulates to the Horner-Makela legacy of actions and propositions and brings it forward into the Maiasaur Project display. ROM #4470 stands between and so connects the Horner-Makela maternalist Maiasaur with the pewter Maiasaur. By this practice, we begin to follow a winding, branching array of factishes – an unruly, rhizomatic map, so to speak – articulating and extending fossils from Montana sedimentary rocks to ROM collections and preparation work, to the display of a particular specimen, to a pewter reconstruction, to digitized forms animating walking Maiasaurs in the "Meet a Maiasaur" theatre, to Tippett's stop-motion mass nesting sites and loving maternal Maiasaurs. Scores of human actors or animateurs are also articulated in this stream of factishes. With *T. rex* the killer carnivore potently haunting the possibility of how one might imagine this time-space, what precipitates as imaginative accretion is this kinder, gentler saurian living in a Mesozoic time-space that is perfect "for raising a family," and offered sympathetically within a museum setting that itself can be experienced as a perfect time-space "for raising a family" – that being the "family" which is the modal audience constituting ROM's expressed target market.

The moment of the Maiasaur Project's emergence contrasts with the Osborn moment at the start of the twentieth century, when saurian natures were more directly enunciated through presumptively insular, authorial, and hierarchic declarations issuing from small networks of scientists and technicians – and more emphatically by an imperious museum director and palaeontologist – in their undertakings with allied fossils and exhibit development. It is vital to recall that the Maiasaur Project display was produced at the end of the twentieth century in the wealthy northern nation of Canada, a moment marked by an increased participatory and collaborative ethos in mainstream museums in wealthy states. This was also a moment when museums had become – and still continue to be in the 2010s – all the more driven to respond to their visiting publics as "consumer markets." When market concerns seek to reconstitute dinosaurs in line with expected audience in-filling, as was the case in this exhibit, the natures resulting are always at risk of congealing into a market-oriented politics-nature – a movement away from the robustness flowing from the vitally intimate, situated engagements of palaeontologists with the fossil and geological material.

Rather than downplaying or simply ignoring the struggles over whose phantasies count most in conditions of mounting market pressures at the ROM, one is compelled to ask speculative questions: might this exhibit have had even greater effect to allow public engagement in the dynamic processes sought by curator Henson if it had spoken more extensively of the scope of polemics, concerns, and techniques which informed it? Was it adequate to the fossil specimen, ROM #44770? Henson knew the contingency of many of the truth claims and iconized stories from the outset: *Maiasaura*'s naming as the "good mother lizard"; maternal nurturance of hatchlings and juveniles; social behaviour in family groups eighty million years ago situated in a time and place when flowering plants flourished.

If one takes seriously that these are the iconic stories at play – and certainly Henson and the exhibit team members knew this explicitly to be the case – and that these stories express concerns of moment, then what would have been the risks in allowing visitors to share in the ramifying actions that brought those concerns to bear in the articulate animalians they would encounter in the exhibit halls? Rather than bringing forth that which animated the exhibited animal's autonomy as once-living being – a movement further interrupted by the loss of the in-gallery preparation lab and the omission of herding-behaviour videos – the museum left these two factishes, ROM #44770 and the model Maiasaur, to speak all the more for themselves, and so they did.

Of course it was not Henson, or any exhibit team member, or a single planning decision along the way, but rather the complex apparatus and intra-actor exchanges that put limits on the potential to fully realize Henson's wish to engage publics in the meticulous and rich "process to get to that final mounted skeleton." This is the very point and challenge of politics/natures in action. Henson's wishes and fears – and arguably those of all the exhibit team members – resolved on being able to respond adequately and reliably to the specimen and the fossils, and therefore to the call for responsible, scrupulous attention to the dynamic process of crafting the scientific autonomy of *Maiasaura*. Had this thoroughness and attention been realized, one then begins to wonder whether a child's furtive, dangerous touch of a little pewter dinosaur could have led to so much more, that is, to an all the more articulate *Maiasaura* nature.

It is in the intricate passage then, from *mirabile contactu* to *mirabile fabricatu*, that we begin to glean what lies so richly behind the scientist's capacity and authority as expert in achieving the final declaration: *mirabile dictu!*

16 "Just Trying to Be a Scientist": Another Mesozoic Is Possible

[Y]ou basically get away from the notion of dinosaur as monster to dinosaur as animal, capable of complex behaviour ... So, from primitive brute, you get to this sort of civilized, developed organism ...

Andreas Henson, 1998, speaking figuratively
of the transformation of dinosaurs

The question is not ... "does it make a difference?" but "does it make an 'interesting' difference?" that is, does it articulate in interesting new ways ...?

Bruno Latour[1]

Finally, we can speak of the performative materialization and being of the potent little pewter Maiasaur, ROM #44770, and *Maisaura herself* by a remarkable articulation of intricate histories, networks, and actions: specimens, film animation, scientific illustration, palaeontological reconstruction extrapolating from skeletal remains, baby and adult Maiasaurs, displays of the Statue of Liberty, the artistic will of a sculptor, the choice and value of precious materials, digitizing work, the fall-out of carnivore/herbivore tropes, Edward Drinker Cope's descriptive studies of a century ago, the finding of a fossil skeleton on Blackfeet Reservation lands, the entry of that skeleton into the competitive museum collections and acquisitions marketplace, the interpretive savvy of an exhibit planner, the normative dichotomies of gender, the harnessing of the public's desires to touch and so make intimate what is otherwise prohibited from their touch, and the institutional drive to be identified with massively popular dinosaur media spectacles such as *Jurassic Park*.

All of this prompts me to recall the point I made in chapter 5:

> What becomes important is which technical workings and materializations in display and scientific work become the enduring, layered-up nexuses of natural/political relations, achieving dominant inter-performative effects.

ROM #44770 – a.k.a. Henrietta – and its pewter counterpart conjoin into a nexus of articulations, articulating alternate relations in the ebb and flow, being at least partially the consequence of the intricate, intimate, yet sometimes fortuitous choices made by the ROM dinosaur collections curator, Andreas Henson. We know that the historical flows and swerves in the flux of the specimen-spectacle apparatus have conditioned much – and certainly far more than the fairly obvious gendering swerves contoured in this book. It is also clear that scientist-curators have played a vital part in the work of translation and articulation. In this final chapter, then, it is to the actions of the curator, this individual human actor, that I turn my attention, not to suggest the exceptionalism of the curator-scientist as the privileged authority, but rather to consider the complexity, vulnerability, and promise of this role in articulating dinosaurs well.

In one interview, during which we discussed the stresses which massive public awareness of dinosaurs produces in the work life of dinosaur palaeontologists, Henson lamented, "I'm just trying to be a scientist!" His entreaty spoke of the increasing demands upon working as a palaeontologist, especially in a museum, where directly serving public concerns is a mandated aspect of the job – confounded by volatile institutional changes often around market potential – and where the amount of work in managing those concerns in ever more complex institutional arrangements becomes overwhelming. I take from his lamentation and his efforts in the Maiasaur Project some optimistic signs – Henson's recognition that being a scientist entails far more than just technical practices, and extends, as we saw in the last chapter, to public responsiveness and an obligation to speak honestly and honourably about the material and knowledge under consideration in one's research ambit. In fact, that concern aligns with the classical notion of what it is to be a "curator" in a museum. So, in this chapter, I attempt to recover the hope and promise in curatorial work, however modest that may be, and notably in regard to the fashioning of Mesozoic possibility.

In the effort of just trying to be a scientist, both Henson and the fossils were subjected to a vexing bifurcation: the narrative and theatrical model of good mothering and the marketing, branding model of CGI techno-spectacle hijacked the speculative and generative model of engaging visitors in an adequate praxis of the dynamic process of palaeontological investigation. Henson paid the price of being situated in an institutional set-up – one component of the consumer capitalist milieu in which he lived and worked – that continued to prize market-sensitive spectacle productions to which the honest practice of engaging fossil specimens as curator-palaeontologist would have to be accommodated. Henson's "sense of a dynamic process" survived by its yoking to good mothering analytic models – tired and normative as they may be – as contrasts to fearsome hunters that were already captured into the specimen/spectacle complex.

On the proposition that factishes such as ROM #44770 and the pewter model Maiasaur exist through the assembling together of humans with non-humans, and that they work so well as such for being so effectively fabulated, I want to move now to the question of *how they are situated*, in what milieus, and what they might tell us about natures, politics, and yes, Mesozoic *futures* from within those milieus. In so doing I will also be situating and calling forward the critical role of the exhibit curator and other folks involved in the development of the Maiasaur Project exhibit. This is also to ask the question, rather in the manner Bruno Latour has asked in his proposition of political ecology:[2] how are such dinosaurian natures and their factishes brought into society, or better, into public, civil life? An important point of distinction of my terms is required here to properly consider this question: by "civil" I am following the dictionary meaning, "of or relating to ordinary citizens and their concerns," and not some progressive, universalist notion as is often problematically signalled by the word "civilization."[3]

I will follow two related coordinates in considering this matter of the civil life of dinosaurs and their allies: first, that of the museum as an institutional apparatus in which scientists and specimens are often at risk of being captured – or, alternatively, where they might work as hopeful projects of civil engagement. The second coordinate traces the interesting assemblage of curators, specimens, displays, and communications specialists, and the promise of choice conferred upon the curator in generating potential articulations – that is, I consider how to situate the curator in relation to the apparatus, the dispositif. Karen Barad's proposition of the "agential cut," which I discussed in chapter 2, has clear purchase in this milieu, as much as it does in the choice-making of quantum physicists in their experimental set-ups. This, then, foregrounds the efforts and plight of the curator as scientist, and the tension of seeking to act responsively in an apparatus that works over and over to reconcile, mend, and produce plausible articulations between that which it otherwise has separated and divided up: specimens and spectacles, collections and displays, the scientific and the public, the natural and the political. To follow once more on Adrienne Rich's call and responding to Henson's lament, I ask: "What then can be spoken honestly and responsibly by the scientist caught in this perplexing tension?" It is clear that simple utterances of the truth of this or that dinosaur would be irresponsible to the way that evidence comes together and is articulated to other evidence, within many milieus of knowledge production. On the other hand, one can be responsive to the conditions of possibility in which the process of knowing of these creatures, the fossils, and palaeontological matter has been situated and has come about. The full responsibility lies in the compelling challenge to move with, learn with, and offer this increasing complexity that has allowed dinosaurs to exist so convincingly among us as public-scientific animated beings – real, virtual, and otherwise, yet always political.

Finally, and resolving much that has been said from the outset of this book, I consider directly the potential in articulating and rearticulating dinosaurs. I take up a proposition advancing from the discussions of Mesozoic performativity, between the moments of Osborn and Henson, trusting by now that it comes as a far less challenging one. It is the proposition of how in dinosaurian natures *another Mesozoic ought to be possible*, and with that I offer a modest address to the promise such a move holds.

1: The Civility and Incivility of Museums

The apparatus in which the Maiasaur Project and its factishes came to be articulated is, *prima facie,* a bureaucratic and quasi-governmental one, but one located in the larger genealogy of the specimen-spectacle complex. Since its inception in 1912, the Royal Ontario Museum's baseline operational funding has come from the provincial government of Ontario, at times via the University of Toronto, although development and expansion of galleries in recent decades has typically been yoked to large-scale private fundraising efforts through a Board of Governors made up primarily of prominent figures from central Canada's corporate elite. These are the institutional-political coordinates for the playing out of the specimen-spectacle tension faced by curator Henson, for the actions of the exhibit team, and so for the emergence of the Maiasaur Project and its factishes.

The administrative officer for the Palaeobiology Division at the ROM, Beth Jameson, pointed out to me how the history of power relations between the curatorial and the display functions of the museum began to shift significantly in 1968, about the time she joined that section of the museum.[4] Until then, display had been largely a matter of placing objects in showcases, composing them, and affixing descriptive labels. In the four decades that followed, concerns over public visitation and public communication increased.[5] Various struggles saw the museum separate from the University of Toronto, come under the jurisdiction of several provincial government ministries, agree to the unionization of its labour force, go through a major architectural expansion, close to the public for an extended period, and reopen with a royal visit. By the 1980s, interpretation, education, and exhibit planning departments drew the majority of budget allocations, and curatorial departments the minority share – a reversal of the 1960s situation, but one seeking to draw visitors into more of a participatory engagement with the displays and the institution overall. In the course of this, the demand for more articulation between curatorial and public concerns increased, in many ways lessening the divide between exhibits and curatorial functions, though granting increasing agency to the exhibit functions and drawing curatorial interests gradually in their slipstream. All the same, Jameson expressed to me her view that the ROM would always be guided by the curators, as they were, in her words, "the only stable group in the institution."

In terms of the conditions of bureaucracy, however, the struggle over whose accounts and which practices should take precedence in the exhibit was further complicated by the conventional but divisive organizational structure, as Jameson pointed out:

> This is the problem with the administration … it's always been vertical … we had a curatorial *division*, a service *division*, an exhibit *division* … people at the top of each were building their empires … they're *divisions*, the very word means you don't talk to one another across the divisions … It's exactly the same today … [Emphasis Jameson's]

At the core of Jameson's remarks was a very resilient dichotomy in museum functioning: the relation between scientists + collections repeatedly set against the relation between exhibitors + publics, what I have called the specimen-spectacle divide, but dissected further by the dividing up of marketing, interpretation, design, construction, fund-raising, evaluation functions, and associated institutional units.[6] Divisions between functionally organized sections of the museum had arisen continually in the Maiasaur Project. Recall again the precautionary note the ever-pragmatic administrator Jameson had struck:

> getting a dinosaur from the ground to the public, that all needs to be in one area … you can't have it divided … And yes, the Maiasaur Project was like that because it was a team … But they also had Andreas, who was new, who didn't have any baggage … who didn't know of previous problems … he wanted to make his mark … It would never have come off if it hadn't been for that …

In other words, Henson didn't know what he couldn't do, or that the dinosaur specimen and its story would eventually have to run through so many hands, and so he tried things otherwise untried, borrowing on the residues of authority still residing in the role called "curator." To be sure, the conventional – one might even say arcane – role of a curator was that which Henson had seen as his prerogative: *to select, organize, and look after items in a collection or exhibition*. What he had to contend with now was that this role had been far more distributed across a team of specialists, many of whom were untrained in palaeontology but highly trained in distinct aspects of exhibit communications.

Still, because or in spite of the "team" approach to exhibit curation, the baggage Jameson referred to is precisely what ended up generating the altered divide in the Maiasaur Project. This divide played itself out poignantly in the struggle over representational priority between curator Henson and media producer Tomasenko in attempting to appeal to "audience" and "market." The work to bridge the gap was taken up valiantly by interpreter Ross, acting as mediator and animateur, but who also admitted to slipping up in those mediation and articulation obligations when she opposed the use of "slave monitors." These monitors, seemingly insignificant as mere technical devices, appear now as potently crucial tools for knowledge articulation, and for animating and materializing palaeontological expertise and knowing within this specimen-centred exhibit.

Writing of the Smithsonian Institution, Steve Allison-Bunnell witnessed a similar struggle:

> The division between exhibition and research has served to dissolve the unique characteristic of natural history museums as places where scientific research and public representation interlock ... there is indeed no particular reason why either science or exhibition must continue to coexist under the same roof, and the identity and mission of natural history museums will be decided by struggles between these two programmes, rather than the negotiation of a symbiosis.[7]

To a limited extent at least, the Maiasaur Project signals a constrained effort to work as a team towards collective public natures. Nonetheless, the Maiasaur Project illustrates how consumer-oriented marketing concerns have become part of the action of the collective, as well as a feature of the exhibitionary outcome.[8] The audience-market, consumerist move had directly affected curator Henson from the very outset. To ensure that the Working Lab could be developed for the display, Henson knew "that only by putting in an over-abundance of interactive technology which somehow made it hip and modern ... was I able to sell that very simple idea." He did so because he understood that "Museums have become a leisure time activity. So it's very important to compete with other leisure time activities and to provide entertainment." These moves would haunt the exhibit and its project team throughout the development. Henson had also noted that the move to the "family" story was not *his* choice, but rather something "that just happened" – in spite of the fact that it was he who had enrolled Horner's family-structure account in promoting the acquisition of the Maiasaur specimen. Both the domestic family lifestyle account and the high-tech multimedia were oriented to audience and market for their effects as consumer fetishes.

So, the curator's intent was gradually swept up by other institutional tides. But he did not act alone in facing the currents. Interpreters and technical staff made choices that also generated these currents, allowing the specimen and the model Maiasaurs to act in a way that could mirror the concerns of the market. What took place in the lab became, ultimately, not what informed the surrounding exhibits, but rather the display of legitimating palaeontological preparation techniques as the crucial means for scientists and publics alike *to be fully responsive to the specimens* themselves. Its articulation with the rest of the exhibit resided mostly in its being interactive and visual. As I learned from those who worked on the displays – and the preparators who worked *in* the display lab itself – the lab did captivate visitors. Indeed, one guide at the museum blogged, some fifteen years after the fact, of how the lab would come to transform his entire sense of nature and museums:

> This one visit in particular was really influential because one of the scientists brought me into the lab so I could see what he was studying. It was ridiculously cool. There were fossils everywhere, and he showed me the tools he was using and where the bones came from and why science is so fun. I remember also being really impressed that they got to sit on spinning chairs all day. Needless to say I never forgot this experience. It was so special for me to meet the people who actually put the galleries together that I love so much, collect the specimens and artifacts that I'm in awe of, and do the research behind some of the greatest discoveries.[9]

Yet, apart from the lab, the "science" of the specimen itself was articulated suggestively rather than explicitly in the majority of media elements in the display. Visitors were pushed all the more to absorb the "science" in the story that was relayed in the display: the virtualized Maiasaur and the familial life and times of the good mother.

Of any person in the exhibit development process, curator Henson exercised the greatest influence over the materializations in the Maiasaur Project. His respected guidance and input was felt across the network of interactions: with the collections, with the purchase of the specimen, the display media, the storyline, in budgeting, in the specimen preparation process, in model-making, scene-making, and dinosaur animations. He acted as a constant authority to planners, designers, managers, technicians, board members, sponsors, news reporters, script writers, marketing staff, web page designers, etc. He was seen and heard in a video in the galleries describing the field locality where the specimen was found. In research museums like the ROM in the 1990s, scientists and curators were arguably still the most "stable group in the institution," as Beth Jameson suggested, positioned at many of the points of juncture (or rupture) between the technical and public practices of museums.

In sum, the civilizing project of the exhibition and its emergent being was one wrought through great struggles to resist and overcome division after division, to defy and work around those divisions, while sustaining the democratic role and commitment of curatorship. Once more, such "civilizing" works not in the tongue-in-cheek, progressive evolutionary sense of the term that curator Henson was playing upon to get his transformative aims across. Instead, it arises as the action of struggling to bring something into civil life, that is, life with people, with humans – literally, the bringing of nature articulately into society, into the *demos* – where, curiously, it always does get articulated, disarticulated, and rearticulated, as Latour has proposed.

Hypotheses out of Unholy Alliances

Alliances make a difference in nature. Institutions such as the ROM exist in the persistent tension between dynamics of *what can possibly be known* and those aligning to *what kind of knowing will attract paying visitors*. In a volume entitled *Museums and the Paradox of Change*,[10] Dr Robert Janes, an archaeologist and former director of the Glenbow Museum in Calgary, Alberta, attempted to come to terms with the place of Canadian public museums in times of diminishing – even vanishing – government funds for their operation as natural and cultural history institutions. The Glenbow, once recipient of large government grants to support its operations, had throughout the 1990s and into the 2000s become what Janes calls an "autonomous museum," relying on private contributions and sponsorships, admission revenues, and entrepreneurial projects for the lion's share of its budgetary needs. Janes's somewhat uneasy commentary about the shifts in museums towards increasing levels of collaboration with corporations, or in joint ventures with anything from fast food restaurants to international consulting, speaks of these relationships as "unholy alliances." The curator's alliance to particular specimens, to closely studied collections, offers the counter to such problematic institutional alliances.

In an effort to carve out a role for museums as they end up less and less linked to government through grant funding, Janes situated them in relation to his working definition of "civil society" as being

> the space between the individual and the state, and … the realm of autonomous institutions that are not run by governments, but that act as agents of the will of the people … Predictably, in a country where the vast majority of professional museums are owned and operated by government, the role of museums in civil society remains unexplored.[11]

However, the hopeful work that goes on in the space which Janes discusses is similar, I would argue, for government-funded agencies like the ROM, which have also come to ally themselves with corporate supporters and partners, and which also vie with autonomous museums and other leisure attractions for public attention.[12] As Tony Bennett notes, that work still takes aim at the visitor in order to "induct her or him into new forms of programming ... aimed at producing new types of conduct and self-shaping," which of course will be more prone to refracting the ethos of private corporate sponsors and capital-sensitive boards of directors.

Yet Janes has a vital point. Museums of all sorts may be aligned in relation to this project, and those who steward these museums can advance or upset the ongoing drift towards market orientations, choosing their alliances accordingly. Put another way, those choices can lead to a pernicious and ongoing capture of the institution and its knowledge-generation by articulation to markets, or they can interfere with and slow down that capture – or even, one might think, generate alternative articulations and alliances that counteract the market drift. That is, they can interrupt what Deleuze and Guattari note when they say that market orientation in performative action is akin to what they call "strata," which work as "acts of capture ... like 'blackholes' or occlusions striving to seize whatever comes within their reach."[13]

In the 2000s, the ROM had a chance to do this when it embarked on a different development project, known as the "Crystal" – an architectural expansion offering a new set of galleries that began to be opened in the summer of 2007, including a showpiece gallery fashioned around a highly aestheticized presentation of dinosaur skeletal mounts. Still, arguably, it did no more than generate new forms of alliance to the market in this specimen-spectacle move, ones predicated not on an immersive experience in the potential dynamics of science and natural process as sought in the Maiasaur Project with its potent factishes, but on the appeal (a market-responsive one at that) of museums as publicly accessible, elite sites of the precious, the exquisite, the superlative, the beautiful. A very different adventure in the specimen-spectacle complex, and yet one that still mobilizes around the veracity of the real, the didacticism of the mediated. The necessity of a curator is increasingly removed in this form of display. The ROM Crystal made what was suggested would be a new move, but which in fact was fully based on the familiar pattern of the modern pretension: aestheticize the specimen as spectacle, almost as art-piece; displace the knowing that allows those specimens to stand as autonomous creations.

This only braces up what happens in the space of which Janes speaks, and raises many additional questions. Janes speaks of the "will of the people," but just who are the people for whom such displays and selections are acting, whether in the case of the Crystal or the Maiasaur

Project? What is the complex set of everyday agencies which are mobilized in museum sites and which serve to reconfigure what Janes refers to as the individual and the liberal state? More directly in relation to the work of dinosaur world-making, for cultural institutions in particular, what is it that precipitates uniquely, diversely, or repetitively in that heterogeneous space in between?

Janes, sounding much like Henson, eventually suggests that the work of museums "is really a hypothesis, and every day we frame our questions and assumptions in accordance with what we know at the time," something that tomorrow could be rather different.[14] A more compelling claim could hardly be made. Sharon Macdonald echoes Janes, using "theory" rather than "hypothesis":

> Any museum, or exhibition is, in effect, a statement of position. It is a theory: a suggested way of seeing the world. And, like any theory, it contains certain assumptions, speaks to some matters and ignores others, and is intimately bound up with – and capable of affecting – broader social and cultural relations.[15]

And, I would add here, "socio-natural" relations. A more pointed term than *hypothesis* with which to think about museums is *proposition*, as used by Alfred North Whitehead, as "something which matters" and so which sooner or later can stand as autonomous, the way *Maiasaura* has become autonomous.[16] Museums, dinosaurs, displays are propositions, yet complexly articulated ones. The Maiasaur Project traded on something that mattered publicly, though along two significant vectors. One vector runs through the dynamics of science as practice. The second runs through the extension and tension of the tropic gendering in masculinized hunter-carnivory and feminized maternal-herbivory. The latter vector extends on a now-normative story. The former indicates what might be opened up by civil engagement in the dynamic practices of sciences.

Knowing so well now that dinosaurs and dinosaur exhibits do indeed matter as propositions of natures, the museum-located Maiasaur Project also affirms what Tom Mitchell suggested more generally concerning "what is to be done" with the dinosaur today:

> "[K]eep an eye on it"... pay attention to what is happening to it, try to make sense of it. The creature has an uncanny capacity for working both symptomatically and diagnostically. It expresses the political unconscious

of each era of modern life, manifesting collective anxieties about disaster and extinction, epitomizing our own ambivalence toward our collective condition.[17]

But clearly, there are more possibilities for action than just watching the dinosaur. What of the choices that are made concerning which fossils to search for, to collect, to describe, to make theory around, to bring to the public? What of the opportunity offered by dinosaur fossil diversity for civil engagement, that is, to generate civil saurians, civil natures? What of the obligation, when situated in the messy apparatus of the museum, to speak honestly? For the dinosaur, or rather *dinosaurs*, plural, bring their own agency resulting from scientists' choices to pursue one kind or another of fossil material. We are more than merely a market for dinosaurs. Rather we participate with their being at every turn, our questions, dreams, hopes for them (and so for us) rebounding into the flows constituting them, their constitution then rebounding to us. As Barad has noted, there is a mutuality of the agency and choices of the person with that of the entity constituted.

2: Caring Curators, Orthogonal Museums, Articulate Dinosaurs

Science and democracy are co-extensive to the social fabric.

Bruno Latour, "Re-enacting Science," lecture at Science Gallery, Trinity College, Dublin, 2012

Bruno Latour put the question of articulation that I have been exploring in very direct terms:

> What would happen to the collective understanding of a discipline, if scientists were no longer trying to extirpate themselves from the sin of being connected, but accepted the vascularization as so many positive features that would turn their science into a well articulated one?[18]

As if in answer to Latour's question, in his interview with me, palaeontology technician Phil Thomm pointed to the salient possibility of articulating through the Maiasaur Project, notably through its specimen, in the manner Henson had been seeking:

> Palaeontology is a really good example of this point … it is usually done behind closed doors … it's sort of mysterious that way. And in the time of

> cutbacks, and Mike Harris,[19] and growing unsureness about learning institutions and the museum, and where it was going, curatorially anyway ... it's about accountability I would say, and trying to let people know that there's nine floors joined to the gallery space that maybe they don't know about ... I guess that was part of what they're [trying] to get from the specimen ...

Thomm, much like Latour, suggested how ROM #44770 had a "vascular" connection to the cutbacks of the Ontario government. Set against those flows, which had the effect of *disarticulating the scientist's practices of knowing*, were the vascular connections that Henson sought to enable, a means to bring the fuller dynamics *articulating scientific techniques, specimens, biological and ecological actions* to bear in what the public would come to engage. Henson and the team, while always understanding the contingency of their own knowing and the milieus in which it unfolded, retained the hope that they could provide an alternate and actually more robust accountability to the natural dynamics animating the creature, model, and specimen of *Maiasaura*. They had to settle on a compromise in the end – something that gave the "sense" of this process – but their hope and concern did not falter.

My recurrent point in this book is that dinosaurs prove themselves over and over to be natural/political agents, literally and materially, establishing their agency, the power to act upon us and our engagement with what is known as natural, an agency we have in part bestowed upon them. Sometimes this agency is stated lightly, other times it is embedded deeply. Thomm's comment is signal, yet again:

> If you were to go up to a child and ask them what kind of dinosaur they wanted to see, it would be a *T. rex*, right ... There's no doubt ... And if you were a business ... you just look at Sue ... If you're a McDonalds or a Walt Disney and you're big, you want to strike a lot of people, what else would you go for but a Tyrannosaur ... And maybe that relates to capitalism somehow ... the ferociousness of it ...

If Thomm is correct, then *Maiasaura* – the factish creature in general as well as the specimen and the pewter model, in alliance with curator Henson and the exhibit team – would have to be seen as interrupting the extreme workings of capitalism. This dinosaur, in some measure, moved away from the ferocity and destructive force of capitalism embodied by the giant carnosaurs.

A cynical analyst might say that Maiasaur is simply the passive, "kindlier" face of capital, a counterpart in the neoliberal struggle set out in terms of good mothers protecting nests from tyrant kings. The claim is certainly compelling. However, such cynicism also works as a capitulation to the inevitability of neoliberalism, a projection of that project. More soberly and modestly, in the face of such conditions, scientists, curators, and museum workers of all sorts have taken up the challenge of bringing their concerns forward, seeking the small or large points of responsiveness that might, in effect, slow the specimen-spectacle apparatus in materializing and capturing natures so thoroughly infected by consumer interest. In moving from anxiety to action, curators and curatorial exhibit teams contribute to the often unpredictable articulation of appropriately responsive dinosaurian civil natures. The conduct of these civil society actors, then, enhances the conduct of civil natures.

Whereas cultural theorist Tom Mitchell concentrated on the distributed and notably visual fetishism of dinosaurs – leading him to call dinosaurs the "totem animal of *modernity*"[20] – Robert Janes has chosen to concentrate on this *civil* function of museums as a space for the people and their engagement with the natural, the cultural, a space of connection and care. The Maiasaur Project aids in showing how these two actions (Mitchell's modernist fetishism and Janes's civil society) are complexly yoked together, uneasy, often fumbling conversants in late liberal projects to produce hopeful, viable articulations of nature. Janes, full of uncertain hope, suggests,

> Museums, especially if they are autonomous agents, are an important counterbalance to the aspirations of free-marketeers. As such they have an enduring role to play in the civil society, by demonstrating the need for balancing the forces of the marketplace with the perspective that no one group or ideology possesses the sole truth about how society should develop.[21]

When he wrote again on this topic in 2009, Janes's worries had deepened, as he noted how many of those who work within museums face institutional pressures disconnecting them from public concern, apart from those that might enhance marketing missions of the institutions.[22] His worries have been that museums are on the verge of irrelevance as a consequence. Instead, and responding boldly to crumbling dreams of unlimited growth in capital and to biospheric destruction resulting from the resource over-consumption of growth-desperate capital, Janes has now begun asking those who steward museums to consider their responsibility, their obligations:

> All museums have the responsibility to … empower and honour all people in the search for a sustainable and just world – by creating a mission that focuses on the interconnectedness of our world and its challenges and promotes the integration of disparate perspectives.[23]

Janes also remarks how "thought and action are largely uncoupled in the museum world – a primary cause of the drift into irrelevance."[24] He goes further by pointing to the routinized "fallacy of authoritative neutrality" that those who steward museums are and must somehow remain disengaged from the full and complex worldly milieus that they have been charged to study and bring genuinely to the public. They maintain this disengagement in the very moment that museum leaders ask them to answer to market competition for a consumer public, rather than to be responsive to the citizen public.

With Janes, we arrive at the action of disarticulation, which in turns triggers questions of *obligations to articulate and rearticulate well*, to generate political natures that are adequate to the world unfolding, what we might call cosmopolitical obligation. In museum vertebrate palaeobiology, it remains with curators to respond well when the moment of decision arises as to which dinosaur, what story, which process of the natural ought to be brought into public conversation. Andreas Henson sought in the Maiasaur Project to build and foreground that which would overcome and rearticulate the many divides of the specimen-spectacle complex: the genuine engaging of the visiting public in the dynamic process of palaeontological investigation.

However we might judge curators, their curatorial intentionality in the history of public palaeontology is pro forma, a resilient feature of curatorial praxis. Recall again the 1850s when Benjamin Waterhouse Hawkins and Sir Richard Owen conceived of their "Secondary Island" of Iguanodon and Megalosaurus models that could bring publics into coordination with the then-understood natural history and sedimentology of the newly described *Dinosauria*. Theirs was a project in support of the improvement of the English people, in line with the progressive civilizational will of Queen Victoria and her consort Prince Albert, echoing the utilitarian social philosophies of Jeremy Bentham. Recall how, in the early decades of the 1900s, Henry Fairfield Osborn sought to extend into the revised maps and moments of the Mesozoic those selective specimen-creatures and modes of palaeontological reanimation that best lined up with and supported his own will to ascendancy that he shared with the wealthy white New York fraternity with whom he consorted. Through *Tyrannosaurus* and other creatures, he would be able to assemble a modified political nature for his delineated publics at the American Museum of Natural History, one that would meet well with Galtonian eugenics programs of the day.

What we will recognize now is that Osborn had decisive say over his allied specimens and key articulator *T. rex* within the AMNH apparatus of public natures, directing those who would do the work of constituting the mounted specimens, displays, handbooks, and public interventions. Henson, on the other hand, though working with similar modalities and techniques of fossil preparation and reanimation, was positioned in shared decision-making with the exhibit team, in a far more democratic fashion, though he certainly retained a significant measure of authority.

In a historical sweep, *Maiasaura* may seem superficially but a small augmentation to Mesozoic natures. But even the challenging intervention of Henson in his attempts to relay dynamic process amounts to a reorienting of politics/natures. However partial or limited the specimen engagement might have been – especially with the loss of the lab and the slave monitors – a deeper engagement with the fossils in the accounts and experiences was offered in the exhibit. And as discussed in the previous chapter, it was here that ROM #44770 and the pewter Maiasaur were articulated into being, articulated as autonomous figures of life, and here that they indeed became articulated to each other as civil saurians, civil natures. These were the real fruits of Henson's struggle.

Henson's lament, a reflection on the action of exhibit-making, about "just trying to be a scientist" makes full sense now as a response and resistance to the pressures of being captured by an institutional apparatus gone awry. He was attempting to speak honestly as a scientist, as a curator. Isabelle Stengers remarked,

> When an operation of capture succeeds, then, one will learn instead to yell, to cry, to find words that rise up like lamentations to speak this disgrace, to transform it into a force that obliges one to think/feel/act.[25]

Stengers is reclaiming these lamentations as assertions of the will to articulate responsively – a will that is available to, if not the principal *modus operandi* of, committed scientists, the will to respond adequately to "the question of the obligations proper to their practice."[26]

The power of scientists, Stengers points out, resides in their capacity to constitute "phenomena as actors in the discussion, that is, not only of letting them [e.g., specimens] speak, but of letting them speak in a way that all other scientists recognize as reliable."[27] Henson achieved this to some extent, after all, in his effort to *remain* an articulating scientist – and *not* a high priest rendering truths – within the more or less democratically configured space of science museum palaeontology and exhibit development. Stengers's remarks also gesture to the multiple milieus that condition scientists' *obligations to care for phenomena*. They allow me at this juncture to tie together many threads around the idea and practice of care in curatorial and museological activity – that is, to reclaim the promise in those practices.

First, what are the operating principles animating the very idea of curation? Etymologically, the Latin root of the word "curator" is "curare," literally to cure, to "take care of" something or someone, or to act as a guardian. It is a word denoting responsibility and obligation within a milieu. Its ancient roots refer to care of souls, so the milieu of sacred life, whereas in Middle English the reference moves to the application of medicine, literally of a cure, moving to milieus of healing others. By early modern times, English usage of the term "curator" began to be associated with those who might carry pastoral responsibility, as in the cleric's tending to a congregation, derived from the pastoralist's tending to a flock, so the milieu of beings gathered into localized assembly. Next, in a secularizing museum world of the late nineteenth century, the idea of the collections and museum curator as keeper came into being – the care for an assemblage of things in encounter with, and on behalf of, a public, a very different sort of "flock."[28]

A propos, to curate in a museum has conventionally been to care, meticulously and in an orderly and professional way, for that tangible and intangible stuff which museums gather together on behalf of their publics (which they also gather together), and which they call their collections. Rather than acting in the manner of a high priest, curatorship is aligned to an obligation to act in a caring way for *both* this material and the publics who would avail themselves of the knowledge-benefits of those collections. The role of a museum curator is, therefore, to be a consummate and caring articulator of public natures (or cultures, or creative expressions), working *in the meso*, by means of acquisitions and management of collections, research in one's own discipline (in

Henson's case, as a vertebrate palaeobiologist), writing about the collections, exhibiting them, and creating public as well as scientific community connections to them. This is political natural articulation, *tout court*, and very much in line with Stengers's observation on grassroots scientific obligations.

To continue, curatorial science is also museological praxis, taking place within a bureaucratic milieu. The challenge to curators in such a setting is to ensure that their responsibilities to their science, to exercising care, are not captured by an apparatus that diminishes the possibility of applying the mindful, even meditative stewardship role that curators have agreed to accept.[29] Robert Janes has called this mode of responsible, responsive curation in market-driven museum milieus an act of "becoming orthogonal," noting the emergent meaning of *orthogonal*:

> Although museums cannot meditate, they can become orthogonal, a rather awkward mathematical word that means "intersecting or lying at right angles." Its meaning[,] however, has been broadened to include a rotation in consciousness – orthogonally, or at right angles to conventional reality. It is a matter of what one is willing to see or ignore, and to what extent one is able to ignore perceptions and remain habitually inattentive to what is really going on. Orthogonal thinking is the antidote to the constraints of conventional thinking and conditioned views ...[30]

Though at times corralled by institutional arrangements and conventional images and stories of dinosaurian life, at least in a very modest sense these were the sorts of moves curator Henson sought to make; the new curator who began with no baggage, as Beth Jameson noted. Janes, citing Jon Kabat-Zinn, proposes how orthogonal engagement "can admit possibilities of freedom, resolution, acceptance, creativity, compassion, and wisdom." Janes adds that orthogonal practice "acknowledges the need for stewardship of the common good, commits to determining what this might require, and then acts to take care of what needs to be done."[31] Given the propositions I have offered in regard to the precipitation of political natures, such caring stewardship – or the lack thereof – will have consequences not just for the publics engaging museums, but for the public natures and factishes that are assembled in the action of curation. While Henson's gestures were perhaps quiet, working within the apparatus rather than against it, they had a slow, subversive effect, concerned at the end with the possibility of sustaining concern.

In the more collective production space of contemporary museums today, such impulses and acts of care are rarely contained in the singular person of the curator-scientist, but rather are shared out among many actors, and those communicative "things" that make up museum collections, exhibits, and the structures in which they act. There is a "power with" – to extend on Hannah Arendt's formulation – that is drawn upon in this sharing among many agentful persons, and many agentful things put into motion together. Arendt smartly notes that "power corresponds to the human ability not just to act but to act in concert."[32] Here, *human ability acts in concert with the ability of specimens and other "things"* to act upon each other, and upon all of us, in the ever-unfolding actions of articulation.[33]

The individual and assembled struggles of the exhibit team to respond caringly to the specimen, and out of dismay with marketing directives, were manifest in many ways, notably in their multiple approaches to rendering *Maiasaura* as a living, breathing, social creature – indeed, a caring creature – whether through ecological and anatomical reconstruction, through animated movement, through time-space localizing, through scenario- and narrative-building, or through the use of familiar human kin terms and references (e.g., "families," "cousins," "neighbours").

Notwithstanding other missed opportunities to provide an even more *careful* articulation of scientific practices with public cultures as public natures, the exhibit team clearly made genuine efforts to animate ROM #44770 more complexly. Their action together, and Henson key among them, suggest to me three provisional points which mark a modest political promise for curatorially based museums. I propose these as three particular questions concerning *articulating power* – or *power-with* – that have risen to the fore in this ethnographic study of the actions in the Maiasaur Project.

(1) What is the articulating power in selectivity? In choosing to ally with a particular specimen for the display of the dynamics of science and its techniques, a curator is also selecting, launching, and modifying the right "factish" for the job of providing more complexity in natural processes, and with that, in society. That very selection can open or foreclose the valences for public in-filling and counter-imagining of that which comes to be presented in exhibitions. Selectivity in historical flux shows that *Maiasaura* was refashioned by society as much as the reverse. In their choice of what should be studied, collected, or displayed, it is clear that curators do – or at least can – have special power in refashioning nature/politics.

(2) What is the articulating power of the collective? To overstate the responsibility and privilege of the curator alone risks ignoring the collective which brought this dinosaur into physical, exhibitionary being, over a time of historic unfolding. The curator was part of a collective of specimens, museum workers, and exhibit fashioners, of collections, buildings, procedures, instruments, the fossil market, and many other resources – all of whom *and* all of which acted together to reconstruct and animate dinosaurian life as science and culture and nature. They acted as well to find the democratic connecting points to their audiences, although admittedly this exposed them to the risks that could arise as over-orientation to market-thinking. This complex, performatively engaged collectivity is what ultimately drew from and reanimated ROM #44770, *Maiasaura*, and its Mesozoic life and times. Turning the privilege, prerogative, and care of the curator-scientist to support the collectivity and the specimen itself, rather than simply heeling to market impulses, was vital to a responsive, modestly reliable means of performatively articulating phenomena for scientists and publics alike.

(3) What is the articulating power of phantasizing? Scientific work, exhibit design, communication practices – as much as visitor engagements with museum exhibits – necessarily entail phantasizing.[34] Throughout the collective action in the making of a curated exhibition, phantasies were continually brought into play. Everyone (museum staff, visitors, consultants, sponsors) and everything (fossils, interactive consoles, animations, videos, diagrams, the pewter Maiasaur model) impinged upon shared and particular senses of this dinosaur and its world moment. Those phantasies were potencies of the performative reality of the display – phantasies like "Maiasaur family life," "complex biology," "friendly dinosaurs," a "living, breathing environment," "morphing dinosaurs," a "pettable Maiasaur," a "pewter" Maiasaur, "co-evolution," and technical science performed both as theatre and *as nature*. None of these would have worked without a faith that visitors could comprehend and participate in these phantasies when encountered in the materialized form of an exhibition of dinosaurs, with its concomitant chronotopes of the Mesozoic.[35]

Thinking of how these phantasies enter and exit the collectivity (of humans and non-humans) gives greater texture to the discussion offered in chapter 7 ("Phantasmatics in the Systematics of Life") of the points of Deleuze and Guattari, and of Emily Martin, in their theorizing of "rhizomatic" movements in and through the world, society, personal life, objects, productions, actions, and organizations. In other words, the Mesozoic acts much in the same way as Deleuze and Guattari's "strata" for capturing dinosaurian possibility, allowing for alternative phantasies to reconstitute the Mesozoic, materially and figuratively.

Phantasy, in its unexpected, often idiosyncratic assembling of ideas and practices, is always on the verge of setting something different in place, often anticipating another new set of relations by the collision of things, impulses, metaphors, programs, and scientific action. One consequence of these actions working together can be, literally, the *fabulating* of new possibilities of socio-natural being and relations, pressing new politics/natures, and, more consequentially, a resolute cosmopolitics.[36]

3: Rearticulating the Mesozoic, Dinosaurs, Museums, and Planetary Futures

This book has been about different modes of *civilizing* of two different dinosaurs – *Tyrannosaurus rex* and *Maiasaura peeblesorum* – in two different yet historically confluent moments and milieus. The original specimens collected of these dinosaurs were by any account marvellous finds, and their subsequent relay into scientific and public imagining and materialization made them both scientific and spectacular. The contingently articulated natures that each freighted with them were also in many ways stunning, so much so that now they can be thought of as quotidian natures of dinosaurs, however much they are debated. Acknowledging their multiple manifestations and transformations, we can say that they instance two predominant formulations of dinosaurian natures and the situated, curatorial/museological regimes that have enabled their formulation.

It is worth briefly revisiting the contrasts of the two specifically examined performative dinosaurs in this book. Crucial to *T. rex*'s impressiveness was its exceptional anatomical form – an enormous beast at some twelve metres, with formidable-looking teeth in enormous jaws, massively powerful-looking rear limbs, and diminutive forelimbs. No association with a particular milieu or history was needed to impress. In contrast, what makes *Maiasaura* impressive, spectacular if you like, is less its form than its attachment to particular worlds and stories.

Indeed, this contrast is telling: *Maiasaura* thrives on attachments, stories, lives lived, associations with young dinosaurs, other dinosaurs;

T. rex on detachment, a dictatorial truth, needing association with little more than lives preyed upon, so long as they were of flesh. Osborn channelled his imagining via energy and vitality, and placed the animal in an *a priori* world. Horner made worlds and associations from the creature, no doubt with some *a priori* considerations, but more responsively to them. Then along came Henson, who deftly borrowed on the distinctions wrought through the specimen-palaeontologist interchanges of both Osborn and Horner!

We then discern at least three provisional approaches, each of which increases in its articulations, all of which work by connecting to publics: (1) find the spectacle and simply animate its being based on one's reading so as to amplify the public spectacle and reinfuse how the specimen is known (Osborn); (2) find the specimen and look for the spectacle in its attachments (Horner); (3) juxtapose an attached spectacle (*Maiasaura* via Horner) against a detached one (*Tyrannosaurus* via Osborn) to make people attend to the attachments, so as to make new, different, and more extensive dynamic relations apparent (Henson via the Maiasaur Project). In effect, Henson was seeking, in sympathy with Horner, to refresh the discipline, to surpass the tired and presumptive Mesozoic/lost world practices aligned with Osborn and his masculinist, apical, killer *T. rex*.

Yet, as Henson would learn in the course of team display work, the more one recedes from the kind of direct intervening in animation as exampled by Osborn, or the direct articulating that Horner did, the more others in the team will participate in the articulation work. In this way, other arrays of thinking and matter enter into the composition, and the curator all the more redistributes her expertise in the composition of resulting natures – towards other politics/natures. To caricature the performative move from Osborn to Henson: a dictatorial, progressively apical dinosaurian nature is displaced by a democratic, dynamically relational dinosaurian nature. But of course dictators and democrats thrive in different political environments, different scientific/public milieus – and so do the natures that are animated in their respective moments.

Given what has been observed in the contrasts and unfoldings of the Osborn *T. rex* nexus and the Horner/Henson *Maiasaura* nexus, the two parts of this book also provide an illustration of an important shift between the formulations. These can be thought of as paradigmatic shifts, and some might even say they are ontological shifts, but certainly they are shifts in political natures. The juxtaposition between these two time periods, each generating new dinosaur forms within common sorts of setting, has served to "lay bare the permanent possibility of alternative configurations," as Annemarie Mol has so aptly put this potential.[37]

As I have already suggested, the promise and challenge in such alternative configurations lies in a larger potential of participating in the reclaiming and rearticulating of Mesozoic natures and their dinosaurian inhabitants and ecologies. Stengers, mindful of our shared planet and the precarious ecological situation pressing itself upon us ("the intrusion of Gaia"), put the challenge this way:

> If our practices have to play a part in reclaiming the capacity to answer the consequences of the intrusion of Gaia, I will propose that they have not to just give up the idea of a purely human history of progress and conquest, which is precisely what this intrusion challenges. They have to reclaim a different, positive, definition of themselves and of civilization, in order to regain relevance and become able to weave different relations with peoples and natures.[38]

Uttering a concern similar to that of Robert Janes in regard to the "mission" of museums, Stengers's propositions about reclaiming what it means to "civilize" nature speak directly to *the power of fabulating with natures* so as to "regain relevance." Such statements of concern are helpful in thinking of the generative potential of committed efforts on the part of scientists like Henson, and those who might risk going even farther than Henson. Rather than being totally captured by the closed circuits of consumer-generated natures, curatorial action can be inserted in more resolute ways to attach alternative adventures by way of participation in Mesozoic performativity. Dinosaurs in their Mesozoic natures stand to be shaped though our situation as caring, responsive earth-residing beings, especially now in this very vulnerable moment, facing a very public nature and the scientific problems flowing from humanly articulated, planetary environmental change. The historically changing Mesozoic and its many ecologies yet have a part to play, now that we can give a positive answer to the question "Is another Mesozoic possible?"

From where I stand, this is a most liberating possibility: to know that how we participate politically with the earth-borne matter of palaeontology allows natures to be recomposed, as Bruno Latour has put it.[39] It returns power both to the things assigned as nature and to those who engage such things. Hannah Arendt's propositions on power are befitting once more:

> Power is actualized only where word and deed have not parted company, where words are not empty and deeds not brutal, where words are not used to veil intentions but to disclose realities, and where deeds are not used to violate and destroy but to establish relations and create new realities.

Continuing on this optimistic potential, in undertaking this research I have been struck intensely by the simultaneous sophistication, hope, and anxiety of scientists and museum workers as they engage with and respond to their collections of specimens, and as they attempt then to study, promote, animate, imagine, educate, and entertain with dinosaurs – their collective effort to "disclose realities." They know as well as Mitchell, Osborn, Horner, and Henson that dinosaurs are partial, so revisable beings that can be recomposed. They know that dinosaurs, encountered in part by empirical knowing, embody and effect socio-natural change. This same productive hope and anxiety – a collective edge if you like – emerged in almost every quarter of the exhibit development process. It echoes the same troubled but hopeful searching expressed by museum directors like Robert Janes. Young palaeobiologists are beginning to express such hopes as well. Matthew Bonnan writes:

> [H]istory's lessons ... are very much an open book for us. If we can anticipate what the future will bring, we can act on it. If we decide to put our imagination to good use, we can create positive change in the world. The non-avian dinosaurs could not learn from their past, but perhaps we can learn from them ...[40]

We should very well expect that the humble but deliberative and imaginative engagement Bonnan proposes, in positively affecting our possible futures by learning from our shared pasts/presents with other creatures, will intensify as worldly conditions demand it of us.[41] Yet, hope and care are likely to be mobilized in multiple and intricate ways. To have a positive effect, such moves will have to be more than *pro forma* ones that would be captured and stripped of their potency within the apparatus. Striking a precautionary note, Marilyn Strathern suggests that instrumentalist, technocratic incorporation of *hope* into research should not be prescriptive:

> It is a short step to asking how best to manage [hope] ... or to build such hope into research protocols. I do not think that we should necessarily do

either. Re-engagement needs to be re-engagement, a matter for the future that the present should leave undefined.[42]

Pragmatically, it is most likely that what we are and will be witnessing is a responsiveness that moves between non-prescription, engaged projection, and what Strathern has called "ecological necessity" to propose and speculate upon hopeful possible futures, *and* pasts, out of our shared present. It is also the complex of human and non-human actions and histories that always and already collide, and from which we can more deliberately learn – the anthropological purchase, one might say. In our museological case, might it be possible for museum workers, professionals, exhibit developers, managers, scientists, and visitors as well to understand the extension of this complex flux that affects their actions – including the sorts of objects, instruments, and finished displays which they work with or help bring into being? Presumably, it will take more than a faith in counterbalancing the free-marketeers if, as it appears, the market logics are to be found shot through the network, from drawing the fossils from the ground to the ultimate forms of life produced at the nexus of actions. Actions and alternative entities will have to be inserted, immanently and at times by design, throughout the entire complex to recompose worlds, effecting shifts in other, generative directions, responding to each and all of the articulate factishes that come into being.

Such a potent move is that of which Stengers and Latour have been writing for two decades and longer in their propositions concerning cosmopolitics, the participation of humans and non-humans in the collective composition of the "common world," one that is not given, but enacted together as sharable, shared, and liveable.[43] One sign that some significant shift has occurred will not be in the appearance of a more pure or true or scientifically accurate dinosaur exhibition, but rather in the emergence of different sorts of dinosaur factishes, ones that move considerably beyond notions of good mothers as counterparts to king tyrants, incorporating broader, altered, more inclusive collectivities into their effective constitution. A second sign will be that the *dynamic processes* will not simply "give a sense" of the labour, techniques, and tools of extracting fossils from encasing matrices, or of the limited work of palaeontological analyses applied to fossils and the sites in which they are found, but will extend to include the more heterogeneous complex of human and non-human agencies, instruments, and indeed phantasies which bear upon the outcomes in emergent natures and beings.[44]

A third sign will be if the emergent creatures and Mesozoic time-space worlds and relations take on the aspect of the worldly conditions and forces that beset us unpredictably and at times shockingly in our current moment, calling us into action. We are still slow to witness this significantly, even if this potential exists, as the tendency to cleave to banalities persists. For instance, in 2006, palaeontologists Robert Bakker and John Long assigned the name *Dracorex hogwartsia* to an unusual pachycephalosaurine (i.e., "bone-headed") dinosaur from South Dakota in celebration of J.K. Rowling's massively popular *Harry Potter* novels.[45] While playful and escapist engagements such as this will no doubt arise as a cathartic relief from the increasingly dire findings presented to us in climate, ecological, and a host of other sciences concerning future prospects for biospheric health on this planet, presumably we can do something more *interesting, articulate, connected, and disruptive* than assigning pop culture references to (possibly) new dinosaur species.

Dinosaurian Responsiveness, Worldly Change

A more interesting and important move will be the emergence of Mesozoic worlds, ecologies, and beings that speak back to or in concert with the imposing conditions we face, conducting natural possibilities of what it means to live well together on a planet, amplifying a collective effort of recomposing natures well, together. If we know that our choices and our capacity for responsiveness have purchase in bringing about different political natures, then we should not be surprised to see an obliging turn in attention to selecting and highlighting fossil-based creatures and formulations of the Mesozoic that articulate with pressing questions of our times, creatures that are relevant to the politics of our current predicaments – such choices would come about since the ecological demand obliges a response.

The familiar example, raised by museum scholar Ben Dibley as by Latour, Stengers, and many others, is climate change and its political and ecological ramifications. We have seen signs of such a transformed Mesozoic as well as a constituent fauna that is subject to – or resistant to – drastic climate cooling or warming, or to catastrophic extinction events. This came very close in the 1980s, when the propositions of Walter and Louis Alvarez on the atmospheric dissemination of sun-blocking particulate matter from an asteroid impact at the end of the Mesozoic (the K-T or Cretaceous-Tertiary boundary) became grist for dinosaur extinction scenarios. Such scenarios found currency in turn with Carl Sagan in regard to human extinction scenarios, when he and his colleagues drew parallels with the climate-changing effects of sun-blocking atmospheric plume blankets that would arise after multiple massive nuclear explosions in the late Cold War, encapsulated in what was known as the "nuclear winter hypothesis."[46] Did we learn from this? When a meteorite blasted through the atmosphere in February 2013 just east of the Ural mountains in Russia, in the same moment that a comet passed between the moon and the earth, garnering worldwide media attention and comment, were not long-recurring fears of imminent cosmic catastrophe brought forward yet again, with thoughts often turning to the proposition of a Late Cretaceous "large-body" impact that coincided with the end of the dinosaurs?

Apart from the single catastrophic event scenarios in Mesozoic performativity, we are now beginning to witness alternative extinction concerns as attention moves to consider the variability of climate parameters in late Mesozoic terrestrial ecosystems,[47] climate-prompted movement in dinosaur populations,[48] temperature-dependent sex determination in terminal Cretaceous dinosaur populations,[49] and even digestive methane gas production from mega-populous plant-eating dinosaurs and the effects on Mesozoic global warming.[50]

Such a set of turns, with such emergences, is signalling a fuller shift in recognition that we are always and already connected in the nexus of relations that are constitutive of the worldly conditions in which we live. The move to *political/natural relationalism*, a move to cosmopolitics, is an approach linking relevance to knowing, a move past the problems of positivism and relativism – the former of which *rejects concerns* as a source of deformation to knowing, the latter of which risks *forfeiting reliability* of knowing by relegating all action to the status of contingency.

A recomposed *civil Mesozoic* is simultaneously concerned, reliable, and still open to subsequent recomposition. Following the pragmatist thinking of Stengers, Deleuze, James, and Whitehead, participation in the recomposition of the Mesozoic promises *lines of flight* from previous, erstwhile "disinterested" figurations of the Mesozoic as an "order of nature," and from the previous colonizing state as a time-space irrevocably designed to support hierarchical, tyrannical natural forms. It is a way of undoing certain political natures, by rejoining the Mesozoic to the Cosmos, which as Stengers notes "must include the thinker. Thinking itself becomes part of the Cosmic adventure."[51]

So, at the last, I can return to our own intimate participation in the Cosmos, through and with dinosaurs. In this foray into practices of Mesozoic performativity and articulation, it has become clear to me that when we articulate dinosaurs, there is an inexorable boomerang effect of those articulations back *into* the workings, back through circulating matter, back through the people and things that allowed the dinosaurs to be articulated in the first instance. We don't just *articulate the creatures* and their time-space worlds. We also *articulate ourselves* as we articulate to these creatures and our own time-space worlds, knowing we are always and already connected. This raises the call for care. As Deborah Bird Rose has observed: "There is no position outside of connection … What happens to one has effects on the well-being of others … to care for others is to care for oneself."[52]

This is what I have learned from the action of dinosaur articulation. The resolve of "just trying to be a scientist" becomes less an expression of exasperation, less a last gasp to hold onto truth. Instead it becomes a cry of reclamation, a statement of protection, of resilient hope that human engagement, the power of selection, care, intimacy, and collectivity to live in concert with rather than apart, will thrive.[53] It is a cry for articulation – especially now in our newly affirmed geochronotope: the Anthropocene.[54]

I return to where I began, with young Maiasaur Project visitor Amy, who uttered the matter of concern so poignantly, thoughtfully, and simply:

> the more you learn about them, and where they came from and what they lived like, it tells the story about us … It is our history too.

The stories, lives, and histories of dinosaurs in their worlds are ours as well. Could there ever be a more compelling reason to articulate them well, and to do so slowly, thoughtfully, collectively, *and* with utmost care?

Notes

1 Can There Really Be an Anthropology of Dinosaurs?

1 Interview text from ROM, July 26, 1998.
2 See Bonnan's webpage: http://matthewbonnan.wordpress.com/2012/12/07/dead-dinosaurs-and-reasons-for-hope, accessed March 11, 2013. Bonnan is currently associate professor of biology at The Richard Stockton College of New Jersey.
3 Some years after the findings of the research informing this book were posted as a University of Alberta dissertation, i.e. 1999, geoscientist David Fastovsky (2007) offered a most helpful and and directly cognate discussion of the matter of how what we might call the "cultural gendering of paleontology" needed to be taken far more seriously. Written from the stance of a specialist in palaeoenvironmental reconstruction, his comments are all the more supportive of what is presented in ethnographic and ethnohistorical detail in this book. In Fastovsky's cogent analysis, he signals social and political "contexts," whereas in this present volume I argue that the very knowledge domain divisions of text/context, fact/value, science/society repeat a common fallacy that there is some pure knowledge of the natural, vs. outward sources leading to distortions. If there is something approaching "fact," it is that both domains are at work together, so really this is a contingent separation into domains (see chapter 2 this volume on performativity, for example). This, quite precisely, is how anthropology has long pursued its understandings, by allowing for all that is at play to be considered, and then quite potently by turning attention to the relations between (cf. Strathern 2005).
4 Martin 1997, 145–6.
5 Kroetsch 1975, 118.

6 Gupta and Ferguson 1997, 39.
7 Franklin 1995a, 179–80.
8 Latour (2004, 18) refers to politics in a particular, ecological and eco-social manner as "the progressive composition of a common world."
9 Law and Urry 2004, 395.
10 Cf. Ortner 1984.
11 Snow 1993 [1959].
12 Strathern 1992a, 177. See Raymond Williams 1985 on the understood complexity of the term "culture" underscoring how such questioning arises.
13 See Segal and Yanagisako 2005. Stanford University offers a rich and fascinating story of disciplinary anthropology's vexing trials in keeping naturalism-scientism and culturalism-humanism in accord. See "A Forced Anthropology Merger," *Inside Higher Education*, February 6, 2007.
14 In recent years, anthropologists Descola (2009) and Viveiros de Castro (2012) have engaged in discussions around "multi-naturalism" and "cosomological perspectivism," both of which trouble restrictive notions of a universal nature as against relative cultures. Blaser (2009) and de la Cadena (2010), who also engage with the cosmological distinctiveness of Indigenous peoples notably in South America, have sought yet another move, which is to speak to the even messier political complexes that arise if one takes seriously the proposition of multiple radically distinctive ontologies, multiple natural/cultural worlds, so to speak.
15 Kirksey and Helmreich 2010.
16 Landecker and Panofsky 2013; Pickersgill et al. 2013; Meloni and Testa 2014; Rose 2013.
17 Franklin 2007; Raffles 2011; Haraway 2007; and see the theme issue of *Critique of Anthropology*, edited by Smart (2014).
18 Strathern 1992a, 184.
19 Latour 2011, 72–3.
20 Latour 1993, 7.
21 See Slack (1996) for a genealogical review of these late Marxist approaches to the idea of articulation.
22 See my master's thesis (Noble 1994), which was generously cited by Mitchell (1998). Papers relating to this have been presented at a number of scholarly venues: Noble 1997a, 1997b, 1998, 1999.
23 Rudwick 1992.
24 Cf. Said 1979.
25 See various discussions of monstrous geographies and monstrous races in Wittkower 1942; Park and Daston 1981; Céard 1977, 1991; Braidotti 1996; Warner 1994.

26 See Pomian 1990. For detailed examinations of histories of notable English fossils and popular natural history in the nineteenth century, see O'Connor 2008 and Lightman 2007.

27 See Noble 1994. Secord (2004) makes a somewhat interesting historicist point that the Crystal Palace creatures should not be seen as part of the history of dinosaur fascination, in part since they were referred to by the generic term "antediluvian animals" rather than the term "dinosaur," and in part because they were the endpoint of a curious history fusing "commerce, education and reason." However, when examined as an apparatus of island-born creatures meant to present the "Secondary Period," which later would be dubbed the "Mesozoic," the creatures can be situated as part of the lineage of Mesozoic performativity that I discuss in this volume.

28 See Hawkins 1854.

29 Bennett 1995, 46–7. Bennett's formulations about training civility through a progressive telos of history are also informed by Fabian (1983). Bennett (1995, 39) notes: "The most crucial development concerned the extensions of time produced by discoveries in the fields of geology and palaeontology, especially in the 1830s and 1840s, and the reorientations of anthropology which this production of a deep historical time prompted in allowing for the historicization of other peoples as 'primitive.' While important differences remained between competing schools of evolutionary thought throughout the nineteenth century, the predominating tendency was one in which the different times of geology, biology, anthropology and history were connected to one another so as to form a universal time. Such a temporality links together the stories of the earth's formation, of the development of life on earth, of the evolution of human life out of animal life and its development from 'primitive' to 'civilized' forms, into a single narrative which posits modern Man (white, male, and middle class, as Catherine Hall would put it) as the outcome and, in some cases, *telos* of these processes." Also see Bennett 1998.

30 Bennett 1995, 35, after Pomian 1990. Also see the interesting work of Kirby (2011), who looks at how scientists over the last one hundred years have worked as consultants with filmmakers – adjuncts to the spectacle sector, if you like – to reshape, advise upon, and edit the scientific content of otherwise fiction-based films.

31 Bennett 1995, 59–88, and cf. Altick 1978.

32 See Agamben 2009.

33 At the time of this volume going to press, another richly contoured ethnography of dinosaur exhibition-making and museological communication had recently been completed as the doctoral work of anthropologist

Diana Marsh. There are notable resonances, as one might expect given the common ethnographic focus on dinosaurs, though set in different North American museums. See Diana Marsh (2014), "From 'Extinct Monsters' to Deep Time: An Ethnography of Fossil Exhibits Production at the Smithsonian's National Museum of Natural History."

34 Mitchell 1998, 261.

35 Mitchell 1998, 186. See Adorno 1984 [1951], 115.

36 Lévi-Strauss 1963, 16.

37 Mitchell 1998, 58.

38 Fatovsky 2007, 239.

39 See for instance Debus 2006; Debus and Debus 2002; Glut 1997.

40 I use "imaginary" in its noun form in the texts, to signal the sharable, transferable character of imagining, as in the sense of a "cultural imaginary" or a "political imaginary." In this sense, the imaginary, while extremely plastic, is an immanent "thing" in that, with the aid of material things, language, and people, it may be exchanged and in due course hardened into things. There is then a sliding between imagining and material outcomes. My point is (i) that people do imagine, reflect, phantasize; and (ii) they then act to translate such imagining into things, while also exchanging translations with other people. The entire fabric, then – material and intangible – may be thought of as an "imaginary." A very important dimension of generative human agency (especially) is lost if imagining is left out of the tracing of, for instance, actor networks. Latour in particular seems often to remove such action from his considerations, but the case of dinosaurs very much troubles his arguments (cf. Latour 1999, 146ff).

41 Mitchell 1998, 254. See his discussion of dinosaurs as "transitional objects," a term he borrows from British psychoanalyst D.W. Winnicott (256–62).

42 Haraway 2007, 244. Haraway asks specifically: "How is becoming with a practice of becoming worldly?" (2007, 3), and this volume offers another take on that question.

43 One could argue, in echoing elements of Kuhn's general point around "paradigms" (1962), that this constitutes a cultural-scientific shift, though the intricacy of the actions is in many ways far more complex than Kuhn had probably contemplated.

44 Franklin 1995a.

2 Materializing Mesozoic Time-Space

1 Whyte 1988, 141.

2 Mitchell 1998, 58, 107ff.

3 *Washington Post* (1999) web site report, sourced from Associated Press film statistics: "Exhibitor Relations; The Associated Press," http://www.washingtonpost.com/wp-srv/style/daily/movies/100million/article.htm, accessed January 19, 2009. The original 1993 *Jurassic Park* film, based on Michael Crichton's 1990 novel of the same title, grossed $356 million at the box office, the highest year-of-release earnings for any film up to that time. Crichton (1995) also relayed Doyle's title in his second book in the series. All the *Jurassic Park* films (1993, 1997, 2001, 2015) were distinguished by advanced use of computer-generated imagery (CGI) techniques, which have become the hallmark of many subsequent box office record breaking films, including *Shrek*, *Batman: The Dark Knight*, *Spider-Man*, and *Lord of the Rings*, and indeed are the basis of the platforms undergirding the multi-billion-dollar video gaming industry. See Abbott 2006.

4 Note that, although the lost geography of dinosaurs may be bounded and separate and so apparently disentangled from human lives, this separation is a rhetorical, practised one which achieves just the inverse. Any ethnographic situating of this figure – as space, time, or materialization – shows that the work it does is that of entangling experts and non-experts in biological, evolutionary, museological, entertainment, and multiple other sorts of contemporary and socially salient knowledges and practices. As a critical part of the scientific-public trading apparatus, it is therefore fully entangled with the human within the scope of wealthy techno-scientific nations and communities. For a fuller consideration of the notion of "entanglement" as a process effectively drawing entities into relations, see Callon 1998a; Callon 1998b, 1–57; Thomas 1991.

5 See Sommer 2007 for a remarkably cognate discussion of the Lost World figurations of Edgar Rice Burroughs, and the action of Henry Fairfield Osborn. I first took up Doyle's and Osborn's work and figurations in the early 1990s (Noble 1994). That an independent identification of these circulations arrived in other quarters is not surprising, given the ubiquity of these figures in literature and science. Of course, the work of Jules Verne was germinal in this genre: Verne 1877, 1965.

6 Definition of "orient" from *New Oxford American Dictionary*; for a more substantive discussion, see Said's (1979) germinal text *Orientalism*.

7 Cryptozoology is the academically marginalized study or endeavour aimed at finding evidence or living specimens of creatures otherwise thought to be extinct, or of creatures spoken of in folk tales or "legend" for which there is reasoned potential of finding empirically verifiable, living samples. On the other hand, it had its counterparts in legitimate nineteenth-century natural history and classification procedures associated

with European colonial expansion into "new" terrain, as eloquently discussed by Harriet Ritvo (1997). See Heuvelmans 1995.

8 Doyle 1994 [1912], 59–60.

9 Bacon 1852 [1605], 69. Also see Bacon 1952, *Novum Organum*.

10 For further discussions of teratology and monster geographies, see, for instance, Wittkower 1942; Warner 1994; Park and Daston 1981; and Braidotti 1996.

11 Bakhtin 1981, 84.

12 Bakhtin 1981, 245.

13 Bakhtin 1981, 250.

14 Bakhtin 1981, 258.

15 For more on the term "phantasy," see Whyte 1988; Butler 1990; Laplanche and Pontalis 1986.

16 Bakhtin 1981, 254.

17 I would make an even stronger claim than Bakhtin, venturing that chronotopes allow reality and representation to emerge as a structural opposition where they constitute the existence and differentiation of each other.

18 Kirby 2011, 142.

19 The Germanic spelling of the root "phantasy" throughout this chapter acknowledges the nuanced distinction from the English usage, spelled as "fantasy." Laplanche and Pontalis remarked (1986, 141): "Fantasy in German 'Phantasie', is the term used to denote the imagination, and not so much the faculty of imagining … as the imaginary world and its contents, the imaginings or fantasies into which the poet or the neurotic so willingly withdraws." My use suggests the intergrading of "imaginary" (adj.) and "the imaginary" (n.).

20 Dinosaur scientists are hardly insensitive to the way that fossil time-spaces are constituted through palaeontological practice. See discussion in Dodson and Dawson 1991.

21 Butler 1990; Laplanche and Pontalis 1986.

22 For a related discussion, see Barad 2003.

23 A notable exception is the outright contestation posed in discourses of Christian scientific creationism. See Gish 1979. Typical in that counter-imaginary practice is the constituting of biblically endorsed moral other-worlds like "Eden," or the lands East of Eden, which, as I have discussed elsewhere (Noble 1994), maintained the oppositionality of serpentine, saurian, and monstrous alter-beings. In otherworld-making, mainstream organic evolution and scientific creationism both converge in chronotopic figures, and indeed, it is in the otherworld that the contests for legitimation of these canonically opposed discourses are waged.

24 Things like palaeobiogeographic maps or drawn stratigraphic profiles are the translational devices that circulate across networks, changing hands with scientists, technicians, publishers, etc., producing the effect of the Mesozoic as a materiality. Both Latour and Haraway borrow from Alfred North Whitehead's proposition of "misplaced concretism" when they discuss similar effects.

25 Butler 1993, 9. See her earlier work on gender perfomativity as well: Butler 1988.

26 See Callon 2010; Butler 2010; Callon 2007; Callon 1998a, 1998b; MacKenzie 2007; MacKenzie, Muniesa, and Siu 2007; and for similar extensions to how things become "public" see Latour 2005; or for the complexity in stabilizing "bodies" see Mol 2002, or the work of Karen Barad (2006 and 2003), who discusses Nils Bohr's intra-active performativity inherent to the quantum model of the atomic.

27 Esterhammer (2000) gestures to intimations, and versions of illocutionary acts in German romanticism.

28 Butler 1993, 13. For these points she references J.L. Austin's (1955) *How to Do Things with Words*, as well as his *Philosophical Papers* (1961), especially 233–52. Butler also notes Felman 1983; Wittgenstein 1958, part 1; and Pratt 1977.

29 Derrida 1988, 18.

30 Butler 1993, 225–6.

31 Deleuze and Guattari 1988, 75–82.

32 Deleuze and Guattari speak extensively of "lines of flight" in *A Thousand Plateaus*, but in regard to order words remark upon "the K.-function," which "designates the line of flight or deterritorialization that carries away all of the assemblages but also undergoes all kinds of reterritorialization and redundancies" (1988, 88–9). Put another way, order words (e.g., "the Mesozoic") stabilize only to generate alternatives which may either undo that stability or become subject to becoming stabilized themselves.

33 Butler 1993, 12–13.

34 Butler 2010, 147–8. This article also debates some of the ways in which Callon takes up performativity of markets, where he appears to grant an excess of agency to economists themselves as ones who "perform the economy." See Callon (2010) for his friendly response. In addition, Butler (1997) has written of the interruptive dimensions of counter-normative gender performativity. While, in my accounts, palaeontologists and fiction writers do exercise agency, it proves to be quite attenuated, always dependent upon the conditions in which the performative is imparted and received, so akin to the perlocutionary formulation from Austin and Butler.

35 Interestingly, Xerox Corporation responded with public relations campaigns to slacken the vernacularization – the "genericide" – of their corporate name, seeing such moves to normative or generic usage as a threat to proprietary interest in their corporate trademark. See Ingram 2003.

36 For a summary of the dense arguments posited by Butler see the review article by Pheng Cheah. Cheah (1996) summarizes: "Butler's account of productive historical forms and her theory of performative agency take the notion of phantasmatic identification – the assumption of the material mark of sex or the intelligible outline of a body through imaginary and symbolic ingestion – as the paradigm for oppression and subversion. Her immediate frame of reference is, of course, the field of gender, sexuality, and desire. Generalized into a political theory, this notion of phantasmatic identification promises to democratize contestation through the interminable proliferation and destabilization of provisional cross-identifications: 'the contemporary political demand on thinking is to map out the interrelationships that connect, without simplistically uniting, a variety of dynamic and relational positionalities within the political field.'"

37 On a technical point, a concern which might be raised by students of Lacanian theory relates to my conjoining of the "phantasmatic" with the "performative." The Mesozoic might better be taken as an expression of the "social symbolic" rather than the "imaginary" – the latter being associated with the phantasmatic in Lacan's tripartite scheme of the Real/Imaginary/Symbolic. Trish Salah, personal communication July 10, 2004; cf. Bowie 1991, 88–121.

38 For a discussion of the "phantasmagoria" see Altick 1978, 217–19.

39 An interesting case of the conjoining of these two ideas is presented in Wendy Lesser's (1987) book on the underworld, *Life below the Ground: A Study of the Subterranean in Literature and History*.

40 Barad 2003, 815.

41 Barad 2003, 814–15. One can think of dinosaurs as the outcome of an "agential cut" where "the agential cut enacts a local resolution within the phenomenon of the inherent ontological indeterminacy. In other words, *relata* do not preexist relations; rather, relata-within-phenomena emerge through specific intra-actions. Crucially then, intra-actions enact agential separability – the local condition of exteriority – within-phenomena" (815). Put another way, things only become apprehensible as separate things through our ontological engagement. We, and the stuff we work with/upon, are co-agents in apprehending phenomena and, in a sense, bringing those phenomena into being.

42 For film animation articulations, see especially chapters 6 and 14 of this volume.

43 I discuss only three cases here. Some other notable examples of explicit phantasy engagement by palaeontologists include Robert T. Bakker's (1995), George Gaylord Simpson's (1996), and William Sarjeant's involvements in writing speculative fiction novels; Jack Horner's consulting on Spielberg dinosaur films; and Donald Baird's longstanding interest in 3D dinosaur imagery. The list goes on, and I'm sure that most dinosaur paleontologists would admit some current or past involvement in speculative or imaginative world-making – even something as simple as dinosaur model collecting – in addition to those which they unavoidably engage in while undertaking technical work.

44 Currie 1993.

45 See Teilhard de Chardin 1976 [1959].

46 For an interesting online discussion of the evolutionary modelling that could inform thought experiments such as Russell's, see http://scienceblogs.com/laelaps/2007/10/23/troodon-sapiens-thoughts-on-th/, accessed January 5, 2009.

47 Personal communication with the author, May 8, 1986, while on expedition in the Gobi Desert. Some of Russell's writing and correspondence relevant to UFO encounter reports is available and compiled with the "Barney and Betty Hill Files" at the University of New Hampshire Library. See http://www.library.unh.edu/find/archives/collections/betty-and-barney-hill-papers-1961-2006, accessed January 8, 2009.

48 Although Russell has distanced himself from discussions of the Dinosauroid over the years, there is at least one highly lauded palaeobiologist at Cambridge, Simon Conway Morris, who has in the 2000s taken the thought experiment to what for many in palaeobiology are thought to be very controversial extremes, following what many believe is a morally questionable inevitablism. See http://news.bbc.co.uk/2/hi/science/nature/6444811.stm, accessed January 6, 2009; see also Morris 2003.

49 The quotes presented here are from Russell 1987, 127–8.

50 Russell 1987, 127.

51 Wittgenstein 1969, 52e (#411).

52 McLaren 1970. As I will discuss in chapter 8, boundary work is also cognate to what Deleuze and Guattari (1988) speak of in the procedure of "articulating" forms from substance (and substance from forms, conversely), using their geology-derived metaphor of "strata."

53 Rudwick 1976, 200. See Farlow 1997; Gradstein et al. 1994.

54 See Atran 1992.

55 Simpson 1984, 14. Simpson was referring not only to Mesozoic strata but to all the geological strata of South America.

56 Molnar 1997, 581.

57 The contours of the relation of domination wrought of travel and knowledge are discussed in Nicholas 1994.
58 Kielan-Jaworowska 1969, 176.
59 "Fearfully great lizard" was the original gloss for "dinosaur" intended by Richard Owen in 1842 when he coined the term for this group of animals. Farlow and Brett-Surman (1997, ix) emphasize this point as a corrective to the more typical gloss, "terrible lizard"; they insist that Owen's intention is that which should be used, as Owen used the superlative form of the Greek root *Deinos*, meaning "fearfully great." Dutiful dino-philes as they are, they add: "Dinosaurs are not lizards, nor are they terrible. They are, instead, the world's most famous 'living' superlative!"

3 Land of the Fear, Home of the Bravado

1 Doyle 1994 [1912], 58.
2 "Homosociality" is a term used in critical gender studies and usually refers to the generalized sense of male fraternal solidarity and social bonding.
3 See Fabian (1983) for his complementary discussion of how anthropology produces its object, "the other," by means of temporal distancing.
4 Doyle 1994 [1912], 135. Note that several editions of Doyle's novel have been used as sources. The 1994 edition noted here is used as a textual source, one which for all intents and purposes matches the original editions. Images cited and used in this essay are from one of two 1912 (i.e., first) editions of the novel, one being a London edition, the other New York.
5 See discussion of the prevalence of the hybridity theme within colonial discourse in Young 1995.
6 Doyle 1994 [1912], 171, 173.
7 Doyle 1994 [1912], 232, 233.
8 Doyle 1994 [1912], 177.
9 Kestner 1997, 212.
10 Jaffe 1987, 99.
11 Jaffe 1987, 99.
12 George Mosse, *Fallen Soldiers: Reshaping the Memory of the World Wars* (1996), 76, 78, 79, as quoted in Kestner 1997, 7.
13 Doyle 1994 [1912], 179.
14 Jaffe 1987, 98.
15 Doyle 1896. His comment in a speech published in the *Critic*, August 1, 1896: 78–9.
16 Doyle 1994 [1912], 192.
17 For the discussion of Osborn, I have drawn on several sources, but in particular the social history work of Ronald Rainger in his 1991 book,

An Agenda for Antiquity: Henry Fairfield Osborn and Vertebrate Paleontology at the American Museum of Natural History, 1890–1935. For an account of the imperialist metaphors in dinosaur palaeontology, and Osborn's place in promulgating them, see Semonin 1997.

18 Haraway 1989b [1984].

19 Rainger 1991, 85.

20 The 1998 3D IMAX film *T-rex: Back to the Cretaceous* presents Brown in these terms. Among palaeontological and historical accounts see Colbert 1968; Spalding 1993; Farlow and Brett-Surman 1997, 712; and also see Mitchell 1998, 143.

21 Rainger 1991, 62.

22 Rainger 1991, 62.

23 Cf. Miller 1918. For an uncritical summary account, see Rexer and Klein 1995, 163–71.

24 Higham 1987, 240.

25 Preston 1986, 98.

26 Andrews 1922, 5.

27 Mitchell 1998, 168.

28 Quoted in Rainger 1991, 102. Original in Andrews 1932, 9–10.

29 Doyle 1912a, 19.

30 Quoted in Haraway 1989a, 57. For another discussion of the connection between dinosaurs and Osborn's racism, see Mitchell 1998, 149–152.

31 Semonin 1997, 176.

32 See Glut 2008; 1997; 1980; Glut and Brett-Surman 1997; Czerkas and Glut 1982.

33 In that image, Osborn had directed Knight to depict *Tyrannosaurus* facing off with ceratopsian dinosaurs, his image captioned as "Fig. 102. Offensive and Defensive Energy Complexes."

34 Rainger 1991, 163.

35 Stephen Czerkas, personal communication; and in general see Czerkas and Olson 1987.

36 I've found no mention of this in primary or secondary literature. Historian of Osborn's life Ronald Rainger informed me that he had not come across any indication of their having met or corresponded during his research into Osborn's correspondence (personal communication, November 1998).

37 Noted in Jaffe 1987, 99; Higham 1987, 234; Orel 1991, 165.

38 Doyle 1912a, 39; cf. Lankester 1905, 209, fig. 150.

39 Lankester 1905, 191–2.

40 Dingus and Rowe 1998; Norell 1998.

41 See chapter 2, note 54.

42 Foucault (1978, 100), in his famous formulation of "biopower," discusses how bracketing the human, as a species, was crucial to modernity and

especially to the emergent project of governmentality, commencing in the late eighteenth century, and continuing to the current moment.

43 Osborn 1897; and see Ballou 1897.

44 Lankester 1905, 209, fig.150; Doyle 1912a, frontispiece illustration, facing title page.

45 Stephen Czerkas (Czerkas and Olson 1987) contours in considerable detail the reconstructional history of *Stegosaurus*, attempting in particular to sort out interpretations of plate arrangements in different illustrations. Most importantly, however, he is persuasive in arguing that Knight's life restoration of 1897, plate and spike arrangement notwithstanding, has become the visible template most drawn upon in the conceptualizing of *Stegosaurus*.

46 Franklin 1995b.

47 Stein 1983.

48 Elsewhere (Noble 2000b), in reference to the public/scientific actions associated with the primatology of Jane Goodall and Dian Fossey, I have referred to a "nexus of mediations," which oriented more to the second dictionary gloss of "nexus," being a series of connections. "Nexus" refers to the collectivity of interacting agents and includes the primatologists, the apes, the instruments of scientific practice (e.g., vehicles, binoculars, notebooks, money, etc.), the local environmental elements, field assistants, camps, film crews, and so on. "Mediation" is what takes place in each engagement between two or more of these "agents."

49 Butler 1993, 12–13. *Dispositif* is Michel Foucault's term for an apparatus, or assemblage of techniques and forms, that both produces and is reconstituted by its own effects. As he noted in this 1980 (1977) interview, *dispositif* is "firstly, a thoroughly heterogeneous ensemble consisting of discourses, institutions, architectural forms, regulatory decisions, laws, administrative measures, scientific statements, philosophical, moral and philanthropic propositions – in short, the said as much as the unsaid. Such are the elements of the apparatus. The apparatus itself is the system of relations that can be established between these elements." Akrich and Latour (1996), writing in a similar vein, but generalizing and abstracting the elements more, suggest that *dispositif* is a "set up" that assembles humans and nonhumans, and distributes competencies.

50 Squier 1999.

51 Squier 1999, 132–3.

52 Latour 1987.

53 Squier 1999, 144.

54 Doyle 1994 [1912], 161.

55 Osborn 1905.

4 Animating *Tyrannosaurus rex*, Modelling the Perfect Race

1 Osborn 1917, ii.
2 Brown 1915, 271.
3 Alvarez 1997. The jacket cover reads, "How was '*Tyrannosaurus rex*' toppled from the apex of creation?"
4 Lessem 1992.
5 For competing views, see Horner 1994 and Horner and Lessem 1994, 203–20, versus Holtz 2008. This then is linked to earlier debates on endothermy in dinosaurs, as discussed by Bakker 1986, 1972, Desmond 1977, and Ostrom 1969.
6 I take up this point in more detail in chapter 5; see also Squier 1999.
7 See Osborn 1898 for an early formulation of how he understood animalian "models."
8 Singer 1931, 542–7.
9 Franklin 2000; Rose 2006.
10 Osborn 1917, 11. And see his discussion of "Tetraplasy" (Osborn 1912).
11 Osborn 1917, 279.
12 Rainger 1991, 150.
13 For a survey of the rise of American eugenics and its infiltration of policy, see Kevles 1985, and a broader review of the history in Marks 1995.
14 Grant 1922 [1916], viii.
15 Davidson 1997, 169–82.
16 Osborn 1897, 14.
17 Osborn 1917, 147 (on psychic and sub-psychic cues), 223 (on hormones).
18 Osborn 1917, 285, introduction to "Appendix."
19 Osborn 1917, 23.
20 See Larson 2008; Breithaupt, Southwell, and Matthews 2008. Larson documents a fourth 10 per cent complete partial skeleton collected by the Carnegie Museum in Montana, CM 1400.
21 See comment in Larson 2008, 10.
22 Chapters by Breithaupt et al., N. Larson, and Glut in P. Larson and Carpenter 2008 have been especially helpful in my unpacking of these associations.
23 Osborn 1905, 259. This, the type specimen, was later sold to the Carnegie Museum, accessioned as CM 9380.
24 The "Dynamosaurus" specimen was later sold to British Museum of Natural History, designated BMNH R7994. The armour plates were later described as deriving from a herbivorous dinosaur, *Ankylosaurus magniventris*, on which the Tyrannosaurid may have lunched (see Carpenter 2004).

25 Osborn 1906.
26 See Daston 2004, for a rich discussion of the hermeneutics and politics of type specimens and their naming.
27 *New York Times*, December 1906, "The Prize Fighter," front page, http://query.nytimes.com/mem/archive-free/pdf?res=9E0CEEDB1731E733A25753C3A9649D946797D6CF, accessed April 6, 2009.
28 Fieldnotes of Barnum Brown, Hell Creek, 1908, http://paleo.amnh.org/notebooks/brown-1908/14.html, accessed April 6, 2009. The AMNH Annual Report (text at http://paleo.amnh.org/photographs/1908-montana-cretaceous, accessed April 6, 2009) offers a summary of the collecting work for AMNH 5027.
29 Of the forty-five major skeletal specimens in scientific collections noted by N. Larson, only three, including the privately collected "Sue" and "Stan," were more complete than AMNH 5027. See table 1.1 in Larson 2008, 3–5.
30 Myers 2006, 25.
31 Here I work with the same proposition as Kelty and Landecker (2004, 32) concerning the status of model and animation "*in relation to knowledge*; in particular, in relation to the systematized knowledge of the biological sciences of the nineteenth and twentieth century" (emphasis added).
32 Griesemer 2004, 437.
33 For a technical discussion of the history of a similar AMNH mounted specimen in a life pose, see William Diller Matthew's notes on *Allosaurus* (Matthew 1915, 19–23).
34 Osborn 1917, iv.
35 Osborn 1913.
36 Also see discussion in Horner and Lessem 1994, 80–3.
37 Osborn 1913, 91.
38 See Krows (1941) for a brief overview of Ditmars's engagements with and advocacy for educational motion pictures.
39 Osborn 1913, 92.
40 Osborn 1913, 92.
41 Brown 1915, 272. In his forthcoming biography of Barnum Brown, Lowell Dingus (unpublished manuscript) suggests, as well, that there was insufficient space in the galleries to complete such a mount. Breithaupt, Southwell, and Matthews (2008, 59) suggested that costs were the limiting factor.
42 Cf. Brown 1915, 270. Barnum Brown's exposé on the proposed life group included a photograph of the battling *T. rex* model, shot from the opposite side (i.e., rear view behind the standing figure), compared to the image in Osborn's article, which is shot from the rear of the "crouching figure."
43 Brown 1915, 271.

44 Marilyn Strathern (2005) provides an extended discussion of how, in social anthropology, the concept of "relationship" is regularly used to understand or uncover lived relationships, and vice versa. Annelise Riles has done the same in pointing to how we use models of property from European law to consider ethnographic encounters with transaction, or how models of the "global" constrain the way we think about how things are actually connected. In one setting, a theory is a concept; in another, the concept is, quite literally, practised (Riles 2004, 2000).
45 Kelty and Landecker 2004, 57.
46 Jordanova 2004, 447.
47 Such generative in-filling of gaps in knowledge offered to visitors to display spaces, "the beholder's share," is explored more thoroughly in discussions of the Maiasaur Project exhibit in the second half of this book.
48 Osborn 1916.
49 Osborn 1917, 214–15.
50 Osborn 1916, 762.
51 Osborn 1916, 763.
52 Osborn 1917, 225.
53 Daston 2004.
54 Haraway 1989b, 55.
55 Osborn 1917, 214.
56 Osborn 1926, 129.
57 Osborn 1917, 225.
58 See Wonders 1993; Rainger 1991.
59 Osborn 1917, 238.
60 Chase 1977, 349.
61 For a concise review of Osborn's racist commitments as they articulated with his social and scientific activities, see Rainger 1991, 147–51.
62 Rainger 1991, 150.
63 Osborn 1917, 223.
64 Osborn 1917, 283–4.
65 Osborn 1924, 238.

5 Politics/Natures, All the Way Down

1 Haraway 1998, 205.
2 Latour and Weibel 2005, jacket text.
3 See Latour 1987, as well as Harding 1991, 10. Shapin and Schaffer (1985) demonstrated the intimacy of the projects of science and politics in their consideration of Boyle and Hobbes in their seventeenth-century historical

moment. Latour (2005, 2004) develops a strong thesis on admitting the political agency of what counts as nature into political collectivities, and likewise acknowledging that this has ever been so in science, even when it is denied by science practitioners. For a specific account of science's "publics," see Hayden 2003. Ground-breaking work on social emplacement was undertaken by Traweek 1992, discussing Japanese and European particle physicists. On the imbricated politics of the co-constitution of knowledge, publics, or populations, as well as the operation of governmentality, see Foucault 2007.

4 Brown 1915, 272.

5 See Hayden's (2003) work. Also see Braun and Whatmore 2010.

6 See Regal 2002, 69; also quoted in Bennett 2004, 116.

7 Trefethen 1961, 258.

8 See, for example, Sternberg 1917; Brown 1919 (the earliest feature article on dinosaurs published by *National Geographic Magazine*); Gilmore 1929, 7–12; L.S. Russell 1967; Kielan-Jaworowska 1969; Spalding 1993; Psihoyos 1994.

9 Quoted in Rexer and Klein 1995, 170.

10 Lankester 1905, 294–5.

11 Haraway 1989a, 27, 31.

12 Haraway 1997, 235; see also Haraway 1989b [1984].

13 Doyle 1994 [1912], epigraph page.

14 Kestner 1997, 6.

15 I note playfully the curious, historiographic puzzle, at the very least a coincidence, in the fact that Roosevelt's initials mirror those of Osborn's great carnosaur, *Tyrannosaurus rex*.

16 Haraway 1989b [1984], 27.

17 It is worthwhile to note that they were both enmeshed in practices that melded nature and empire, but at a moment when European imperial expansion was being challenged by American ambitions as never before. Nonetheless they had inherited the mantle of nineteenth-century natural historical explorations, witnessed in the natural history voyaging of Darwin, Hooker, and Falconer (Browne 1996).

18 Rexer and Klein 1995.

19 Rexer and Klein 1995, 18.

20 Rexer and Klein 1995, 19–20.

21 Also see Lutz and Collins (1993), who discuss the colonial othering inherent in photojournalism and tourist photography as rendered in the *National Geographic Magazine*.

22 See http://www.amnh.org/amnh-expeditions, accessed December 8, 2009.

23 Haraway 1989b [1984].

24 Haraway 1989b [1984], 31.

25 Haraway is explicit that this is the intent and the conformation of the displays, not necessarily the audience reception, which she did not study in her work. As I discuss later in this chapter, Michael Schudson (1997a) misreads Haraway on this point, in his wish for a technical approach that is more in line with highly positivist sociological methodologies.
26 Haraway 1989b [1984], 55. Also see Rainger 1991, 117–22.
27 Wonders 1993, 148.
28 Wonders 1993, 170.
29 Wonders 1993, 9.
30 Wonders 1993, 223.
31 Haraway 1997, 236.
32 Haraway 1991.
33 Haraway 1991, 188.
34 Here, I refer both to Benjamin's formulations (1968 [1936]) and to Latour's discussions of the "factish" (1999, 266–92). I discuss these points on the collapsing of fetishes and facts through dinosaurs in the concluding chapter (chapter 15) of this volume.
35 See Foucault 1980 [1977], 194–228; Agamben 2009; Beuscart and Peerbaye 2006; and Cohen, Walsh, and Richards 2002, for discussion of "socio-technical dispositif."
36 Haraway 2004, 255
37 Osborn (1924, 12), quoted in Bennett (2004, 133).
38 See for example Aaron Glass, "On the Circulation of Ethnographic Knowledge," http://www.materialworldblog.com/2006/10/on-the-circulation-of-ethnographic-knowledge, accessed November 8, 2014.
39 See Boas 1920; Pinkoski 2011; Saunders 2004; Jacknis 1988. Anthropologist Michael Asch, following Trautmann and others, discusses such universalist staged histories as remanifestations of "conjectural histories," notably from Scottish Enlightenment Stadial theory: see Asch 2014, 34ff.
40 See the discussion in *Anthropology News* by Kehoe et al. (2006) of Boas's full self-awareness and profound respect for the peoples in dioramic representations.
41 See Boas's letter of 1905 to M. Jesup in Stocking (1974), 297–300; and, for a more general view of the political machinations in American ethnological circles of the time, see Stocking 1968, 277–84.
42 Conn 1998, 102–9.
43 Young (1992) clearly demonstrates the limits of arguing from and privileging the abstraction of "ideology," inspired specifically by Haraway's semiotic-material practices and situated knowledges.
44 Val Dusek (November 1997) in *Lingua franca*, http://linguafranca.mirror.theinfo.org/Discussion/letters.html, accessed December 8, 2009.

45 Schudson 1997b.
46 Helmreich 2009, 6–9.
47 Barad 2003, 803. I acknowledge that Barad was pressing for something arguably even more intensive than the order of action I present, in the sense of an ontological intra-activity, where the physicist-agent and the constituted particle as agent come together.
48 Haraway 1989a, 30.
49 Haraway 1997, 235.
50 Fabian 1983, 1.
51 Rainger 1991, 180. This echoes Richard Altick's (1978) history of London's public spectacles and "pleasure palaces" of the early part of the nineteenth century.
52 Law 2004a, 2004b.
53 Sarah Franklin, summarizing fellow anthropologist Marilyn Strathern, aptly remarked that "the delightful feature of cultural reproduction is its very reliable tendency never to reproduce itself exactly." Franklin 1997a, 2.
54 Latour 1999, 192.
55 Haraway 2000, 406, and see Haraway 1997, generally.
56 Agamben 2009, 2–3. For the specific text of Foucault's interview, see Foucault 1980 [1977].
57 Agamben 2009, 14.

6 Vestiges of the Lost World: Recirculating the Tyrant Nexus

1 Rovin 1977, 78.
2 Karen Barad (2003) uses the word "intra-active" to signal the co-constitutive relation between people and things. In paleontology this means all things from fossils and collections to instruments of knowledge production, to wider technical and institutional collectives, to the scientists, technicians, animators, writers, etc.
3 For some particularized accounts of these, see the various works of Debus (2010, 2006), and Debus and Debus 2002.
4 I use the notion of near-object along the lines of Star and Griesemer (1989) on "boundary objects" or Latour's (1993) "quasi-objects" and "immutable mobiles." See discussion later in this chapter.
5 Haraway 1997, 236.
6 Farlow and Brett-Surman 1997.
7 Two other encyclopaedic volumes on dinosaurs and dinosaur palaeontology were published in 1997 – something of a banner year for this – providing an enormous quantity of detail on the constitution of dinosaur worlds, biologies, and evolutionary histories. The other two volumes include one

edited by two leading paleontologists, Currie and Padian (1997), and a compilation by Glut (1997).

8 Farlow and Brett-Surman 1997, v. Harryhausen comes first, holding a special position in the order. After that the names are listed alphabetically, suggesting an equal value placed on each of these other figures. Note that in the 2012 second, expanded edition of the book, which added Thomas Holtz to the editorial group, this dedication was omitted.

9 The film saw a blockbuster CGI animation remake, released in 2010.

10 With more of a taste for the dinosaur as icon, rather than as occupant of a geography, aspects of this discussion have been outlined in a different manner by Mitchell (1998), who in turn borrowed liberally from an earlier outline of my own (Noble 1994).

11 See Rainger 1991, 152–81 generally, and 163 on Gilmore.

12 Glut 1980, 50. Ape-friends raise and protect Tarzan in Edgar Rice Burroughs's novels, including *Tarzan the Terrible*, *Pellucidar*, and *At the Earth's Core*, and in several of these the feral white man also confronts monsters drawn after dinosaurs. Burroughs's trilogy, *The Land That Time Forgot*, *The People That Time Forgot*, and *Out of Time's Abyss*, designed a world containing several populations of primitive humanoids, "a veritable assembly line of evolution" (Glut 1980, 59), living in fear of ferocious saurians. "Civilized Man" is present or implied in all of Burroughs's productions, providing a reassurance to the reader that the primitive people are not truly us – only potentially us, lost, however, in some more romantically pure past – and that dinosaurs are wholly, unequivocally *not us*.

13 Glut 1980, 81.

14 Muybridge 1969 [1887].

15 Rovin 1977, 8.

16 This discussion draws on the review of O'Brien's career in Archer 1993.

17 Glut 1980, 82.

18 See Czerkas and Olson 1987; Czerkas and Glut 1982.

19 Glut and Brett-Surman 1997. Stephen Jay Gould (1992, 12) has also remarked on Knight's influence in this history of image-making: "Knight created the canonical picture of dinosaurs for professionals and the public alike." Nonetheless, Gould tends to represent imagery as "social context" and externalities, and stops short of looking at the complex texture of imaginistic circulation in fuller social and material terms.

20 George Lucas honoured the memory and contribution of Willis O'Brien to Hollywood's animation history by borrowing O'Brien's nickname, "Obie," and incorporating it into his famed wizard-like *Star Wars* character, "Obi-wan Kenobi." Those who know Lucas's virtual icons of Hollywood will also notice his ample use of monstrous aliens. Indeed, the Tatooine bar

scene in *Star Wars* is peopled with otherworldly beings which appear to be modelled after the monster-bestiaries illustrated in Rudolf Wittkower's (1942) "Marvels of the East: A Study in the History of Monsters."

21 Cf. Mitchell 1998, 171.
22 See Czerkas and Glut 1982, 12.
23 Kirby 2011, 120. Glut (2008) discusses the history of pictorial, model, and cinematic representations of *T. rex* as two-digit vs. three-digit forelimbed creatures.
24 See, for example, Mulvey 1985, or Hansen, Needham, and Nichols 1991.
25 For a critical psychoanalytic appraisal, see Sippi 1989. And for a masculinist portrayal of ape-woman sexual intrigue, see Hayes 1986; and cf. Noble 2000b.
26 See Noble 1994, 102–3, and subsequently Mitchell 1998, 170–3, for further discussion of *King Kong* and the monster imaginary of difference and spectacle in which it participates.
27 Glut 1980, 91.
28 Phil Tippett's animation work would make its way into "The Maiasaur Project" video displays, as noted in chapter 11 of this volume.
29 In *One Million Years B.C.*, there are no speaking roles, no narration, which would bring even more attention to the body as iconic form. As such, Raquel Welch as a sex symbol is exaggerated as the memorable iconic form in the film, alongside the dinosaurs.
30 Glut 1980, 113.
31 Glut 1980, 107.
32 See http://patrickgarone.blogspot.ca/2010/02/monster-movie-of-week-black-scorpion.html, accessed January 20, 2015.
33 Glut 1980.
34 Cf. Wilford 1986; Desmond 1977.
35 Ostrom 1969.
36 See Desmond (1977), which discusses the so-called "Dinosaur Renaissance," revolving most famously around the work of John Ostrom and his student Robert T. Bakker, who is the late twentieth century's most outspoken vertebrate palaeontologist to advocate for the behavioural interpretation and "point of view" of large and small carnivorous dinosaurs alike. See, for instance, his 1998 novel *Raptor Red*, which is a third-person account of the life and times of a rather fierce "top predator" akin to *Deinonychus*, known as *Utahraptor*.
37 Star 1989, 21.
38 Fujimura 1992.
39 Latour 1993, 89. Also see the discussion of "black boxes" in Latour 1987, 139–40.

40 Star (ed.) 1995, 118.
41 Ross, Duggan-Haas, and Allmon 2013.
42 Farlow and Brett-Surman 1997, 674.

7 Phantasmatics in the Systematics of Life

1 Ward (1992, 207) made this comment about how palaeontologists like himself work persistently with time and history. In this instance he had been inspired by the character Billy in Kurt Vonnegut's novel *Slaughterhouse-Five*.
2 Russell 1987, 115.
3 *The Random House Dictionary* (Stein [ed.] 1983) defines mathematical and physical "approximation" as "a result that is not necessarily exact but is within the limits of accuracy required for a given purpose."
4 Kelty and Landecker 2004, 58.
5 Russell 1987, 117.
6 Personal communication with palaeoartists Mike Skrepnyk, Brian Cooley, Steve Czerkas, Jan Sovak, and Manfred Tolman.
7 Mitchell 1998, 108.
8 Henderson 1997, 167.
9 Head 1998. For an example, of which many could be cited, and following the same format as described in the Head article, see Evans and Reisz 2007, which describes and situates Lambeosaurine dinosaurs from the Late Cretaceous Dinosaur Park Formation of southwestern Alberta.
10 Two very succinct articles discussing cladistics are those on dinosaurian cladistics by Sereno (1990) and Patterson (1980).
11 Palaeontologist Andreas Henson has told me that these two highly popularized dinosaurs, quite remarkably, remain relatively poorly studied from a cladistics standpoint! It is other members of Theropoda which have been the subject of intense investigation, including such creatures as *Baryonyx*, *Allosaurus*, *Deinonychus*, *Dromaeosaurus*, and *Tröodon*.
12 Wiley and Lieberman 2011, xiii.
13 Cf. Gould 1993
14 The AMNH has issued educational materials on cladisitics to extend upon the cladistics schema used to organize their fossil galleries. See http://www.amnh.org/explore/curriculum-collections/dinosaurs-ancient-fossils-new-discoveries/understanding-cladistics, accessed December 12, 2014.
15 See the various discussions of revisionings of *T. rex* in Larson and Carpenter (eds.) 2008. Also see Van Valkenburgh and Molnar 2002.
16 Norell, Gaffney, and Dingus 1995, xii.
17 Sereno 1990, 9–20.

18 Sereno 1990, 17. Sereno continued to stake out a lead role in shaping of phylogenetic systematics; see Sereno 1999.
19 Synapomorphies are used to define monophyletic groups, groups whose members share the biological (e.g., morphological) character, and who in turn share that character with a common ancestor, thus indicating a phylogenetic lineage.
20 Interview with A. Henson, August 8, 1998.
21 Strathern (ed.) 1995, 11.
22 There is at least one book bearing "phantasmatic kinship" as its title: Alberto Eiguer's *Phantasmatic Kinship: Transference and Countertransference in Psychoanalytic Family Therapy*. The volume is a family counselling book by a psychotherapist, drawing upon the Lacanian notion of "phantasmatics of identification."
23 See http://paulsereno.uchicago.edu/about, accessed December 29, 2014.
24 Sereno's own webpage lists these accolades quite unabashedly.
25 Gould 1993.
26 D. Brinkman, personal communication.
27 My colleague at the ROM, Dr Andreas Henson, was studying this specimen at the time of writing. Henson relayed to me many times his delight in these practised jokes in systematic palaeontology, a source of relief, no doubt, to the superficially drab and serious landscape of systematic science.
28 O'Hara 1992, 135–60.
29 Norell, Gaffney, and Dingus 1995, xiii.
30 Deleuze and Guattari 1988, 15 (see also 117).
31 Martin 1997, 138.
32 See her extensive findings in Martin 1994.
33 Martin 1997, 145.
34 The recent work of Jensen and Rödje (2010) is a promising exception in this regard as a rejoinder of anthropology, science and technology studies, and the work of Deleuze.
35 Posadas, Grossi, and Ortiz-Jaureguizar 2013.
36 Upchurch, Hunn, and Norman 2002, 277.
37 Posadas, Grossi, and Ortiz-Jaureguizar 2013, 234.

8 Articulating *Maiasaura peeblesorum*: The Life, Times, and Relations of ROM #44770

1 Pachner 1995, 42.
2 I use the word "delimited" rather than "ascertained" or "established," since the former refers to something knowable in terms of the limits, the

boundaries within which it is intelligible, and the latter two terms suggest something far more fixed and immutable.

3 Callon 2010; 1998a; and see MacKenzie, Muniesa, and Siu 2007.

4 Latour makes a pointed distinction between the correctness of statements and the articulation of propositions, the latter of which open up possibilities for recomposing nature/politics: "Statements are different from propositions. This is especially clear if we consider how we make judgments about their quality. Statements are true or false depending on whether or not there is a state of affairs corresponding to the statement – with all the difficulties outlined by the philosophy of language. I will propose to say, however, that propositions are good or bad depending on whether they are articulate, or inarticulate" (Latour 2000, 375).

5 Ames (1999) writes of the shift in renegotiation of authority in museums, when communities of interest are brought into direct curatorial participation. This trend has continued especially in cultural community engagements. But, as this book has examined it, participation of the "natural" is a far more distributed set of publics and experts.

6 For a cognate disciplinary case of how gender attention shifts natural science possibilities and praxes, see Strum and Fedigan 2000, and within that my own discussion on this, Noble 2000b.

7 The complexity of Mesozoic recomposition is dynamic, and I would never, could never claim to contour it *in toto*. As this is a partciular ethnography following a particular dinosaur and exhibit, I will not be addressing in any kind of detail the many other interesting shifts in the action of dinosaur palaeobiology, such as the intensive interest in dinosaurian/avian evolution, or the rise of multidisciplinary research teams.

8 See L. Russell 1967; Evans and Reisz 2007; Psihoyos 1994; Spalding 1993.

9 Strathern 1995, 183. Strathern's comment is a response to Donna Haraway's influential article on feminist cultural studies of science, "A Game of Cat's Cradle: Science Studies, Feminist Theory, Cultural Studies" (Haraway 1994).

10 Law 2004a.

11 Stengers 2008d.

12 Stengers 2009.

13 Jacket copy, Stengers 2010a.

14 Latour 1999, 303, and see the slow case study on soil science in the same volume, 24–80.

15 The way I use "articulation" moves from Latour's <human+nonhuman> usage, by pressing home the importance of phantasies in moving between humans and non-humans, where the usage of articulation comes to reside in things or communicative actions, and so resolving the further dichotomies to which Latour points. Latour's discussions are compelling but also

imaginative, that is, enabled by phantasmatic possibility inhering in the scholars' thought, and in the texts they share.

16 See Slack 1996, on "The Theory and Method of Articulation in Cultural Studies." Also see Laclau and Mouffe 2001.

17 Fujimura 1992. Also see Fujimura 1987; Star and Griesemer 1989.

18 Mol 2002, 153.

19 Deleuze and Guattari 1988, 41.

20 Deleuze and Guattari 1988, 41.

21 Deleuze and Guattari 1988, 40.

22 As quoted in Slack 1996, 116.

23 Latour 2005, 127.

24 Stengers 2008c.

25 See the many works of Latour, Callon, and Law on tracing socio-technical networks, and the semiotics of entities.

26 By *relationality* I mean the multiplicity of possible ways of enacting or making relations between humans and humans, humans and nonhumans, etc.

27 I use the term "reliable" to signal that what is presented in these pages is a confident proposition with attendant case material on the heterogeneity of public science, public natures as practice.

28 During the ethnographic documenting, and in keeping with my training, I attended as best I could to gender, age, and other social-positioning cues of the anonymous visitors witnessed engaging with the exhibit.

29 Pignarre and Stengers 2011, 23–30.

30 Stengers 2008d.

9 "A Real Sense of a Dynamic Process"

1 This quote is from Wayne Hunt's *Urban Entertainment Graphics* (1997, 194). Copies of the pages in this special book profiling international graphic design work were given to me by the project's 2D designer, Sam Enright.

2 Interviews with Andreas Henson, January 8, 1998, and August 4, 1998.

3 Interview with Beth Jameson, March 1999.

4 This interesting articulation of *Maiasaura* and the ROM with Blackfeet and the Flamand family, while rich, is one I cannot pursue here, as it opens up many more complex trajectories of concern to consider, especially around relations of ownership and transaction between Indigenous peoples and institutions they engage with; cf. Noble 2007.

5 Latour 2000, 375–9; and see Fedigan's remarks in same volume, Strum and Fedigan 2000, 511–12.

6 The question of herding behaviour had long been rising as an issue of intense scholarly investigation, especially in relation to Horner's work with Maiasaura (discussed in several subsequent chapters). See also Varricchio et al. 2008. Placed in paleo-ecological context of a particular fossil rich site and geological sequence, see Currie and Koppelhus 2005.
7 See Callon 1986, and the encompassing volume in which his chapter appears.
8 Cf. Barry 1998; Bearman 1993.
9 While many different accounts of the costs were reported to me, the direct costs of the exhibit were roughly $550,000 Cdn; approximately $200–250,000 for the specimen and $300,000 for construction, media elements, hardware, lab installation, etc. An additional $250,000 was donated by sponsors in animation software, labour, and hardware. These figures are approximates based on interview statements and budget documents prepared during the planning phase.
10 The Burgess Shale of the Canadian Rocky Mountains has been widely popularized in the writing of Stephen Jay Gould's book *Wonderful Life* (2000). The ROM houses the largest and most representative collection of Burgess Shale fossils in the world.

10 A Really Big Jurassic Place: When Specimens and Chronotopes Meet

1 Pachner 1995, 42.
2 This text is from the invitation to the exhibition launch for ROM members. In addition to my own transcriptions of texts in the display area, various copy material for invitations, signs, display text, etc. was kindly provided to me by ROM designer Sam Enright, by the interpretive planner, Jennifer Ross, and by the marketing and promotions coordinator, Brenda Mikelsen, all of whom were members of the Maiasaur Project development team.
3 Report of the then Marketing and Communications Department on promotion of "The Maiasaur Project," November 1, 1995.
4 Interview with Brenda Mikelsen, July 7, 1998.
5 ROM "Event" announcement, May 31, 1995.
6 This point is notable in that the ROM has had a sometimes troubled history of relations with various cultural communities whose interests, histories, and identities have been inadequately or unevenly represented, and whose participation in the making of those representations has on occasion been peripheralized. See Riegel 1996.

7 As will be noted in the next chapter, the block of sediment (matrix) containing the Maiasaur adult skeleton was thought to have also included remains of an infantile or juvenile skeleton, but this turned out not to be the case.

8 That said, fossils have long been associated with the miraculous in European histories, as in the seventeenth century when it was well considered by natural philosophers that they were wrought by the hand of God and placed in the earth to test our faith in his presence and power. See discussion in Rudwick 1976, 1–48.

9 Horner and Gorman 1988; and Horner and Makela 1979.

10 Interview with Andreas Henson, August 4, 1998.

11 Horner and Makela 1979, 297.

12 Horner and Makela 1979, 298.

13 Dahlberg (ed.) 1981; Geller 2008.

14 Palaeontologist David Norman suggested a slightly different etymology (Norman 1991, 181–3), where the name refers to "an old Roman mother goddess also known as the Bona Dea or 'good goddess.'" Norman adds: "The name was chosen deliberately" in keeping with the "strong evidence that the parents looked after their young." As usual, the assigning of gender to the parenting role is left unexplained.

15 CityTV production MediaTelevision feature on the Maiasaur Project exhibition (first broadcast date: July 25, 1995). The point was confirmed repeatedly by several members of the exhibit development team.

16 ROM "Event" announcement, May 31, 1995.

17 I have participated in dinosaur exhibit planning and development with palaeontologists during the early planning stages for galleries of the Royal Tyrrell Museum of Palaeontology in Drumheller, Canada (1983); for a large, temporary exhibition at Spain's Museo Nacional de Ciencias Naturales (1989); and for the Ex Terra Foundation's "Dinosaur World Tour" resulting from the joint scientific research of the China-Canada Dinosaur Project (1987–8).

18 I discuss these wire frame elements and their visual play in the exhibit media in chapter 14.

19 CityTV, then owned by media "innovator" Moses Znaimer, was well known in Canadian broadcast circles for its "hip" approaches to programming, along with a special savvy for new media. Presumably, the new media elements of the Maiasaur Project attracted the attention of the producers for its thematic affinity to their "MediaTelevision" series. Given the market orientation of Znaimer's CityTV, such a series could only be advanced as a viable sort of programming if it held particularly great public currency and appeal for prospective television advertisers. In other words,

media-in-the-making was considered a marketable story-product for which there was an audience base.

20 CityTV production MediaTelevision feature on the Maiasaur Project exhibition (first broadcast date: July 25, 1995). The story was re-broadcast at least once, according to the ROM Marketing and Communications section's report on the campaign.

21 Allison-Bunnell 1998; Bennett 1995; Altick 1978. Allison-Bunnell (1998) discusses the tension in the early 1960s activities of the Smithsonian Natural History Museum, when curatorial scientists had more of an integral role in exhibit development, and with that came to embed their technical interests in the making of exhibitions and educational films. He also points out how that moment has now passed, as institutional shifts have seen an increasing specialization of the researcher, while exhibit development has in parallel become highly professionalized. According to Allison-Bunnell, the terrain between the two activities of the Smithsonian, in this instance, has been overtaken by the exhibition professionals, while the researcher has become little more than an advising consultant.

22 Barry 1998, 112. Also see discussions in Janes 1995; Witcomb 2006.

23 Silverstone 1992, 41.

24 Barry 1998, 101, quoting Kirby.

25 Interview with Brenda Mikelsen, July 7, 1998. All subsequent quotes by Mikelsen are from this interview.

26 That polarization is discussed in the dichotomy of "wonder" and "resonance" by Steven Greenblatt (1991).

27 Other poster slogans have included "Dragon Food in a Tin" for a medieval armour poster; "Visitors Velcome" [*sic*] for a bat exhibition poster; "The body is 2/3 water. Come see the other 1/3" for an Egyptian mummy poster; "Hockey Night in Upper Canada" for a poster with selected eighteenth- and nineteenth-century Canadian skating artefacts; "Life before Nintendo Games" for a poster with antique toys; and "Some Constrictions Apply" on a poster depicting a boa constrictor.

28 Harvey 1996, 150; see also Macdonald 1993.

29 Strathern 2005, 82–91.

30 See Latour 1993, 121, and Strathern's dense discussion (1995, 177–85) on the "local," the "global," and location.

31 The Czech artist Zdenek Burian was probably the most prolific painter of the ancient life-worlds of dinosaurs in Europe (see Augusta 1964). One of his major influences in the genre was Charles Knight. Burian's paintings appeared in many books, museum displays, and television shows, and next to Charles Knight's, his work may have been one of the most highly

circulated sources of reconstructed scenes and images of dinosaurs up until the 1990s.

32 Also see Bakhtin's commentary in "Discourse on the Novel," where he writes: "The word in language is half someone else's. It becomes 'one's own' only when the speaker populates it with his own intention, his own accent, when he appropriates the word, adapting it to his own semantic and expressive intention. Prior to this moment of appropriation, the word does not exist in a neutral and impersonal language … but rather it exists in other people's mouths, in other people's contexts, serving other people's intentions: it is from there that one must take the word, and make it one's own … Language is not a neutral medium that passes freely and easily into the private property of the speaker's intentions; it is populated – overpopulated with the intentions of others. Expropriating it, forcing it to submit to one's own intentions and accents, is a difficult and complicated process" (Bakhtin 1981, 294).
33 Bakhtin 1986, 170.
34 Here, by borrowing from Barad, I am modifying the "zone of contact" proposition of James Clifford and Mary Louise Pratt. Clifford 1997, 188–219; Pratt 1992.
35 Cf. Adorno 1993 [1978].
36 For the reader's sake, the da Vinci object was one of his notebooks with mirror-writing, while the Lacroix and Westwood objects were exclusive designer shoes. Shoes, of course, have long been taken as exemplary sexual fetishes.

11 Need to Say, Need to Know: Planning to Articulate Specimen and Spectacle

1 Interviews with Jennifer Ross, March 30 and 31, 1998. All subsequent interview quotes from Jennifer Ross are from those two dates.
2 I must add a note on the necessary contingency of reports. As the interviews drawn upon mostly took place in 1998, the quoted commentaries on the documents and actions have to be recognized as retrospective. Some reimagining of what took place on the part of those interviewed has to be acknowledged. Nonetheless, the correspondences between the various accounts, documents, and the exhibit result give a reliable sense of the agencies at play, the complexity of interactions and mediations, and the emerging effects.
3 As is discussed below, "Henrietta" was a name used occasionally by staff for the specimen, and originated with the company that collected the specimen, Bearpaw Palaeontological, Inc.
4 Personal communication with the president of Bearpaw Palaeontological, Inc., August 10, 1999.

5 The matter of family relations is taken up in some depth in chapter 13, "A Perfect Time for Raising a Family."
6 Interview with Sam Enright, June 16, 1998.
7 I found that different individuals used names of documents inconsistently, mixing up interpretation statements, exhibit briefs, and concept summaries. The "Concept Design Summary" usually contains outlines of (1) the administration of the project, including team members, schedule, and budget; (2) the interpretive approach; (3) 2D and 3D design approaches, floor plans, and sketches; and (4) related activities such as donor recognition, special events, and marketing and publicity plans. In the case of the Maiasaur Project, a further section was included to outline multimedia and audio-visual programs, this being something of a new venture in this kind of programming for the ROM in 1995.
8 See Riles (ed.) 2006 for a full discussion of the instrumental agency of documents; and Ferraris 2009.
9 December 3, 1993, memorandum, "Re: Curatorial proposal concerning an exhibit featuring the preparation and mounting of the skeleton of the Cretaceous duckbilled dinosaur *Maiasaura*," from Andreas Henson, associate curator, Department of Vertebrate Palaeontology, to Sylvia Fleming, associate director, Public Programs; March 1, 1994, memorandum, "Revised Version ..." from Andreas Henson to Lydia Chisolm, manager, Exhibit Programming.
10 The "mess" being discerned out of the careful work of planning coincides with and recapitulates the empirical observations of John Law in his discussions of the challenge of methods and messiness in everyday practices of science and social science. See Law 2004a.
11 "Concept Design Summary," 7. n.b. Emphases with bold and font sizes approximate those in the original.
12 Rainger 1991.
13 Additional articles related to *Maiasaura* or hadrosaur nesting behaviour include Horner 1982; 1983; 1984; 1987. See also Carpenter 1982.
14 Only one other palaeontologist has received a MacArthur Fellowship, that being Harvard scientist, historian, and science popularizer Stephen Jay Gould.
15 See the collection of essays edited by Strum and Fedigan (2000), which traces aspects of this double unfolding in the field of primate ethology, as witnessed by both animal scientists themselves and a number of science studies scholars, including Donna Haraway, Bruno Latour, Greg Mittman, Alison Wylie, Evelyn Fox Keller, and this author as well. The Mesozoic, then, is also an excellent machinery for capturing new resources, narratives from the human realm as it acquires new material from the non-human realm.

16 Quote from versions of curatorial proposal of A. Henson, December 3, 1993 and March 1, 1994.
17 Following the points on enframing from Callon (1998a), the Mesozoic is a "metrological device," which allows the capturing of the right matter, and its hybridizing with the right human matter, to advance science, scientists, and nature in particular directions. In true colonizing fashion, it incorporates material according both to human and non-human exigencies. In this way, it works very much like capitalist expansion, always adding in new resources, and justifying that process as it goes.
18 The term "herd," though more familiarly applied to mammals, is used consistently by dinosaur palaeontologists in both technical and non-technical discourse. Dinosaur palaeontology can be quite flexible in freely reapplying zoomorphic (or, as I've been discussing, anthropomorphic) terms in this manner. This is all rather dualistic and mimics contemporary mammal behavioural terms quite directly: the term "herding dinosaurs" is only applied to herbivorous dinosaurs who are thought to have amassed gregariously in large numbers, whereas carnivorous dinosaurs, to my knowledge, have not been described as "herding" creatures. Instead, quite remarkable inferences about small carnivorous dinosaurs such as *Velociraptor*, *Dromaeosaurus*, and *Deinonychus* have led palaeontologists to describe them as "pack-hunting" – a term typically associated with the social predatory behaviour of living mammals, such as wolves and hyenas.
19 This point on the collapsing of time in the film *Jurassic Park*, which mixed together dinosaurs known to have lived both in the Jurassic (e.g., *Brachiosaurus*, *Compsognathus*) and the Cretaceous (e.g., *Tyrannosaurus*, *Triceratops*, *Velociraptor*), caused no end of consternation for professional and amateur palaeontologists, who in Internet chat groups, media interviews, and film reviews – most notably one by Stephen Jay Gould (1993) – have pointed out their annoyance with such anachronisms.
20 For this section, I draw on both the Interpretation Statement and the Multimedia/Audio-Visual Statements of the Concept Design Summary.
21 Interview with Manfred Tolman, August 20, 1999.
22 Cf. Jay 1988.

12 The Difference a Lab Can Make

1 Latour 1983. In the title of his highly cited article, Latour parodies the words of Archimedes: "Give me a place to stand on, and I will move the Earth." Also see the canonical work of Latour and Woolgar 1986.
2 Interviews with Sam Enright, July 29, 1998.

3 Quoted statements from Phil Thomm in the following discussion relating to the Working Lab are drawn from an interview of March 31, 1998.

4 Note that this illustrates only the lab section. More complete schematics are presented in chapters 13 and 14, which address the other two major sections of the exhibition.

5 Andreas Henson emphasized the technical and safety complexities as well in his curatorial proposal (December 3, 1993): "The preparation area would have to be completely enclosed in Plexiglas or a similar translucent material to prevent the spread of dust and chemical fumes while at the same time facilitating public viewing. It needs to be suitably ventilated to protect the technical personnel as required and specified by federal and provincial laws."

6 See Meyer (2011) for a discussion of the very recent turn to build labs into displays. It should be noted that the ROM had, over the preceding two decades, held annual events during which technical and curatorial staff would bring artefacts, specimens, and technical matter into the museum galleries. The major event of this sort was "March Break at the ROM," which takes place during the spring break in the school year. At that time, enormous numbers of school children would descend upon the museum with parents, guardians, friends, and relatives to participate in this fair-like event. Tables would be distributed throughout the museum for technical demonstrations, talks, participatory events, and artefact-handling opportunities. Performances would also take place. This tremendously popular event would allow the visitors a rare opportunity to participate in and "catch a glimpse" of what takes place "behind the scenes" at the ROM.

7 Though a professional mounted cast of the skeleton with the skull in place was placed in the gallery, the original Maiasaur skull itself from ROM #44770 was never actually put on display. In fact, it was still in the process of being prepared "behind the scenes" in the Department of Veterbrate Palaeontology preparation lab in April of 1999 when my research visit came to its end. According to Ian Morrison, January 11, 2000, the original was never displayed. In fact, the cast made used both ROM and to a lesser extent MoR material. Moreover, no technical description had been published on the specimen. Astonishingly, Paul Sereno, on viewing ROM #44770, felt uncertain that the creature was even a Maiasaur! The ROM specimen's skull had an intact crest, while the holotype (Princeton U.) of Horner's had a crushed crest. No proper comparative study had been undertaken to ascertain the relations.

8 Note that this illustrates only the lab section of the exhibit.

9 Curatorial proposal, second version, March 4, 1994. Henson had also remarked on his particular interests in these topics in one interview: "I very much like the notion of co-evolution ... which is a notion that, even though it has some antiquity within evolutionary biology, having been proposed in the early 1960s ... it's not something people really think about ... and I think it's a very important concept to think about in terms of our current challenges, in terms of the dramatic environmental changes we are going through, that yes, animals relate to plants and that the evolution of plants influenced the evolution of animals and vice versa." Compare these comments with the article review and discussion article "Mesozoic and Early Cenozoic Terrestrial Ecosystems," by Wing et al. (1992).
10 Pignarre and Stengers 2011, 14.
11 Latour 2000.
12 Strathern 2004.
13 Winsor (1991, 121ff) points out how public and systematic collection spaces were separated in nineteenth-century museums of natural history. In line with this division, the active work of the curator and the curator's work bench would also be removed to "behind-the-scenes" spaces hidden away from the areas of public exhibition.

13 A Perfect Time for Raising a Family: Kinship as New Syntax for Dinosaurian Natures?

1 Although this next section is primarily based on one slow, deliberate walk-through, I have added in fragments from several other visits (and from visitor "follows") to give the reader a fuller sense of the diversity of visitor experiences.
2 For a rather laudatory account of the ROM's history and the emergence of the gallery layouts into which this exhibit was positioned, see Dickson 1986.
3 Law 2004a.
4 See Schneider 1980.
5 "Family" is a remarkably contested signifying category in anthropological discourse. Its meaning is immanent in the worldly examples that give it shape, such as the use of "family" as a target audience at the ROM, or the Maiasaur "family" of the exhibit video. For a discussion of the category of "family" in anthropological discourse, see Collier, Rosaldo, and Yanagisake 1992.
6 Stengers 2011, 387.
7 The comments in this section refer to an ethnographic walk-through that took place in early June of 1998.

8 I drew the main schematic plan prior to looking at the designer's plans and drawings. It is based on the exhibit's layout as I had understood it over several visits. A few modifications in the titling of display elements have been made to correlate this plan to the "official" exhibit floor plan presented in the preceding chapter.
9 Terms from Barry 1998.
10 As noted in the previous chapter, both preparator Phil Thomm and sculptor Manfred Tolman had told me that Henson had encouraged them to review Cope's article. Curiously, the unusual beak associated with the ROM specimen was not depicted in the model. According to Tolman, he had been instructed by curator Henson not to portray this feature (personal communication with artist).
11 Latour 1999, 140.
12 Interview with Jennifer Ross, March 30, 1998.
13 Interview with Sam Enright, July 9, 1998.
14 From "Concept Design Summary" document.
15 I was told by several staff that provincial statute requires the ROM to present its materials in both official languages of Canada.
16 The French and English glosses do not match precisely, but already there are small but significant indexes to the scope of meaning which this exhibition entailed. *Maiasaur* becomes *Maiasaura* in French, matching the Linnaean scientific term for the genus. French linguistic rules assign gender to every noun; *Maiasaura* is a feminine form.
17 Visitor interview, July 20, 1998, with a woman from Ohio (approximately fifty years old).
18 Interview with Andreas Henson, August 4, 1998.
19 Interview with Andreas Henson, August 4, 1998.
20 Macdonald 1997, 167.
21 Philip Currie, personal communication, December 28, 2011.
22 Macdonald (1997, 158–9) pointed out an exactly parallel situation in the Science Museum in London. She notes that the programming team for the exhibition "Food for Thought" were all women, adding, "Team members were relatively low-graded to be charged with the task of making an exhibition … Their gender and status as 'non-experts' seemed to become meshed together as markers of difference from what they mostly conceptualized as a traditional and conservative museum establishment in which 'science' itself was masculinely gendered … Their identification with the public, then, was an identification with a public which had been disregarded by the museum and scientific establishment; and part of the rhetoric during the making of the exhibition was not just about getting more

science to that public, but about challenging some of the establishment high ground."

23 Horner and Makela 1979; Desmond 1977.

24 Interview with Wendy Madsen, April 7, 1998.

25 Visitor interviews, July 24, 1998.

26 Biography may well be the adult translation of this genre, signalled typically by phrases like "the life and times" of the noted person. For a sampling of genres of children's literature see Lukens 1999, chapter 2, "Genre in Children's Literature," 13–40.

27 Visitor interviews, July 26, 1998.

28 Franklin 1997b.

29 Cf. Harris 1990; Benjamin 2006, 9–16. Benjamin speaks of Baudelaire's Parisian "flâneur," one who strolls through the built environment as part of everyday experience – a figure that would count as readily for a shopper in a shopping arcade, or a museum-goer in a museum. Harris writes of the historical relation of the built spaces of consumption, and those of public spectacle and museums. Also see Altick 1978.

30 Macdonald 1998a, 136.

31 Most major museums now insert shops into the very spaces of temporary exhibits to market theme-specific or niche merchandise, as would be the case with the V&A exhibit at the ROM.

32 Strathern 2005, 67. Whereas Strathern offers a methodical discussion of "emergent properties" of kinship in procreative discourse and human genetic technologies, we can see a parallel process here in the emergence of dinosaur relatedness in public palaeontology.

33 Mitchell (1998) moves much of his argument around the psychoanalytic part-for-whole relation of dinosaurs, going so far as to suggest that they are, in their very excessive and obsolescent embodiment, a Freudian "totem" for America as a whole, and in this sense the "totem animal of modernity" where dinosaurs and America become correlated exemplars of modernity, its excesses, its limits, its delusions of truth and grandeur. The intimate stories of *Maiasaura* and *Tyrannosaurus* offer up a more textured set of possibilities, interesting and disturbing in more particular ways than grand statements of being quasi-religious stand-ins for all of modernity.

34 Macdonald 1998b, 2.

35 See Stengers 2008b, 43, where she gestures to ideas from Deleuze and Guattari of how the most forceful accounts, and indeed forces, are made all the more forceful if and once "captured" into an assemblage. The force of the real is captured, and uttered as authentic, in the assemblages we call

"museums." See also Latour 1999 for his various discussions of the "reality of science studies."

36 See Ames 1986; Bennett 1995, 2004, 2006; Hooper-Greenhill 1992; Impey and McGregor 1985; and Findlen 1994.
37 Gombrich 2000 [1961].
38 Franklin and McKinnon (2001) offer numerous anthropological cases of emergent kinships, wrought in varied milieus.
39 Visitor interviews, July 26, 1998.
40 Visitor interviews, July 26, 1998.
41 I interviewed over forty individuals. Most could recite the name of the dinosaur, that it was a plant-eater, and that it lived "eighty million years ago."
42 Said 1979.
43 Term used by Lockett, Menna, and Walker 1998.
44 Helmreich 1998.
45 Visitor interviews, July 24, 1998.
46 Visitor interviews, July 24, 1998.
47 For an example of how the semiotics of museum exhibits generate expected and unexpected meanings, see Wilson 1988.
48 Visitor interviews, July 25, 1998.
49 Wilford 1986; Desmond 1977; Norman 1991; Bakker 1986; Horner and Gorman 1988; and Horner 1987.
50 Because the same animation technique was used, the visage was reminiscent of the stop-motion style of monsters and dinosaurs from Willis O'Brien and Ray Harryhausen films of the 1930s through the 1960s. See discussion in chapter 6, "Vestiges of the Lost World"
51 Kyla and Allison were seven years old, while Emma was five years old.
52 Visitor interviews, July 26, 1998.
53 Anthropologist Kenneth Little passed on this tale to me in February of 1998, just a couple of days after the exchange with Will had taken place. At the time, Little was professor and chair of the Department of Anthropology at York University, Toronto.
54 Visitor interviews, July 24, 1998.
55 See, for example, Fiorillo and Gangloff 2001; Varricchio et al. 2008; Persons and Currie 2014.
56 Title of draft of paper by Lockett, Menna, and Walker (1998).
57 Habitat groups are a staple of twentieth-century natural science museums, meant to capture, snapshot-like, a moment in the life-world of creatures. See Wonders 1993, 12.

14 Technotheatrical Natures: Maiasaur's World, by Default?

1 Pachner 1995, 47.
2 Stengers quoted the last message of Gilles Deleuze, one of her mentors, who urged readers in his collaborative volume with Félix Guattari: "Lodge yourself on a stratum, experiment with the possibilities it offers, find an advantageous place on it, find potential lines of deterritorialization, possible lines of flight, experience them, produce flow conjunctions here and there …" (Stengers 2010b, 2010c; Deleuze and Guattari 1988, 161). The Mesozoic specimen/spectacle nexus is such a stratum, prone, as it were and at many turns, to emergent and multiple possible lines of flight. *Maiasaura* then, it could be said, opens up a line of flight from the Mesozoic stratum of *Tyrannosaurus*.
3 CityTV production MediaTelevision feature on the Maiasaur Project exhibition (first broadcast date: July 25, 1995).
4 This description continues in the mode of the ethnographic descriptions in the previous chapter. A single walk-through is presented with interjections from interviews, quotes, commentaries, etc. As mentioned previously, the exhibit's layout and elements shifted over the course of the intended run. From May 1995 to June 1997, the lab was operating in the third section of the exhibit. From July 1997 to May 1998, the mounted skeleton of the specimen replaced the lab in the third section. From May 1998 to September 1998, in order to accommodate space needed for the "A Grand Design" exhibit from the Victoria and Albert Museum in London, the third space was completely overtaken, with the skeleton being moved into the "Meet a Maiasaur" theatre space.
5 See chapter 10.
6 Law 2004a, 56.
7 The wire-frame lines were retained in the "transmorphic" Maiasaur used in many of the graphics produced for way-finding and for promotional materials; cf. figure 10.2.
8 Claymation is an animation technique which makes use of clay models. Here she was referring to the "Visit to the Cretaceous" videos showing Maiasaur nesting.
9 Visitor interview, June 14, 1998.
10 Interestingly, the elevated-tail, vs. tail-dragging, vision of *T. rex* was quite probably informing this child's analogizing, as discussed by Bakker's (1987) provocative writing of "dancing dinosaurs," which saw the revival of Cope's vision of his leaping Dryptosaurs, as discussed in chapter 4 of this volume.

11 Visitor interview, July 20, 1998.
12 Stengers, 2012b.
13 Gombrich 2000 [1961], 180ff; Jordanova 2004, 447; Rancière 2009.
14 Cf. Czerkas and Glut 1982; Augusta 1964.
15 ROM interpretation guide in conversation with a member of a guided tour through the "Meet a Maiasaur" theatre. June 5, 1998, gallery observations. This group stayed in the exhibit about four minutes.
16 I should also point out that Disney and Hollywood productions use very similar approaches to technical verification of animated dinosaur reconstructions, consulting with palaeontologists – usually for a fee – and then expending far more extensive resources than most museums to realize the animated outcome.
17 These are the words of Walter Tomasenko, who effectively parented the multimedia developments of this display. A somewhat remarkable footnote to his comment, however, is that in the late 1960s, the communications theorist Marshall McLuhan had worked with painter Harley Parker on an experimental invertebrate palaeontology display project at the ROM which is rarely discussed. Lovat Dickson (1986, 152–3) wrote: "When Parker designed the ... gallery in 1967, complete with a 'total environment' provided by films, stills, tapes, telephones, push-buttons, smells and sounds, the resulting assault on the senses was so great that the display became known as the 'discothèque gallery.'" Reports from long-time staff of the ROM are that McLuhan's use of televisual and interactive components was more or less a public failure, the implication being that the everyday competence in knowing how to engage such media had yet to become widely distributed to the point that visitors could "make the transition" to the displays in a museum context. The McLuhan-Parker project is worthy of in-depth study and comparison in light of the wildfire spreading of interactive, digital media in cultural institutions today. See McLuhan, Parker, and Barzun 1969.
18 Interview with Jennifer Ross, March 30, 1998.
19 Interview with Walter Tomasenko, June 28, 1998. All subsequent quotes from Tomasenko in this chapter are from either that interview or a second one on July 8, 1998.
20 Janes 1995.
21 Cf. Latour's use of this term (2000).
22 Following Michel Callon, Andrew Barry (1998, 112) suggested that interactive media might indeed have this power when he noted: "Certainly, some of those associated with science museums and science centres have hoped that interactive devices could serve both to distribute roles to humans and

to generate certain human capacities." This was the very sort of widely dispersed socio-technical effect which Tomasenko was calculating with, as well – one that for him was just so obviously useful to everyone and for everything.

23 See Gregory (2007), who gives an example of how "Disneyfication" is used instrumentally by all kinds of museums, and notably in the Christian "Creation Museum and Family Discovery Centre," that clearly operate within the specimen-spectacle complex, but according to their own terms and concerns. To understand the robustness of that museum's claims, one could trace the complex of articulations to the specimens that are meant to uphold accounts offered. I would presume the complex is more readily described in tripartite terms: specimen-scripture-spectacle.

24 Pignarre and Stengers 2011.

25 Recall previous discussion in this book on such relations as Russell and Kish, Osborn and Knight, Owen and Hawkins. Also see Tom Mitchell's wide-ranging discussions of dinosaur art (1998, 48–56 and 265–75).

26 Whitehead 2011, 64.

27 As Deleuze and Guattari noted: "From this standpoint, when we talk about 'undecidable propositions,' we are not referring to the uncertainty of the results, which is necessarily a part of every system. We are referring, on the contrary, to the coexistence and inseparability of that which the system conjugates, and that which never ceases to escape it following lines of flight that are themselves connectable. The germ and locus par excellence of revolutionary decisions" (Deleuze and Guattari 1988, 473).

15 *Mirabile dictu*! Factishes All the Way Down

1 Rich 1979.

2 Stengers 2012a, b.

3 The "Joy of Rex" posters and campaign were designed by the Toronto advertising company Roche MacAuley, which had long established itself as the purveyor of tongue-in-cheek ad campaigns on behalf of the ROM.

4 Latour 1999, 274.

5 Stengers 2010a, 113, and see Stengers 2012a, 340–50.

6 Boghossian 2002, 227.

7 *Science Daily* recirculated the "lost world" phantasmatics on this creature in its article "Amazing Horned Dinosaurs Unearthed on 'Lost Continent'; New Discoveries Include Bizarre Beast with 15 Horns" at http://www.sciencedaily.com/releases/2010/09/100922121943.htm, accessed August 1, 2011.

8 Yale News remarked: "When Nicholas Longrich discovered a new dinosaur species with a heart-shaped frill on its head, he wanted to come up with a name just as flamboyant as the dinosaur's appearance. Over a few beers with fellow paleontologists one night, he blurted out the first thing that came to mind: Mojoceratops ... 'It was just a joke, but then everyone stopped and looked at each other and said, "Wait – that actually sounds cool,"' said Longrich, a postdoctoral associate at Yale University. 'I tried to come up with serious names after that, but Mojoceratops just sort of stuck. You're supposed to use Latin and Greek names, but this just seemed more fun,' Longrich said. 'You can do good science and still have some fun, too. So why not?'" http://news.yale.edu/2010/07/08/mojoceratops-new-dinosaur-species-named-flamboyant-frill, accessed August 1, 2011.

9 Haraway 1998 184.

10 Pachner 1995, 44–7.

11 Sources for the following discussion are mostly popular and virtual: Tippett Studios web page: http://www.tippett.com/; http://articles.latimes.com/1997/nov/02/entertainment/ca-49413; Matt Williams, *BFS Newsletter* 21(1), January/February 1997; Raging Polygons Inhabit The Lost World By Ron Magid.

12 Kirby 2011, 130–1.

13 See http://www.cycho.com.au/cycho-site/scenes.html, accessed June 1, 2014.

14 See Benjamin's (1968 [1936]) discussion of both Marx's and his own formulations.

15 See Matt Williams's fan article in *BFS Newsletter* 21(1).

16 Latour (2010) expands on the compositional effort.

17 Michael Sragow, "Building the Perfect Beast," http://www.sfweekly.com/sanfrancisco/building-the-perfect-beast/Content?oid=2134756, accessed March 2, 2015.

18 Those animations have over the course of the last decade and a half become the standardized image sources for popular imagining of Maiasaur behaviour – good mothering has become the norm. In the 1997 *Jurassic Park* film, *Tyrannosaurus rex* had as well become a good mother – indeed, even a good father.

19 See John Rosengrant's commentary on the embodied, speculative mimicry of wearing a "raptor suit" in "JURASSIC PARK – Evolution of a Raptor Suit with John Rosengrant" at http://www.youtube.com/watch?v=jAzQr3Ml0UI, accessed February 21, 2013.

20 The Universal Pictures "behind the scenes" website story outlined the process used by the animation team of ILM during the 1997 production, as

well as the technical implications for thinking palaeozoologically: "artists sculpted scale maquettes of all the dinosaurs; maquettes were then precisely scaled into clay sculptures to create animatronic live-action dinosaurs and then scanned into ILM's computers. The dinosaurs were animated at ILM on Silicon Graphics hardware using SoftImage and Caricature, ILM's in-house proprietary animation software developed for Dragonheart. The dinos and other 3D animations were rendered using Pixar Renderman, Discreet Logic FLINT and FLAME, and ILM's in-house SABRE system ... Dutra, former Tippett right-hand man [notes], 'We have to realize that we're working with tonnage and not just pixels ... A huge part of animating is understanding how an animal negotiates its weight and scale. Both compys and T. rexes are bipedal and have short arms. However, the compys weigh two pounds while the T. rexes weigh tons. This means we have to account for vast weight differences.'"

21 Strathern 1996, 522–4; Macdonald 1998b, 129; Jordanova 1989, 38.

22 Callon 2007.

23 Macdonald 1998b, 129.

16 "Just Trying to Be a Scientist": Another Mesozoic Is Possible

1 Latour, comment in Strum and Fedigan 2000, 315.

2 Latour 2004.

3 See the *Oxford American Dictionary* entry for "civil," accessed April 12, 2012. Many well-known thinkers, including Karl Popper, Johann Gottfried Herder, Dipesh Chakrabarty, and Johannes Fabian, have offered accounts that delink notions of progress from those of civilization, and I am in accord on that point. In this discussion, I am allying with Isabelle Stengers in reclaiming the actions of *civilizing, civilly*.

4 Discussion is based on an interview with Beth Jameson, March 23, 1999.

5 For a sketch review of the transitions from the late 1960s through to 1982, see Dickson 1986, 130–76.

6 Polly Winsor (1991) provided an exquisite discussion of this around Louis and Alexander Agassiz's involvement in Harvard's Museum of Comparative Zoology in the late 1800s, when a transatlantic move was underway to separate systematic scientific collections from teaching (read: "public") collections, signalling with it a new dividing and specializing of the museum labour activities and focus, and a dividing of curatorial activities at the same time.

7 Allison-Bunnell 1998, 94–5. A similar argument about the V&A in London is made by Sandino 2012.

8 The argument supporting a collective approach is that a more plural process of knowledge production is taking place, and that the natural worlds emerging in exhibitionary form will take in far greater arrays of interests depending on how much the various actors in the collective are given agency to act. The contrast here with the Osborn practices of exclusionary world-building is heartening. With Osborn, exhibitionary and scientific outcomes arose through a top-down authoritarian practice of amplifying certain agendas.

9 Posted April 10, 2012, by Kiron Mukherjee: http://www.rom.on.ca/en/blog/archaeology-weekend-meet-your-museum-heroes.

10 Janes 1995.

11 Janes 1995, 232. This is not unlike the problem-solving proposition of Antonio Gramsci in his formulations on civil society.

12 A similar point about the participation of governmentally affiliated museums as sites of participation in civil society is made by anthropologist Ivan Karp (ed. 1992, 4). Following Gramsci, Karp approximately distinguishes "political society" from "civil society" as that which has a strict regulatory, policy enforcement (e.g., "policing"), or statutory function versus that which has a hegemonic function through the complex everyday action of public culture. Also see Bobbio 1971.

13 Deleuze and Guattari 1988, 40.

14 Janes 1995, 258.

15 Macdonald 1996, 14.

16 See Stengers 2008c, for a compelling discussion of Whitehead's "matters of concern" as developed in his volume *Process and Reality*.

17 Mitchell 1998, 261–2.

18 Latour 2000, 380–1.

19 Mike Harris was the notoriously neoconservative premier of the province of Ontario who led an all-out cutback campaign against many public institutions over the 1990s. His finance minister, Jim Flaherty, was at the time of writing (2013) overseeing a similar campaign for Canada's "Harper Government," where he had also become the minister of finance. See comment by Howie West, "Harper's Museum and Art Gallery Policy: Cultural Devolution and Privatization" at http://www.policyalternatives.ca/sites/default/files/uploads/publications/National_Office_Pubs/2008/HarperRecord/Harpers_Museum_and_Art_Gallery_Policy.pdf, accessed October 12, 2015.

20 Mitchell 1998.

21 Janes 1995, 255.

22 See for instance Karp et al. (eds) 2006.

23 Janes 2009, 173.
24 Janes 2009, 166.
25 Pignarre and Stengers 2011, 96.
26 Stengers 2010d, 31.
27 Stengers 1997, 85. See also Disch 2009, 269.
28 Various usages noted in the *Oxford English Dictionary* online, accessed February 8, 2013.
29 While many have written of the "death of the curator," such pronouncements remain premature, if we consider the place of the nominal curator within a curatorial team charged with executing an exhibit and so bringing about public natures. Writing more on art curation, Anjali Gupta has reiterated David Lévi-Strauss's assay of the "curatorial disposition": "His register includes the administrator, auteur, bricoleur, broker, bureaucrat, catalyst, the collaborator, cultural impresario and the diplomat. To this I would add the cat herder, the confidant, the casting director, the enabler, the overly entitled, the magician, the producer and the delinquent. While these are not new incarnations, some are antithetical to the traditional role of the curator as a caretaker of objects. And while most of us discarded that construct long ago, there are those who still function solely within that rubric. I see figures in long robes, scurrying about to plug up holes in the convex surface of a decaying diaphragm, while provocateurs prick its surface from the other side, in the hopes of giving birth to a hybrid creature – an entity unconcerned with titles and semantics – an entity capable of navigating the cultural landscape to come." Where I differ from Gupta and Lévi-Strauss is in their thinking that the only thing curators did in the past was purely the "caring for" function, but certainly it was and remains an honourable action, especially when combined with the second dimension Gupta points to and that palaeontological curators have continually displayed: being provocateurs. However we consider or value their actions, Owen, Osborn, Horner, and Henson were all provocateurs who, through their dinosaur allies, continually aided in the generation of hybrid creatures (dinosaurs are always such), and they were "capable of navigating the cultural landscape to come." See Gupta's (2008) comment on the "Death of the Curator."
30 Janes 2009, 151.
31 Janes 2009, 152, 153.
32 Arendt 1970, 44. See the discussion of Arendt's formulation of "power" by Habermas and McCarthy 1977. For an incisive analysis of putting "power with" into action, see Tully 2011.

33 This hearkens once more to Karen Barad's intra-action that I discussed in previous chapters, and Latour's inter-objectivity; see note following.

34 In relation to actor networks, I reiterate that phantasy should not be thought of as divorced from the actions of humans and non-humans. Rather, it takes place through the active engaging of human with non-human entities. Phantasies move between and come to reside in humans and non-humans through collective actions. Humans, therefore, are privileged to some extent, because they are able to take up or act upon phantasies very rapidly. Specific consideration of how phantasies move between people and things might fruitfully be taken up through ethnomethodological studies. See Latour 1996, to which I am suggesting the simple move of readmitting phantasy to the exchanges, helping along his point that there need not be "a yawning gulf separating the agent from structure, the individual from society" (232).

35 The movement to democratic participatory engagement in contemporary web-mediated and physical museum visitor experience is advancing rapidly in the 2010s. A leader in this movement is Santa Cruz museum director Nina Simon, who writes on her blog "Museums 2.0": "What do I mean by 2.0? 'Web 2.0' is not just a buzzword; it's a definition of web-based applications with an 'architecture of participation,' that is, one in which users generate, share, and curate the content. The web started with sites (1.0) that are authoritative content distributors – like traditional museums. The user experience with web 1.0 is passive; you are a viewer, a consumer. Web 2.0 removes the authority from the content provider and places it in the hands of the user. Now, you are a participant. You determine what's on the site, and you judge which content is most valuable … I believe that museums have the potential to undergo a similar (r)evolution as that on the web, to transform from static content authorities to dynamic platforms for content generation and sharing. I believe that visitors can become users, and museums central to social interactions. Web 2.0 opens up opportunity, but it also demonstrates where museums are lacking. The intention of this blog is to explore these opportunities and shortcomings with regard to museums and interactive design." At http://museumtwo.blogspot.co.nz/2006/12/what-is-museum-20.html, accessed December 9, 2012.

36 The corollary here is to note that when we attend to Mesozoic fabulation it is *to catch movement of the constitution of dinosaurian politics/natures*. Here I am experimenting and turning the notion of *fabulation* from Gilles Deleuze (1995) to incorporate natures, more than just peoples. Drawing upon Henri Bergson, Deleuze had left "natures" out of his discussion when

remarking: "To catch someone in the act of legending [i.e., fabulating] is to catch movement of the constitution of a people." See Deleuze 1995, 125–6. See also the detailed discussion in Mengue 2008.

37 Mol 2002. Mol also launches what has become a larger conversation in the Science Studies discourse over the last decade, on whether such alternatives are in the order of paradigmatic or ontological distinctions, a matter which some readers might consider through the extended case studies of this book on dinosaur articulation.

38 Isabelle Stengers, Situsci Public Lecture, "Cosmopolitics: Learning to Think with Sciences, Natures, Peoples," March 5, 2012, at https://youtu.be/1I0ipr61SI8, accessed April 5, 2016.

39 Latour 2010; Neale Wheeler Watson Lecture, "May Nature Be Recomposed? A Few Questions of Cosmopolitics," Nobel Museum, Svenska Akademiens Börssal, May 11, 2010. Video of lecture at http://www.bruno-latour.fr/node/269, accessed February 1, 2013.

40 See Bonnan's webpage: http://matthewbonnan.wordpress.com/2012/12/07/dead-dinosaurs-and-reasons-for-hope/, accessed March 11, 2013.

41 For one of the more adventurous speculative palaeontologists, see the work of Dale A. Russell (2009).

42 Strathern 2006, 203. See also Miyazaki and Riles, 2005.

43 Adopting Stengers's phrasing directly, museums scholar Ben Dibley (2011, 162) recently wrote of the same concerns and possibilities I have been writing on here, pointing to cosmopolitical potential: "Perhaps, the cosmopolitical museum might come to put forward proposals by which 'we' think of our decisions 'in the presence of' those others once disqualified by the borders of nation, species, and animation, not on the assumption that we nevertheless share a common world, but that we enter into the hard work of its composition."

44 Cf. Latour 1997, where he notes that it is in the very fracturing of such totals that the promise of more adequate articulation lies. Rather than seeking "truth" we should be after truths-in-the-making.

45 See Bakker et al. 2006.

46 See Sagan et al. 1986; Turco et al. 1990; and Raup 1999, on the "Nemesis affair," addressing the perennial fear of large-body impacts with the earth bringing on extinction.

47 Mannion et al. 2012; Morley 2011.

48 Eberth et al. 2013.

49 Miller, Summers, and Silber 2004.

50 Wilkinson, Nisbet, and Ruxton 2012.

51 Stengers 2008a.

52 Rose 2011, 27. I thank my student Amy Donovan (not to be confused with the museum visitor Amy whom I also quote) for isolating this apt statement from Deborah Bird Rose's important book.
53 I acknowledge here a related call for care in praxis of science and its "matters of fact" from de la Bellacasa 2011.
54 See the work of the "Working Group on the Anthropocene" of the Subcommission on Quaternary Stratigraphy at http://quaternary.stratigraphy.org/workinggroups/anthropocene/, accessed December 20, 2014.

References

Abbott, S. 2006. "Final Frontiers: Computer-Generated Imagery and the Science Fiction Film." *Science Fiction Studies* 33 (1): 89–108.

Adorno, T. 1984 [1951]. *Minima Moralia: Reflections from Damaged Life*. Trans. Edmund Jephcott. Repr. New York: Verso.

Adorno, T. 1993 [1978]. "On the Fetish Character in Music and the Regression of Listening." In H. Marcuse, A. Arato, and E. Gebhardt, *The Essential Frankfurt School Reader*, 270–99. New York: Continuum Press.

Agamben, G. 2009. *What Is an Apparatus: And Other Essays*. San Francisco: Stanford General.

Akrich, M., and B. Latour. 1996. "A Summary of a Convenient Vocabulary for the Semiotics of Human and Nonhuman Assemblies." In W. Bijker and J. Law, eds., *Shaping Technology/Building Society: Studies in Sociotechnical Change*, 259–64. Cambridge, MA: MIT Press.

Allison-Bunnell, S. 1998. "Making Nature 'Real' Again: Natural History Exhibits and Public Rhetorics of Science at the Smithsonian Institution in the Early 1960s." In *The Politics of Display: Museums, Science, Culture*, ed. S. Macdonald, 77–97. New York: Routledge.

Altick, R. 1978. *The Shows of London*. Cambridge, MA: Harvard University Press.

Alvarez, W. 1997. *T. rex and the Crater of Doom*. Princeton: Princeton University Press.

Ames, M. 1986. *Museums, the Public and Anthropology: A Study in the Anthropology of Anthropology*. Vancouver: University of British Columbia Press.

Ames, M. 1999. "How to Decorate a House: The Re-negotiation of Cultural Representations at the University of British Columbia Museum of Anthropology." *Museum Anthropology* 22 (3): 41–51. http://dx.doi.org/10.1525/mua.1999.22.3.41.

Andrews, R.C. 1922. *On the Trail of Ancient Man: A Narrative of the Field Work of the Central Asiatic Expeditions*. Garden City, NY: Garden City Publishing.

Andrews, R.C. 1932. *The New Conquest of Central Asia: A Narrative of the Explorations of the Central Asiatic Expeditions in Mongolia and China, 1921–1930*. New York: American Museum of Natural History.

Archer, S. 1993. *Willis O'Brien: Special Effects Genius*. Jefferson, NC: McFarland.

Arendt, H. 1970. *On Violence*. New York: Harcourt, Brace & World.

Asch, M. 2014. *On Being Here to Stay: Treaties and Aboriginal Rights in Canada*. Toronto: University of Toronto Press.

Atran, S. 1992. *Cognitive Foundations of Natural History: Towards an Anthropology of Science*. Cambridge: Cambridge University Press.

Augusta, J. 1964. *Prehistoric Animals: Illustrated under the Direction of the Author by Zdenek Burian*. London: P. Hamlyn.

Austin, J.L. 1955. *How to Do Things with Words*. Ed. J.O. Urmson and Marina Sbisá. Cambridge, MA: Harvard University Press.

Bacon, F. 1852 [1605]. *The Two Books of Francis Bacon: Of the Proficience and Advancement of Learning, Divine and Human*. Ed. T. Markby. London: John W. Parker and Son.

Bacon, F. 1952 [1605]. *Advancement of Learning; Novum Organum, New Atlantis*. Chicago: Encyclopedia Britannica Books.

Bakhtin, M. 1981. *The Dialogical Imagination: Four Essays*. Ed. M. Holquist and C. Emerson. Austin: University of Texas Press.

Bakhtin, M. 1986. *Speech Genres and Other Late Essays*. Ed. C. Emerson and M. Holquist. Austin: University of Texas Press.

Bakker, R. 1972. "Anatomical and Ecological Evidence of Endothermy in Dinosaurs." *Nature* 238 (5359): 81–5. http://dx.doi.org/10.1038/238081a0.

Bakker, R. 1986. *The Dinosaur Heresies*. New York: William Morrow.

Bakker, R. 1987. "The Return of the Dancing Dinosaurs." In *Dinosaurs Past and Present*, vol. 1, ed. S. Czerkas and E. Olson, 38–69. Los Angeles: Natural History Museum of Los Angeles County.

Bakker, R. 1995. *Raptor Red*. New York: Bantam.

Bakker, R., R. Sullivan, V. Porter, P. Larson, and S. Salusbury. 2006. "*Dracorex hogwartsia*, n. gen., n. sp., a spiked, flat-headed pachycephalosaurid dinosaur from the Upper Cretaceous Hell Creek Formation of South Dakota." In *Late Cretaceous Vertebrates from the Western Interior. New Mexico Museum of Natural History and Science Bulletin 35*, ed. S.G. Lucas and R.M. Sullivan, 331–45. Albuquerque: New Mexico Museum of Natural History.

Ballou. 1897. "Strange Creatures of the Past: Gigantic Saurians of the Reptilian Age." *Century Magazine* 55: 15–23.

Barad, K. 2003. "Posthumanist Performativity: Toward an Understanding of How Matter Comes to Matter." In "Gender and Science: New Issues," special issue of *Signs* 28 (3): 801–31. http://dx.doi.org/10.1086/345321.

Barad, K. 2006. *Meeting the Universe Halfway: Quantum Physics and the Entanglement of Matter and Meaning*. Durham, NC: Duke University Press.

Barry, A. 1998. "On Interactivity: Consumers, Citizens, and Culture." In *The Politics of Display: Museums, Science, Culture*, ed. S. Macdonald, 98–117. New York: Routledge.

Bearman, D. 1993. "Interactivity in American Museums." *Museum Management and Curatorship* 12 (2): 183–93. http://dx.doi.org/10.1080/09647779309515356.

Benjamin, W. 1968 [1936]. "The Work of Art in the Age of Mechanical Reproduction." In *Illuminations: Essays and Reflections*, ed. H. Arendt, 217–52. New York: Schocken.

Benjamin, W. 2006. *The Writer of Modern Life: Essays on Charles Baudelaire*. Cambridge, MA: Harvard University Press.

Bennett, T. 1995. *The Birth of the Museum: History, Theory, Politics*. London: Routledge.

Bennett, T. 1998. *Culture: A Reformer's Science*. London: Sage Publications.

Bennett, T. 2004. *Pasts beyond Memory: Evolution, Museums, Colonialism*. London: Routledge.

Bennett, T. 2006. "Exhibition, Difference, and the Logic of Culture." In *Museum Frictions: Public Cultures/Global Transformations*, ed. I. Karp et al., 46–69. Durham, NC: Duke University Press. http://dx.doi.org/10.1215/9780822388296-003.

Blaser, M. 2009. "Political Ontology." *Cultural Studies* 23 (5–6): 873–96. http://dx.doi.org/10.1080/09502380903208023.

Boas, F. 1920. "The Methods of Ethnology." *American Anthropologist* 22 (4): 311–21. http://dx.doi.org/10.1525/aa.1920.22.4.02a00020.

Bobbio, N. 1971. "Gramsci and the Conception of Civil Society." In *Gramsci and Marxist Theory*, ed. C. Mouffe, 21–47. London: Routledge.

Boghossian, P.A. 2002. "Who's Afraid of Social Construction? Reflections on Ian Hacking's *The Social Construction of What?*" In *Normativity and Legitimacy: Proceedings of the II Meeting Italian-American Philosophy, New York, 1999*, vol. 1. New York: Lit Verlag, 214–32.

Bowie, M. 1991. *Lacan*. Cambridge, MA: Harvard University Press.

Braidotti, R. 1996. "Signs of Wonder and Traces of Doubt: On Teratology and Embodied Differences." In *Between Monsters, Goddesses and Cyborgs: Feminist Confrontations with Science, Medicine, and Cyberspace*, ed. N. Lykke and R. Braidotti, 135–52. London: Zed Books.

Braun, B., and S.J. Whatmore, eds. 2010. *Political Matter: Technoscience, Democracy, and Public Life*. Minneapolis: University of Minnesota Press.

Breithaupt, B., E. Southwell, and N. Matthews. 2008. "Wyoming's *Dynamosaurus imperiosus* and Other Early Discoveries of *Tyrannosaurus rex* in the Rocky Mountain West." In *Tyrannosaurus rex, the Tyrant King*, ed. P. Larson and K. Carpenter, 57–61. Bloomington, Indianapolis: Indiana University Press.

Brown, B. 1915. "Tyrannosaurus, the Largest Flesh-Eating Animal That Ever Lived." *American Museum Journal* 15 (6): 270–9.

Brown, B. 1919. "Hunting Big Game of Other Days." *National Geographic* 35: 407–29.

Browne, J. 1996. "Biogeography and Empire." In *Cultures of Natural History*, ed. N. Jardine, J.A. Secord, and E.C. Spary, 305–21. Cambridge: Cambridge University Press.

Butler, J. 1988. "Performative Acts and Gender Constitution: An Essay in Phenomenology and Feminist Theory." *Theatre Journal* 40 (4): 519–31. http://dx.doi.org/10.2307/3207893.

Butler, J. 1990. "The Force of Fantasy: Feminism, Mapplethorpe, and Discursive Excess." *Differences: A Journal of Feminist Cultural Studies* 2 (2): 105–25.

Butler, J. 1993. *Bodies That Matter: On the Discursive Limits of "Sex."* New York: Routledge.

Butler, J. 1997. *Excitable Speech: A Politics of the Performative*. New York: Routledge.

Butler, J. 2010. "Performative Agency." *Journal of Cultural Economics* 3 (2): 147–61. http://dx.doi.org/10.1080/17530350.2010.494117.

Callon, M. 1986. "The Sociology of an Actor-Network." In *Mapping the Dynamics of Science and Technology*, ed. M. Callon, J. Law, and A. Rip, 19–34. London: Macmillan.

Callon, M. 1998a. "Actor-Network Theory: The Market Test." In *Actor Network and After*, ed. J. Law and J. Hassard, 181–95. Oxford and Keele: Blackwell and the Sociological Review.

Callon, M., ed. 1998b. *The Laws of the Markets*. London: Blackwell.

Callon, M. 2007. "What Does It Mean to Say That Economics Is Performative?" In *Do Economists Make Markets? On the Performativity of Economics*, ed. D. MacKenzie, F. Muniesa, and L. Siu, 311–57. Princeton: Princeton University Press.

Callon, M. 2010. "Performativity, Misfires and Politics." *Journal of Cultural Economics* 3 (2): 163–9. http://dx.doi.org/10.1080/17530350.2010.494119.

Carpenter, K. 1982. "Baby Dinosaurs from the Late Cretaceous Lance and Hell Creek Formations, and a Description of a New Species of Theropod." *Contributions to Geology, University of Wyoming* 20: 123–34.

Carpenter, K. 2004. "Redescription of *Ankylosaurus magniventris* Brown 1908 (Ankylosauridae) from the Upper Cretaceous of the Western Interior of North America." *Canadian Journal of Earth Sciences* 41 (8): 961–86. http://dx.doi.org/10.1139/e04-043.

Céard, J. 1977. *La Nature et les Prodiges. L'insolite en France au XVIème siècle.* Geneva: Librairie Droz.

Céard, J. 1991. "The Crisis in the Science of Monsters." In *Humanism in Crisis: The Decline of the French Renaissance*, ed. P. Desan, 181–205. Ann Arbor: University of Michigan Press.

Chase, P. 1977. *The Legacy of Malthus: The Social Costs of the New Scientific Racism.* New York: Alfred A. Knopf.

Cheah, P. 1996. "Mattering." *Diacritics* 26 (1): 108–39. http://dx.doi.org/10.1353/dia.1996.0004.

Clifford, J. 1997. *Routes: Travel and Translation in the Late Twentieth Century.* Cambridge, MA: Harvard University Press.

Colbert, E. 1965. *Dinosaurs, Their Discovery and Their World.* Toronto: Clark Irwin.

Colbert, E. 1968. *Men and Dinosaurs: The Search in Field and Laboratory.* New York: Dutton.

Collier, J., M. Rosaldo, and S. Yanagisako 1992. "Is There a Family? New Anthropological Views." In *Rethinking the Family*, 2nd ed., ed. B. Thorne, 3–48. Boston: Northeastern University Press.

Conn, S. 1998. *Museums and American Intellectual Life.* Chicago: University of Chicago Press.

Cope, Edward Drinker. "On the Characters of the Skull in the Hadrosauridae." *Proceedings of the Academy of Natural Sciences of Philadelphia* (1883): 97–107.

Crichton, M. 1990. *Jurassic Park.* New York: Alfred Knopf.

Crichton, M. 1995. *The Lost World.* New York: Alfred Knopf.

Currie, P.J. 1993. "On Mahars, Gryfs and the Paleontology of ERB." *Burroughs Bulletin* 16: 21–4.

Currie, P.J., and E.B. Koppelhus. 2005. *Dinosaur Provincial Park: A Spectacular Ancient Ecosystem Revealed* (Vol. 1). Bloomington: Indiana University Press.

Currie, P.J., and K. Padian, eds. 1997. *Encyclopedia of Dinosaurs.* San Diego: Academic Press.

Czerkas, S., and E. Olson, eds. 1987. *Dinosaurs Past and Present.* 2 vols. Los Angeles: Natural History Museum of Los Angeles County.

Czerkas, S.M., and D.F. Glut. 1982. *Dinosaurs, Mammoths, and Cavemen: The Art of Charles R. Knight.* New York: E.P. Dutton.

Dahlberg, F., ed. 1981. *Woman the Gatherer.* New Haven: Yale University Press.

Daston, L. 2004. "Type Specimens and Scientific Memory." Critical Inquiry, Online Paper in series "The Arts of Transmission." http://criticalinquiry.uchicago.edu/features/artsstatements/arts.daston.htm. http://dx.doi.org/10.1086/427306. Accessed April 6, 2009.

Davidson, J. 1997. *The Bone Sharp: The Life of Edward Drinker Cope*. Philadelphia: Academy of Natural Sciences.

de la Cadena, M. 2010. "Indigenous Cosmopolitics in the Andes: Conceptual Reflections Beyond 'Politics.'" *Cultural Anthropology* 25 (2): 334–70. http://dx.doi.org/10.1111/j.1548-1360.2010.01061.x.

Debus, A.A. 2006. *Dinosaurs in Fantastic Fiction: A Thematic Survey*. Jefferson, NC: McFarland.

Debus, A.A. 2010. *Prehistoric Monsters: The Real and Imagined Creatures of the Past That We Love to Fear*. Jefferson, NC: McFarland.

Debus, A.A., and D.E. Debus. 2002. *Paleoimagery: The Evolution of Dinosaurs in Art*. Jefferson, NC: McFarland.

de la Bellacasa, M.P. 2011. "Matters of Care in Technoscience: Assembling Neglected Things." *Social Studies of Science* 41 (1): 85–106. http://dx.doi.org/10.1177/0306312710380301.

Deleuze, G. 1995. *Negotiations, 1972–199*. Trans. M. Joughin. New York: New York University Press.

Deleuze, G., and F. Guattari. 1988. *A Thousand Plateaus: Capitalism and Schizophrenia*. Trans. B. Massumi. London: Athlone Press.

Derrida, J. 1988. "Signature, Event, Context." In *Limited, Inc*, ed. G. Graff, trans. Samuel Weber and Jeffrey Mehlman, 1–24. Evanston: Northwestern University Press.

Descola, P. 2009. "Human Natures." *Social Anthropology* 17 (2): 145–57. http://dx.doi.org/10.1111/j.1469-8676.2009.00063.x.

Desmond, A. 1977. *The Hot-Blooded Dinosaurs: A Revolution in Palaeontology*. London: Futura.

Dibley, B. 2011. "Museums and a Common World: Climate Change, Cosmopolitics, Museum Practice." *Museum and Society* 9 (2): 154–65.

Dickson, L. 1986. *The Museum Makers: The Story of the Royal Ontario Museum*. Toronto: Royal Ontario Museum.

Dingus, L., and T. Rowe. 1998. *The Mistaken Extinction: Dinosaur Evolution and the Origin of Birds*. New York: W.H. Freeman.

Disch, L.J. 2009. "'Faitiche'-izing the People: What Representative Democracy Might Learn from Science Studies." Working paper. http://papers.ssrn.com/sol3/papers.cfm?abstract_id=1449734. Accessed November 11, 2014.

Dodson, P., and S.D. Dawson. 1991. "Making the Fossil Record of Dinosaurs." *Modern Geology* 16: 3–15.

Doyle, A.C. 1896. [Untitled speech]. *The Critic*, August 1: 78–9.
Doyle, A.C. 1912a. *The Lost World: Being an Account of the Recent Amazing Adventures of Professor E. Challenger, Lord John Roxton, Professor Summerlee and Mr. Ed Malone of the Daily Gazette*. London, New York, and Toronto: Hodder and Stoughton.
Doyle, A.C. 1912b. *The Lost World: Being an Account of the Recent Amazing Adventures of Professor E. Challenger, Lord John Roxton, Professor Summerlee and Mr. Ed Malone of the Daily Gazette*. New York: George H. Doran, Hodder and Stoughton.
Doyle, A.C. 1994 [1912]. *The Lost World: Being an Account of the Recent Amazing Adventures of Professor E. Challenger, Lord John Roxton, Professor Summerlee and Mr. Ed Malone of the Daily Gazette*. London: Puffin.
Eberth, D.A., D.C. Evans, D. Brinkman, F. Therrien, D.H. Tanke, L.S. Russell, and Hans Sues. 2013. "Dinosaur Biostratigraphy of the Edmonton Group (Upper Cretaceous), Alberta, Canada: Evidence for Climate Influence." *Canadian Journal of Earth Sciences* 50 (7): 701–26. http://dx.doi.org/10.1139/cjes-2012-0185.
Esterhammer, A. 2000. *The Romantic Performative: Language and Action in British and German Romanticism*. Stanford, CA: Stanford University Press.
Evans, D., and R. Reisz. 2007. "Anatomy and Relationships of *Lambeosaurus magnicristatus*, a Crested Hadrosaurid Dinosaur (Ornithischia) from the Dinosaur Park Formation, Alberta." *Journal of Vertebrate Paleontology* 27 (2): 373–93. http://dx.doi.org/10.1671/0272-4634(2007)27[373:AAROLM]2.0.CO;2.
Fabian, J. 1983. *Time and the Other: How Anthropology Makes Its Object*. New York: Columbia University Press.
Farlow, J. 1997. "Dinosaurs and Geologic Time." In *The Complete Dinosaur*, ed. J. Farlow and M. Brett-Surman, 107–11. Bloomington, Indianapolis: Indiana University Press.
Farlow, J., and M.K. Brett-Surman, eds. 1997. *The Complete Dinosaur*. Bloomington, Indianapolis: Indiana University Press..
Fastovsky, D. 2007. "Ideas in Dinosaur Paleontology: Resonating to Social and Political Context." In *The Paleobiological Revolution: Essays on the Growth of Modern Paleontology*, ed. D. Sepkoski and M. Ruse, 239–53. Chicago: University of Chicago Press.
Felman, S. 1983. *The Literary Speech-Act: Don Juan with J.L. Austin, or Seduction in Two Languages*. Trans. Catherine Porter. Ithaca, NY: Cornell University Press.
Ferraris, M. 2009. "Documentality, or Europe." *Monist* 92 (2): 286–314. http://dx.doi.org/10.5840/monist200992216.
Findlen, P. 1994. *Possessing Nature: Museums, Collecting, and Scientific Culture in Early Modern Italy*. Berkeley, Los Angeles: University of California Press.

Fiorillo, A.R., and R.A. Gangloff. 2001. "The Caribou Migration Model for Arctic Hadrosaurs (Dinosauria: Ornithischia): A Reassessment." *Historical Biology: A Journal of Paleobiology* 15 (4): 323–34. http://dx.doi.org/10.1080/08912960210000037327.

Foucault, M. 1978. *History of Sexuality: The Will to Knowledge*, vol. 1. London: Penguin Books.

Foucault, M. 1980 [1977]. "The Confession of the Flesh." In *Power/Knowledge: Selected Interviews and Other Writings*, ed. C. Gordon, 194–228. London: Harvester Wheatsheaf.

Foucault, M. 2007. *Security, Territory, Population: Lectures at the Collège de France, 1977–1978*. New York: Palgrave Macmillan. http://dx.doi.org/10.1057/9780230245075.

Franklin, S. 1995a. "Science as Culture, Cultures of Science." *Annual Review of Anthropology* 24 (1): 163–84. http://dx.doi.org/10.1146/annurev.an.24.100195.001115.

Franklin, S. 1995b. "Romancing the Helix: Nature and Scientific Discovery." In *Romance Revisited*, ed. L. Pearce and J. Stacey, 63–77. London: Lawrence and Wishart.

Franklin, S. 1997a. *Embodied Progress: A Cultural Account of Assisted Conception*. London: Routledge. http://dx.doi.org/10.4324/9780203414965.

Franklin, S. 1997b. "It Takes All Kinds: Actor Network Theory and the Idea of Relation." Discussion paper for the Keele Actor Network Theory and Beyond Conference, Keele, UK, July 9–12, 1997.

Franklin, S. 2000. "Life Itself: Global Nature and the Genetic Imaginary." In *Global Nature, Global Culture*, ed. S. Franklin, C. Lury, and J. Stacey, 188–227. London: Sage. http://dx.doi.org/10.4135/9781446219768.n7.

Franklin, S. 2007. *Dolly Mixtures: The Remaking of Genealogy*. Durham, NC: Duke University Press. http://dx.doi.org/10.1215/9780822389651.

Franklin, S., and S. McKinnon, eds. 2001. *Relative Values: Reconfiguring Kinship Studies*. Durham, NC: Duke University Press.

Fujimura, J. 1987. "Constructing 'Do-Able' Problems in Cancer Research: Articulating Alignment." *Social Studies of Science* 17 (2): 257–93. http://dx.doi.org/10.1177/030631287017002003.

Fujimura, J. 1992. "Crafting Science: Standardized Packages, Boundary Objects, and 'Translations.'" In *Science as Culture and Practice*, ed. A. Pickering, 168–211. Chicago: University of Chicago Press.

Geller, P.L. 2008. "Conceive Sex: Fomenting a Feminist Bioarchaeology." *Journal of Social Archaeology* 8 (1): 113–38. http://dx.doi.org/10.1177/1469605307086080.

Gilmore, C.W. 1929. "Hunting Dinosaurs in Montana." In *Explorations and Fieldwork of the Smithsonian Institution in 1928*, 7–12. Washington, DC: Smithsonian Institution.

Gish, D.T. 1979. *Evolution? The Fossils Say No!* San Diego: Creation-Life Publishers.

Glut, D.F. 1980. *The Dinosaur Scrapbook*. Secaucus, NJ: Citadel Press.

Glut, D.F. 1997. *Dinosaurs: The Encyclopedia*. Jefferson, NC: McFarland.

Glut, D.F. 2008. "*Tyrannosaurus rex*: A Century of Celebrity." In *Tyrannosaurus rex, the Tyrant King*, ed. P. Larson and K. Carpenter, 398–427. Bloomington, Indianapolis: Indiana University Press.

Glut, D.F., and M.K. Brett-Surman. 1997. "Dinosaurs and the Media." In *The Complete Dinosaur*, ed. J. Farlow and M.K. Brett-Surman, 675–97. Bloomington, Indianapolis: Indiana University Press.

Gombrich, E.H. 2000 [1961]. *Art and Illusion: A Study in the Psychology of Pictorial Representation*. Princeton: Princeton University Press.

Gould, S.J. 1992. "Reconstructing (and Deconstructing) the Past." In *The Book of Life: An Illustrated History of the Evolution of Life on Earth*, 6–21. New York: Norton.

Gould, S.J. 1993. "Dinomania." *New York Review of Books*, August 12: 51–5.

Gould, S.J. 2000. *Wonderful Life: The Burgess Shale and the Nature of History*. New York: Random House.

Gradstein, F.M., F.P. Agterberg, J.G. Ogg, J. Hardenbol, P. van Veen, J. Thierry, and Z. Huang. 1994. "A Mesozoic Time Scale." *Journal of Geophysical Research* 99 (B12): 24051–74. http://dx.doi.org/10.1029/94JB01889.

Grant, M. 1922 [1916]. *The Passing of the Great Race or the Racial Basis of European History*. New York: Scribner.

Greenblatt, S. 1991. "Resonance and Wonder." In *Exhibiting Cultures: The Poetics and Politics of Museum Display*, ed. I. Karp and S.D. Lavine, 42–56. Washington, DC: Smithsonian Institution.

Gregory, K. 2007. "Who's Afraid of Dinosaurs? The Creation Museum and Family Discovery Center, Petersburg, Kentucky." *Contexts* 6 (1): 74–5. http://dx.doi.org/10.1525/ctx.2007.6.1.74.

Griesemer, J. 2004. "Three-Dimensional Models in Philosophical Perspective." In *Models: The Third Dimension of Science*, ed. J. Chadarevian and N. Hopwood, 433–42 . Stanford, CA: Stanford University Press.

Grossberg, L. 1986. "On Postmodernism and Articulation: An Interview with Stuart Hall." *Journal of Communication Inquiry* 10 (2): 45–60. http://dx.doi .org/10.1177/019685998601000204.

Gupta, A., ed. 2008. *Art Lies* 59. http://texashistory.unt.edu/ark:/67531/metapth228024. Accessed January 31, 2015.

Gupta, A., and J. Ferguson, eds. 1997. *Anthropological Locations: Boundaries and Grounds of a Field Science*. Berkeley: University of California Press.

Habermas, J., and T. McCarthy. 1977. "Hannah Arendt's Communications Concept of Power." *Social Research* 44: 3–25.

Hansen, C., C. Needham, and B. Nichols. 1991. "Pornography, Ethnography and the Discourses of Power." In *Representing Reality*, ed. B. Nichols, 201–28. Bloomington: Indiana University Press.

Haraway, D.J. 1989a. *Primate Visions: Gender, Race, and Nature in the World of Modern Science*. London, New York: Routledge.

Haraway, D.J. 1989b [1984]. "Teddy Bear Patriarchy: Taxidermy in the Garden of Eden, New York City, 1908–1936." In *Primate Visions: Gender, Race, and Nature in the World of Modern Science*, 26–58. London, New York: Routledge.

Haraway, D.J. 1991. "Situated Knowledges: The Science Question in Feminism and the Privilege of Partial Perspective." In *Simians, Cyborgs, and Women: The Reinvention of Nature*, ed. D.J. Haraway, 183–202. London, New York: Routledge.

Haraway, D.J. 1994. "A Game of Cat's Cradle: Science Studies, Feminist Theory, Cultural Studies." *Configurations* 2 (1): 59–71. http://dx.doi.org/10.1353/con.1994.0009.

Haraway, D.J. 1997. *Modest_Witness@Second_Millennium.FemaleMan©_Meets_OncoMouse*. London, New York: Routledge.

Haraway, D.J. 1998. "Deanimations: Maps and Portraits of Life Itself." In *Picturing Science, Producing Art*, ed. C. Jones and P. Galison, 181–207. New York: Routledge.

Haraway, D.J. 2000. "Morphing in the Order: Flexible Strategies, Feminist Science Studies, and Primate Revisions." In *Primate Encounters: Models of Science, Gender, and Society*, ed. S. Strum and L. Fedigan, 398–420. Chicago: University of Chicago Press.

Haraway, D.J. 2004. *The Haraway Reader*. Oxford: Psychology Press.

Haraway, D.J. 2007. *When Species Meet*. Minneapolis: University of Minnesota Press.

Harding, S. 1991. *Whose Science? Whose Knowledge?* Ithaca, NY: Cornell University Press.

Harris, N. 1990. *Cultural Excursion: Marketing Appetites and Cultural Tastes in Modern America*. Chicago: University of Chicago Press.

Harvey, P. 1996. *Hybrids of Modernity: Anthropology, the Nation State, and the Universal Exhibition*. London: Routledge.

Hawkins, B.H. 1854. "On Visual Education As Applied to Geology." *Journal of the Society of the Arts* 2 (78): 444–9.

Hayden, C. 2003. *When Nature Goes Public: The Making and Unmaking of Bioprospecting in Mexico*. Princeton: Princeton University Press.

Hayes, H. 1986. "The Dark Romance of Dian Fossey." *Life (Chicago, Ill.)* 9 (11): 64–70.

Head, J. 1998. "A New Species of Basal Hadrosaurid (Dinosauria, Ornithischia) from the Cenomanian of Texas." *Journal of Vertebrate Paleontology* 18 (4): 718–38. http://dx.doi.org/10.1080/02724634.1998.10011101.

Helmreich, S. 1998. *Silicon Second Nature: Culturing Artificial Life in a Digital World*. Berkeley: University of California Press.

Helmreich, S. 2009. *Alien Ocean: Anthropological Voyages in Microbial Seas*. Berkeley: University of California Press.

Henderson, D. 1997. "Restoring Dinosaurs as Living Animals." In *The Complete Dinosaur*, ed. J. Farlow and M.K. Brett-Surman, 165–72. Bloomington, Indianapolis: Indiana University Press.

Heuvelmans, B. 1995. *On the Track of Unknown Animals*. London: Kegan Paul.

Higham, C. 1987. *The Adventures of Conan Doyle: The Life of the Creator of Sherlock Holmes*. New York: Norton and Co.

Holtz, T.R. 2008. "A Critical Reappraisal of the Obligate Scavenging Hypothesis for *Tyrannosaurus rex* and Other Tyrant Saurians." In *Tyrannosaurus rex, the Tyrant King*, ed. P. Larson and K. Carpenter, 370–96. Bloomington, Indianapolis: Indiana University Press.

Hooper-Greenhill, E. 1992. *Museums and the Shaping of Knowledge*. London: Routledge.

Horner, J.R. 1982. "Evidence of Colonial Nesting and 'Site Fidelity' among Ornithischian Dinosaurs." *Nature* 297 (5868): 675–6. http://dx.doi.org/10.1038/297675a0.

Horner, J.R. 1983. "Cranial Osteology and Morphology of the Type Specimen of *Maiasaura peeblesorum* (Ornithischia: Hadrosauridae), with Discussion of its Phylogenetic Position." *Journal of Vertebrate Paleontology* 3 (1): 29–38. http://dx.doi.org/10.1080/02724634.1983.10011954.

Horner, J.R. 1984. "The Nesting Behavior of Dinosaurs." *Scientific American* 250 (4): 130–7. http://dx.doi.org/10.1038/scientificamerican0484-130.

Horner, J.R. 1987. "Ecological and Behavioral Implications Derived from a Dinosaur Nesting Site." In *Dinosaurs Past and Present*, vol. 2, ed. S.J. Czerkas and E.C. Olson, 51–63. Seattle, London: University of Washington Press.

Horner, J.R. 1994. "Steak Knives, Beady Eyes, and Tiny Little Arms (a Portrait of T. rex as a Scavenger)." In *Dino Fest Proceedings*, ed. G. Rosenberg and D. Wolberg. Paleontological Society Special Publication 7.

Horner, J.R., and J. Gorman. 1988. *Digging Dinosaurs: The Search That Unraveled the Mystery of Baby Dinosaurs*. New York: Workman Publishing.

Horner, J.R., and D. Lessem. 1994. *The Complete T. rex: How Stunning New Discoveries Are Changing Our Understanding of the World's Most Famous Dinosaur*. New York: Touchstone.

Horner, J.R., and R. Makela. 1979. "Nest of Juveniles Provides Evidence of Family Structure among Dinosaurs." *Nature* 282 (5736): 296–8. http://dx.doi.org/10.1038/282296a0.

Hunt, W. 1997. *Urban Entertainment Graphics*. New York: Madison Square Press.

Impey, O., and A. McGregor, eds. 1985. *The Origins of Museums*. Oxford: Clarendon Press.

Ingram, J.D. 2003. "The Genericide of Trademarks." *Buffalo IP Law Journal* 2: 154.

Jacknis, I. 1988. "Franz Boas and Exhibits." In *Objects and Others: Essays on Museums and Material Culture*, ed. G.W. Stocking, 75–111. Madison: University of Wisconsin Press.

Jaffe, J. 1987. *Arthur Conan Doyle*. Boston: Twayne Publishers.

Janes, R. 1995. *Museums and the Paradox of Change: A Case Study in Urgent Adaptation*. Calgary: University of Calgary Press.

Janes, R. 2009. *Museums in a Troubled World: Renewal, Irrelevance or Collapse?* New York: Routledge.

Jardine, N., J.A. Secord, and E.C. Spary, eds. 1995. *Cultures of Natural History*. Cambridge: Cambridge University Press.

Jay, M. 1988. "Scopic Regimes of Modernity." In *Vision and Visuality*, ed. Hal Foster, 3–23. Seattle: Bay Press.

Jensen, C., and K. Rödje, eds. 2010. *Deleuzian Intersections: Science, Technology, Anthropology*. New York, Oxford: Berghahn Books.

Jones, C.A., and P. Galison, eds. 1998. *Picturing Science, Producing Art*. New York: Routledge.

Jordanova, L. 1989. "Objects of Knowledge: An Historical Perspective on Museums." In *The New Museology*, ed. P. Vergo, 22–40. London: Reaktion Books.

Jordanova, L. 2004. "Material Models as Visual Culture." In *Models: The Third Dimension of Science*, ed. S.D. Chadarevian and N. Hopwood, 443–51. Stanford, CA: Stanford University Press.

Karp, I., ed. 1992. *Museums and Communities: The Politics of Public Culture*. Washington, DC: Smithsonian Press.

Karp, I., C. Kratz, L. Szwaja, and T. Ybarra-Frausto, eds. 2006. *Museum Frictions: Public Cultures/Global Transformations*. Durham, NC: Duke University Press. http://dx.doi.org/10.1215/9780822388296.

Kehoe, A.B., A.A. Wiley, G. Stocking, R. Darnell, N.F. Boas, and C.M. Hinsley. 2006. "Boas as Hamatsa: Appropriate for the Medal for Exemplary Service to Anthropology Award?" *Anthropology News* 47 (2): 4–5. http://dx.doi.org/10.1525/an.2006.47.2.4.

Kelty, C., and H. Landecker. 2004. "A Theory of Animation: Cells, L-systems, and Film." *Grey Room* 17 (Fall): 30–63. http://dx.doi.org/10.1162/1526381042464536.

Kestner, J.A. 1997. *Sherlock's Men: Masculinity, Conan Doyle, and Cultural History*. Brookfield, VT: Ashfield Publishing.

Kevles, D.J. 1985. *In the Name of Eugenics*. New York: Alfred Knopf.

Kielan-Jaworowska, Z. 1969. *Hunting for Dinosaurs*. Cambridge, MA: MIT Press.

Kirby, D. 2011. *Lab Coats in Hollywood: Science, Scientists, and Cinema*. Cambridge, MA: MIT Press.

Kirksey, S., and S. Helmreich. 2010. "The Emergence of Multispecies Ethnography." *Cultural Anthropology* 25 (4): 545–76. http://dx.doi.org/10.1111/j.1548-1360.2010.01069.x.

Kroetsch, R. 1975. *Badlands*. Toronto: New Press.

Krows, A.E. 1941. "Motion Pictures Not for Theatres: Dr. Raymond Lee Ditmars." *The Educational Screen: The Magazine Devoted to Audio-Visual Aids in Education*, February: 61.

Kuhn, T.S. 1962. *The Structure of Scientific Revolutions*. 1st ed. Chicago: University of Chicago Press.

Laclau, E., and C. Mouffe. 2001. *Hegemony and Socialist Strategy: Towards a Radical Democratic Politics*. London: Verso.

Landecker, H., and A. Panofsky. 2013. "From Social Structure to Gene Regulation, and Back: A Critical Introduction to Environmental Epigenetics for Sociology." *Annual Review of Sociology* 39 (1): 333–57. http://dx.doi.org/10.1146/annurev-soc-071312-145707.

Lankester, E.R. 1905. *Extinct Animals*. London: Archibald Constable & Co. http://dx.doi.org/10.5962/bhl.title.102107.

Laplanche, J., and J. Pontalis. 1986. "Fantasy and the Origins of Sexuality: Retrospect 1986." In *Formations of Fantasy*, ed. V. Burgin, J. Donald, and C. Kaplan, 5–34. London: Methuen.

Larson, N. 2008. "One Hundred Years of *Tyrannosaurus rex*: The Skeletons." In *Tyrannosaurus rex, the Tyrant King*, ed. P. Larson and K. Carpenter, 398–427. Bloomington, Indianapolis: Indiana University Press.

Larson, P., and K. Carpenter, eds. 2008. *Tyrannosaurus rex, the Tyrant King*. Bloomington, Indianapolis: Indiana University Press.

Latour, B. 1983. "Give Me a Laboratory and I Will Raise the World." In *Science Observed*, ed. K. Knorr Cetina and M. Mulkay, 141–70. Beverly Hills, CA: Sage.

Latour, B. 1987. *Science in Action: How to Follow Scientists and Engineers through Society*. Cambridge, MA: Harvard University Press.

Latour, B. 1993. *We Have Never Been Modern*. Cambridge, MA: Harvard University Press.

Latour, B. 1996. "'On Interobjectivity.' Special symposium, *Mind, Culture, and Activity*." *International Journal (Toronto, Ont.)* 3 (4): 228–69.

Latour, B. 1997. "A Few Steps toward an Anthropology of the Iconoclastic Gesture." *Science in Context* 10 (1): 63–83.

Latour, B. 1999. *Pandora's Hope: Essays on the Reality of Science Studies*. Cambridge, MA: Harvard University Press.

Latour, B. 2000. "A Well-Articulated Primatology." In *Primate Encounters: Models of Science, Gender, and Society*, ed. S. Strum and L. Fedigan, 358–82. Chicago: University of Chicago Press.

Latour, B. 2004. *Politics of Nature: How to Bring the Sciences into Democracy*. Cambridge, MA: Harvard University Press.

Latour, B. 2005. "From Realpolitik to Dingpolitik – An Introduction to Making Things Public." In *Making Things Public, Atmospheres of Democracy*, ed. B. Latour and P. Weibel, 4–31. Cambridge, MA: MIT Press.

Latour, B. 2010. "An Attempt at a 'Compositionist Manifesto.'" *New Literary History* 41 (3): 471–90.

Latour, B. 2011. "Politics of Nature: East and West Perspectives." *Ethics & Global Politics* 4 (1): 71–80. http://dx.doi.org/10.3402/egp.v4i1.6373.

Latour, B., and P. Weibel, eds. 2005. *Making Things Public, Atmospheres of Democracy*. Cambridge, MA: MIT Press.

Latour, B., and S. Woolgar. 1986. *Laboratory Life: The Construction of Scientific Facts*. Princeton: Princeton University Press.

Law, J. 2004a. *After Method: Mess in Social Science Research*. London: Routledge.

Law, J. 2004b. "Enacting Naturecultures: A Note from STS." Published by the Centre for Science Studies, Lancaster University. http://www.comp.lancs.ac.uk/sociology/papers/law-enacting-naturecultures.pdf. Accessed June 9, 2009.

Law, J., and J. Urry. 2004. "Enacting the Social." *Economy and Society* 33 (3): 390–410. http://dx.doi.org/10.1080/0308514042000225716.

Lessem, D. 1992. *Kings of Creation: How a New Breed of Scientists Is Revolutionizing Our Understanding of Dinosaurs*. New York: Simon & Schuster.

Lesser, W. 1987. *The Life below the Ground: A Study of the Subterranean in Literature and History*. Boston: Faber and Faber.

Lévi-Strauss, C. 1963. *Totemism*. Boston: Beacon Press.

Lightman, B. 2007. *Victorian Popularizers of Science: Designing Nature for New Audiences*. Chicago: University of Chicago Press. http://dx.doi.org/10.7208/chicago/9780226481173.001.0001.

Lockett, C., L. Menna, and E. Walker. 1998. "The Effects of Technology-Based Devices at the Royal Ontario Museum: Observations and Visitors' Perceptions." Unpublished draft article. Toronto: Royal Ontario Museum.
Lukens, R.J. 1999. *A Critical Handbook of Children's Literature*. New York: Longman.
Lutz, C., and J. Collins. 1993. *Reading National Geographic*. Chicago: University of Chicago Press.
Macdonald, S. 1993. "Un nouveau 'corps des visiteurs': Musées et changement culturels." *Publics et Musées* 3 (1): 13–27. http://dx.doi.org/10.3406/pumus.1993.1021.
Macdonald, S. 1996. "Introduction." In *Theorizing Museums: Representing Identity and Diversity in a Changing World*. Sociological Review Monograph Series, ed. S. Macdonald and G. Fyfe, 1–18. Oxford: Blackwell Publishers.
Macdonald, S. 1997. "Authorizing Science: Public Understanding of Science in Museums." In *Misunderstanding Science? The Public Reconstruction of Science and Technology*, ed. A. Irwin and B. Wynne, 153–71. Cambridge: Cambridge University Press.
Macdonald, S. 1998a. "Supermarket Science? Consumers and the 'Public Understanding of Science.'" In *The Politics of Display: Museums, Science, Culture*, ed. S. Macdonald, 118–38. New York: Routledge.
Macdonald, S. 1998b. "Exhibitions of Power and Powers of Exhibition: An Introduction to the Politics of Display." In *The Politics of Display: Museums, Science, Culture*, ed. S. Macdonald, 1–24. New York: Routledge.
MacKenzie, D. 2007. "Is Economics Performative? Option Theory and the Construction of Derivatives Markets." In *Do Economists Make Markets? On the Performativity of Economics*, ed. D. MacKenzie, F. Muniesa, and L. Siu, 54–86. Princeton: Princeton University Press.
MacKenzie, D., F. Muniesa, and L. Siu, eds. 2007. *Do Economists Make Markets? On the Performativity of Economics*. Princeton: Princeton University Press.
Mannion, P.D., R.B.J. Benson, P. Upchurch, R.J. Butler, M.T. Carrano, and P.M. Barrett. 2012. "A Temperate Palaeodiversity Peak in Mesozoic Dinosaurs and Evidence for Late Cretaceous Geographical Partitioning." *Global Ecology and Biogeography* 21 (9): 898–908. http://dx.doi.org/10.1111/j.1466-8238.2011.00735.x.
Marks, J. 1995. *Human Biodiversity: Genes, Race, and History*. New York: Aldine.
Marsh, D. 2014. "From 'Extinct Monsters' to Deep Time: An Ethnography of Fossil Exhibits Production at the Smithsonian's National Museum of Natural History." PhD diss., University of British Columbia.
Martin, E. 1994. *Flexible Bodies: Tracking Immunity in American Culture – From the Days of Polio to the Age of AIDS*. Boston: Beacon Press.

Martin, E. 1997. "Anthropology and the Cultural Study of Science: From Citadels to String Figures." In *Anthropological Locations: Boundaries and Grounds of a Field Science*, ed. A. Gupta and J. Ferguson, 131–46. Berkeley: University of California Press.

Matthew, W.D. 1915. *Dinosaurs, with Special Reference to the American Museum Collections*. New York: American Museum of Natural History. http://dx.doi.org/10.5962/bhl.title.102123.

McLaren, D. 1970. "Presidential Address: Life, Time and Boundaries." *Journal of Paleontology* 44 (5): 815.

McLuhan, M., H. Parker, and J. Barzun. 1969. *Exploration of the Ways, Means, and Values of Museum Communication with the Viewing Public*. New York: Museum of the City of New York.

Meloni, M., and G. Testa. 2014. "Scrutinizing the Epigenetics Revolution." *Biosocieties* 9 (4): 431–56. http://dx.doi.org/10.1057/biosoc.2014.22.

Mengue, P. 2008. "People and Fabulation." In *Deleuze and Politics*, ed. I. Buchanan and N. Thoburn, 218–39. Edinburgh: Edinburgh University Press. http://dx.doi.org/10.3366/edinburgh/9780748632879.003.0011.

Meyer, M. 2011. "Researchers on Display: Moving the Laboratory into the Museum." *Museum Management and Curatorship* 26 (3): 261–72. http://dx.doi.org/10.1080/09647775.2011.585800.

Miller, D., J. Summers, and S. Silber. 2004. "Environmental versus Genetic Sex Determination: A Possible Factor in Dinosaur Extinction?" *Fertility and Sterility* 81 (4): 954–64. http://dx.doi.org/10.1016/j.fertnstert.2003.09.051.

Miller, L.E. 1918. *In the Wilds of South America*. New York: Scribner.

Mitchell, W.J.T. 1998. *The Last Dinosaur Book*. Chicago: University of Chicago Press.

Miyazaki, H., and A. Riles. 2005. "Failure as an Endpoint." In *Global Assemblages: Technology, Politics, and Ethics as Anthropological Problems*, ed. A. Ong and S.J. Collier, 320–31. London: Blackwell Publishing.

Mol, A. 2002. *The Body Multiple*. Durham, NC: Duke University Press. http://dx.doi.org/10.1215/9780822384151.

Molnar, R. 1997. "Biogeography for Dinosaurs." In *The Complete Dinosaur*, ed. J. Farlow and M.K. Brett-Surman, 581–606. Bloomington, Indianapolis: Indiana University Press.

Morley, R.J. 2011. "Cretaceous and Tertiary Climate Change and the Past Distribution of Megathermal Rainforests." In *Tropical Rainforest Responses to Climatic Change*, ed. M. Bush, et al., 1–34. Berlin: Springer Praxis. http://dx.doi.org/10.1007/978-3-642-05383-2_1.

Morris, S.C. 2003. *Life's Solution: Inevitable Humans in a Lonely Universe*. Cambridge: Cambridge University Press. http://dx.doi.org/10.1017/CBO9780511535499.

Mulvey, L. 1985. "Visual Pleasure and Narrative Cinema." In *Movies and Methods*, vol. 2, ed. B. Nichols, 303–15. Berkeley: University of California Press.

Muybridge, E. 1969 [1887]. *Animal Locomotion: An Electro-photographic Investigation of Consecutive Phases of Animal Movements, 1872–1885*. New York: Da Capo Press.

Myers, N. 2006. "Animating Mechanism: Animations and the Propagation of Affect in the Lively Arts of Protein Modelling." *Science Studies* 19 (2): 6–30.

Noble, B.E. 1994. "Dinosaurographies: The Public Politics of Monstrous Fascination." MA thesis, University of Alberta.

Noble, B.E. 1997a. "Nature/Culture, Lost Worlds, Dinosaurs: Fetishes All the Way Down." Presented at the annual meetings of the American Anthropological Association, Washington, DC.

Noble, B.E. 1997b. "Dinosaur Fetishism, Museums and the Lost World: Evolutionism and Mimesis in the Making of the Non-Human Other." Presented in the "Visualizing Differences" session at the annual meetings of the Canadian Anthropological Society, St John's, NL.

Noble, B.E. 1998. "Dinosaur Resurrections: Phantasmatics in the Systematics of Life." Presented at the annual meetings of the Society for the Social Studies of Science, Halifax, NS.

Noble, B.E. 1999. "Saurian Resurrections." MIT/Harvard invited lecturer in Cultural Studies of Science. Cambridge, MA: Center for Literary and Cultural Studies, Harvard University.

Noble, B.E. 2000a. "Between Specimen and Spectacle: Culturing Dinosaurs and Performing Worlds in Museums and Palaeobiology." PhD dissertation, Department of Anthropology, University of Alberta.

Noble, B.E. 2000b. "Politics, Gender, and Worldly Primatology: The Goodall-Fossey Nexus." In *Primate Encounters: Models of Science, Gender, and Society*, ed. S. Strum and L. Fedigan, 436–62. Chicago: University of Chicago Press.

Noble, B.E. 2007. "Justice, Transaction, Translation: Blackfoot Tipi Transfers and WIPO's Search for the Facts of Traditional Knowledge Exchange." *American Anthropologist* 109 (2): 338–49. http://dx.doi.org/10.1525/aa.2007.109.2.338.

Norell, M. 1998. "Review of *T. Rex and the Crater of Doom* by Walter Alvarez." *Quarterly Review of Biology* 73 (3): 343–4. http://dx.doi.org/10.1086/420322.

Norell, M., E. Gaffney, and L. Dingus. 1995. *Discovering Dinosaurs in the American Museum of Natural History*. New York: Alfred A. Knopf.

Norman, D. 1991. *Dinosaur!* London: Boxtree.

O'Connor, R. 2008. *The Earth on Show: Fossils and the Poetics of Popular Science, 1802–1856*. Chicago: University of Chicago Press.

O'Hara, R.J. 1992. "Telling the Tree: Narrative Representation and the Study of Evolutionary History." *Biology and Philosophy* 7 (2): 135–60. http://dx.doi.org/10.1007/BF00129880.

Orel, H. 1991. *Sir Arthur Conan Doyle: Interviews and Recollections*. New York: St Martin's Press.

Ortner, S. 1984. "Theory in Anthropology since the Sixties." *Comparative Studies in Society and History* 26 (1): 126–66. http://dx.doi.org/10.1017/S0010417500010811.

Osborn, H.F. 1897. "A Great Naturalist: Edward Drinker Cope." *Center Magazine* 55: 10–5.

Osborn, H.F. 1898. "Models of Extinct Vertebrates." *Science* 7 (182): 841–5. http://dx.doi.org/10.1126/science.7.182.841.

Osborn, H.F. 1905. "*Tyrannosaurus* and Other Cretaceous Carnivorous Dinosaurs." *Bulletin of the American Museum of Natural History* 21: 259–65.

Osborn, H.F. 1906. "*Tyrannosaurus*, Upper Cretaceous Carnivorous Dinosaur (Second Communication)." *Bulletin of the American Museum of Natural History* 22: 281–96.

Osborn, H.F. 1912. "Tetraplasy, the Law of the Four Inseparable Factors of Evolution." *Journal of the Academy of Natural Sciences of Philadelphia*, special anniversary volume: 275–309.

Osborn, H.F. 1913. "Tyrannosaurus: Restoration and Model of the Skeleton." *Bulletin of the American Museum of Natural History* 32: 91–2.

Osborn, H.F. 1916. "Skeletal Adaptations of *Ornitholestes*, *Struthiomimus*, *Tyrannosaurus*." *Bulletin of the American Museum of Natural History* 35: 761–71.

Osborn, H.F. 1917. *The Origin and Evolution of Life: On the Theory of Action, Reaction and Interaction of Energy. Hale Lectures of the National Academy of Sciences, Washington, April 1916*. New York: Scribner.

Osborn, H.F. 1924. *Impressions of Great Naturalists*. New York: Scribner.

Osborn, H.F. 1926. "The Evolution of Human Races." *Natural History* (January/February); repr. in *Natural History* 89 (April 1980).

Ostrom, J.H. 1969. "A New Theropod Dinosaur from the Lower Cretaceous of Montana." *Yale Peabody Museum Postilla* 128: 1–17.

Pachner, J. 1995. "State of the Art Antiquity." *Electronic Link Magazine* 2 (2): 42–7.

Park, K., and L. Daston. 1981. "Unnatural Conceptions: The Study of Monsters in Sixteenth- and Seventeenth-Century France and England." *Past & Present* 92 (1): 20–54. http://dx.doi.org/10.1093/past/92.1.20.

Patterson, C. 1980. "Cladistics – Pattern versus Process in Nature: A Personal View of a Method and a Controversy." *Biologist (Columbus, Ohio)* 27 (5): 234–40.

Persons, W.S., IV, and P.J. Currie. 2014. "Duckbills on the Run: The Cursorial Abilities of Hadrosaurs and Implications for Tyrannosaur-Avoidance

Strategies." In *Hadrosaurs*, ed. D.A. Eberth and D.C. Evans, 449–58. Bloomington, Indianapolis: Indiana University Press.

Phillips, S. 1859. *Guide to the Crystal Palace and Its Park and Gardens*, rev. ed. Ed. F.K.J. Shelton. London: Robert K. Burt.

Pickersgill, M., J. Niewöhner, R. Müller, P. Martin, and S. Cunningham-Burley. 2013. "Mapping the New Molecular Landscape: Social Dimensions of Epigenetics." *New Genetics & Society* 32 (4): 429–47. http://dx.doi.org/10.1080/14636778.2013.861739.

Pignarre, P., and I. Stengers. 2011. *Capitalist Sorcery: Breaking the Spell*. Basingstoke: Palgrave Macmillan.

Pinkoski, M. 2011. "Back to Boas." *Histories of Anthropology Annual* 7 (1): 127–69. http://dx.doi.org/10.1353/haa.2011.0013.

Pomian, K. 1990. *Collectors and Curiosities. Paris and Venice, 1500–1800*. Cambridge: Polity Press.

Posadas, P., M. Grossi, and E. Ortiz-Jaureguizar. 2013. "Where Is Historical Biogeography Going? The Evolution of the Discipline in the First Decade of the 21st Century." *Progress in Physical Geography* 37 (3): 377–96. http://dx.doi.org/10.1177/0309133313478316.

Pratt, M.L. 1977. *Toward a Speech Act Theory of Literary Discourse*. Bloomington: Indiana University Press.

Pratt, M.L. 1992. *Imperial Eyes: Travel Writing and Transculturation*. London, New York: Routledge. http://dx.doi.org/10.4324/9780203163672.

Preston, D.J. 1986. *Dinosaurs in the Attic: An Excursion into the American Museum of Natural History*. New York: St Martin's Press.

Psihoyos, L. 1994. *Hunting Dinosaurs*. New York: Random House.

Raffles, H. 2011. *Insectopedia*. New York: Vintage.

Rainger, R. 1991. *An Agenda for Antiquity: Henry Fairfield Osborn and Vertebrate Paleontology at the American Museum of Natural History, 1890–1935*. Tuscaloosa: University of Alabama Press.

Rancière, J. 2009. *The Emancipated Spectator*. London: Verso.

Raup, D. 1999. *The Nemesis Affair: A Story of the Death of Dinosaurs and the Ways of Science*. New York: W.W. Norton & Company.

Regal, B. 2002. *Henry Fairfield Osborn: Race and the Search for the Origins of Man*. Aldershot: Ashgate.

Rexer, L., and R. Klein. 1995. *American Museum of Natural History: 125 Years of Expedition and Discovery*. New York: Abrams and AMNH.

Rich, A. 1979. "Women and Honor: Some Notes on Lying." In *On Lies, Secrets and Silence: Selected Prose, 1966–1978*, 185–94. New York: Norton.

Riegel, H. 1996. "Into the Heart of Irony: Ethnographic Exhibitions and the Politics of Difference." In *Theorizing Museums: Representing Identity and*

Diversity in a Changing World. Sociological Review Monograph Series, ed. S. Macdonald and G. Fyfe, 83–104. Oxford: Blackwell.

Riles, A. 2000. *The Network Inside Out*. Ann Arbor: University of Michigan Press.

Riles, A. 2004. "Property as Legal Knowledge: Means and Ends." *Journal of the Royal Anthropological Institute* 10 (4): 775–95. http://dx.doi.org/10.1111/j.1467-9655.2004.00211.x.

Riles, A., ed. 2006. *Documents: Artifacts of Modern Knowledge*. Ann Arbor: University of Michigan Press.

Ritvo, H. 1997. *The Platypus and the Mermaid, and Other Figments of the Classifying Imagination*. Cambridge, MA: Harvard University Press.

Rose, D.B. 2011. *Wild Dog Dreaming: Love and Extinction*. Charlottesville: University of Virginia Press.

Rose, N. 2006. *The Politics of Life Itself*. Princeton: Princeton University Press.

Rose, N. 2013. "The Human Sciences in a Biological Age." *Theory, Culture & Society* 30 (1): 3–34. http://dx.doi.org/10.1177/0263276412456569.

Ross, M.R., D. Duggan-Haas, and W. Allmon. 2013. "The Posture of *Tyrannosaurus rex*: Why Do Student Views Lag behind the Science?" *Journal of Geoscience Education* 61 (1): 145–60. http://dx.doi.org/10.5408/11-259.1.

Rovin, J. 1977. *From the Land beyond Beyond: The Films of Willis O'Brien and Ray Harryhausen*. New York: Berkley Windhover Books.

Rudwick, M. 1976. *The Meaning of Fossils: Episodes in the History of Palaeontology*. New York: Neale Watson Academic Publications.

Rudwick, M. 1992. *Scenes from Deep Time: Early Pictorial Representations of the Prehistoric World*. Chicago: University of Chicago Press.

Russell, D.A. 1987. "Models and Paintings of North American Dinosaurs." In *Dinosaurs Past and Present*, vol. 1, ed. S. Czerkas and E. Olson, 114–31. Los Angeles: Natural History Museum of Los Angeles County.

Russell, D.A. 2009. *Islands in the Cosmos*. Bloomington: University of Indiana Press.

Russell, D.A., and R. Séguin. 1982. "Reconstructions of the Small Cretaceous Theropod, *Stenonychosaurus inequalis*, and a Hypothetical Dinosauroid." Syllogeus 37. National Museum of Natural Sciences, Ottawa.

Russell, L.S. 1967. *Dinosaur Hunting in Western Canada*. Toronto: University of Toronto Press.

Sagan, C., R. Turco, G.W. Rathjens, R.H. Siegel, S.L. Thompson, and S.H. Schneider. 1986. "The Nuclear Winter Debate." *Foreign Affairs* 65 (1): 163–78. http://dx.doi.org/10.2307/20042868.

Said, E. 1979. *Orientalism*. New York: Vintage.

Sandino, L. 2012. "A Curatocracy: Who and What Is a V&A Curator?" In *Museums and Biographies: Stories, Objects, Identities*, ed. K. Hill, 87–101. Ipswich: Boydell Press.

Saunders, B. 2004. "Not a Cultural Relativist. The Boasian Legacy and Burden." In *The Challenges of Native American Studies: Essays in Celebration of the 25th American Indian Workshop*, ed. B. Saunders and L. Zuyderhoudt, 107–23. Leuven: Leuven University Press.

Schneider, D.M. 1980. *American Kinship: A Cultural Account*. Chicago: University of Chicago Press.

Schudson, M. 1997a. "Paper Tigers: A Sociologist Follows Cultural Studies into the Wilderness." *Lingua Franca*, August: 49–56.

Schudson, M. 1997b. "Cultural Studies and the Social Construction of 'Social Construction.'" In *From Sociology to Cultural Studies: New Perspectives*, ed. E. Long, 379–98. Oxford: Blackwell Publishers.

Secord, J.A. 2004. "Monsters at the Crystal Palace." In *Models: The Third Dimension of Science*, ed. S. de Chadarevian and N. Hopwood, 138–69. Stanford, CA: Stanford University Press.

Segal, D., and S. Yanagisako, eds. 2005. *Unwrapping the Sacred Bundle: Reflections on the Disciplining of Anthropology*. Durham, NC: Duke University Press Books. http://dx.doi.org/10.1215/9780822386841.

Semonin, P. 1997. "Empire and Extinction: The Dinosaur as a Metaphor for Dominance in Prehistoric Nature." *Leonardo* 30 (3): 171–82. http://dx.doi.org/10.2307/1576441.

Sereno, P. 1990. "Clades and Grades in Dinosaur Systematics." In *Dinosaur Systematics: Approaches and Perspectives*, ed. K. Carpenter and P. Currie, 9–20. Cambridge: Cambridge University Press. http://dx.doi.org/10.1017/CBO9780511608377.004.

Sereno, P. 1999. "Definitions in Phylogenetic Taxonomy: Critique and Rationale." *Systematic Biology* 48 (2): 329–51. http://dx.doi.org/10.1080/106351599260328.

Shapin, S., and S. Schaffer. 1985. *Leviathan and the Air-Pump: Hobbes, Boyle and the Experimental Life*. Princeton: Princeton University Press.

Silverstone, R. 1992. *Consuming Technologies: Media and Information in Domestic Spaces*. London: Routledge. http://dx.doi.org/10.4324/9780203401491.

Simpson, G.G. 1967 [1949]. *The Meaning of Evolution*. New Haven: Yale University Press.

Simpson, G.G. 1983. *Fossils and the History of Life*. New York: Scientific American Books.

Simpson, G.G. 1984. *Discoverers of the Lost World: An Account of Some of Those Who Brought Back to Life South American Mammals Long Buried in the Abyss of Time*. New Haven, London: Yale University Press.

Simpson, G.G. 1996. *The Dechronization of Sam Magruder: A Novel*. New York: St Martin's Press.

Singer, C. 1931. *A Short History of Biology: A General Introduction to the Study of Living Things*. Oxford: Clarendon Press.

Sippi, D. 1989. "Aping Africa: The Mist of Immaculate Miscegenation." *CineAction* 18: 18.

Slack, J. 1996. "The Theory and Method of Articulation in Cultural Studies." In *Stuart Hall: Critical Dialogues in Cultural Studies*, ed. D. Morley and K. Chen, 112–27. London, New York: Routledge.

Smart, A. 2014. "Critical Perspectives on Multispecies Ethnography." *Critique of Anthropology* 34 (1): 3–7. http://dx.doi.org/10.1177/0308275X13510749.

Snow, C.P. 1993 [1959]. *The Two Cultures*. Cambridge: Cambridge University Press. http://dx.doi.org/10.1017/CBO9780511819940.

Sommer, M. 2007. "The Lost World as Laboratory: The Politics of Evolution between Science and Fiction in the Early Decades of Twentieth-Century America." *Configurations* 15 (3): 299–329. http://dx.doi.org/10.1353/con.0.0033.

Spalding, D.A.E. 1993. *The Dinosaur Hunters*. Toronto: Key Porter.

Squier, S. 1999. "From Omega Man to Mr. Adam: The Importance of Literature for Feminist Science Studies." *Science, Technology & Human Values* 24 (1): 132–58. http://dx.doi.org/10.1177/016224399902400107.

Star, S.L. 1989. *Regions of Mind: Brain Research and the Quest for Scientific Certainty*. Stanford, CA: Stanford University Press.

Star, S.L., ed. 1995. *Ecologies of Knowledge: Work and Politics in Science and Technology*. Albany: SUNY Press.

Star, S.L., and J.R. Griesemer. 1989. "Institutional Ecology, 'Translations,' and Boundary Objects: Amateurs and Professionals in Berkeley's Museum of Vertebrate Zoology, 1907–39." *Social Studies of Science* 19 (3): 387–420. http://dx.doi.org/10.1177/030631289019003001.

Stein, J., ed. 1983. *The Random House Dictionary of the English Language*. New York: Random House.

Stengers, I. 1997. *Power and Invention: Situating Science*. Minneapolis: University of Minnesota Press.

Stengers, I. 2008a. "Whitehead and Science: From Philosophy of Nature to Speculative Cosmology." Seminar paper. http://www.mcgill.ca/files/hpsc/Whitmontreal.pdf. Accessed September 12, 2012.

Stengers, I. 2008b. "Experimenting with Refrains: Subjectivity and the Challenge of Escaping Modern Dualism." *Subjectivity* 22 (1): 38–59. http://dx.doi.org/10.1057/sub.2008.6.

Stengers, I. 2008c. "A Constructivist Reading of *Process and Reality*." *Theory, Culture & Society* 25 (4): 91–110. http://dx.doi.org/10.1177/0263276408091985.

Stengers, I. 2008d. "History through the Middle: Between Macro and Mesopolitics." Interview with Isabelle Stengers, November 25. www.senselab.ca/inflexions/volume_3/.../Stengers_en_mesopolitique.pdf. Accessed November 23, 2010.

Stengers, I. 2009. "William James: An Ethics of Thought?" *Radical Philosophy* 157: 9–17.

Stengers, I. 2010a. *Cosmopolitics I*. Trans. Robert Bononno. Minneapolis: University of Minnesota Press.

Stengers, I. 2010b. "Gilles Deleuze's Last Message." http://www.recalcitrance.com/deleuzelast.htm. Accessed November 15, 2011.

Stengers, I. 2010c. "Experimenting with What Is Philosophy?" In *Deleuzian Intersections: Science, Technology, Anthropology*, ed. C. Jensen and K. Rödje, 39–56. New York, Oxford: Berghahn Books.

Stengers, I. 2010d. "Including Nonhumans in Political Theory: Opening Pandora's Box?" In *Political Matter: Technoscience, Democracy, and Public Life*, ed. B. Braun and S.J. Whatmore, 3–33. Minneapolis: University of Minnesota Press.

Stengers, I. 2011. *Thinking with Whitehead: A Free and Wild Creation of Concepts*. Cambridge, MA: Harvard University Press.

Stengers, I. 2012a. *Cosmopolitics II*. Trans. Robert Bononno. Minneapolis: University of Minnesota Press.

Stengers, I. 2012b. "Cosmopolitics – Learning to Think with Sciences, Peoples, Natures. To See Where It Takes Us." Lecture presented at Saint Mary's University, Halifax, NS. https://www.youtube.com/watch?v=1I0ipr61SI8. Accessed September 15, 2015.

Sternberg, C.H. 1917. *Hunting Dinosaurs on the Red Deer River, Alberta, Canada*. Lawrence, KS: published by author. http://dx.doi.org/10.5962/bhl.title.57209.

Stocking, G. 1968. *Race, Culture, and Evolution: Essays in the History of Anthropology*. New York: Macmillan.

Stocking, G. 1974. *The Shaping of American Anthropology 1883–1911: A Franz Boas Reader*. New York: Basic Books.

Strathern, M. 1992a. *After Nature: English Kinship in the Late 20th Century*. Cambridge: Cambridge University Press.

Strathern, M. 1992b. *Reproducing the Future: Anthropology, Kinship, and the New Reproductive Technologies*. London, New York: Routledge.

Strathern, M. 1996. "Cutting the Network." *Journal of the Royal Anthropological Institute* 2 (3): 517–35. http://dx.doi.org/10.2307/3034901.

Strathern, M. 2004. *Partial Connections*, vol. 3. Walnut Creek, CA: AltaMira Press.

Strathern, M. 2005. *Kinship, Law and the Unexpected: Relatives Are Always a Surprise*. Cambridge: Cambridge University Press. http://dx.doi.org/10.1017/CBO9780511614514.

Strathern, M. 2006. "A Community of Critics? Thoughts on New Knowledge." *Journal of the Royal Anthropological Institute* 12 (1): 191–209. http://dx.doi.org/10.1111/j.1467-9655.2006.00287.x.

Strathern, M., ed. 1995. *Shifting Contexts: Transformations in Anthropological Knowledge*. London: Routledge. http://dx.doi.org/10.4324/9780203450901.

Strum, S., and L. Fedigan, eds. 2000. *Primate Encounters: Changing Images of Primate Societies: Models of Science, Gender, and Society*. Chicago: University of Chicago Press.

Teilhard de Chardin, P. 1976 [1959]. *The Phenomenon of Man*. New York: Harper Perennial.

Thomas, N. 1991. *Entangled Objects: Exchange, Material Culture and Colonialism in the Pacific*. Cambridge, MA: Harvard University Press.

Thomas, N. 1994. *Colonialism's Culture: Anthropology, Travel and Government*. Princeton: Princeton University Press.

Traweek, S. 1992. "Border Crossings: Narrative Strategies in Science Studies and among Physicists in Tsukuba Science City, Japan." In *Science as Culture and Practice*, ed. A. Pickering, 429–66. Chicago: University of Chicago Press.

Trefethen, J.B. 1961. *Crusade for Wildlife: Highlights in Conservation Progress*. Harrisburg: Stackpole Company and Boone and Crockett Club.

Tully, J. 2011. *Challenging Practices of Citizenship: Citizenship for the Love of the World*. Keynote address (draft). Conference on Challenging Citizenship, University of Coimbra, Coimbra, Portugal, June 3.

Turco, R.P., O.B. Toon, T.P. Ackerman, J.B. Pollack, and C. Sagan. 1990. "Climate and Smoke: An Appraisal of Nuclear Winter." *Science* 247 (4939): 166–76. http://dx.doi.org/10.1126/science.11538069.

Upchurch, P., C.A. Hunn, and D.B. Norman. 2002. "An Analysis of Dinosaurian Biogeography: Evidence for the Existence of Vicariance and Dispersal Patterns Caused by Geological Events." *Proceedings of the Royal Society of London. Series B, Biological Sciences* 269 (1491): 613–21. http://dx.doi.org/10.1098/rspb.2001.1921.

Van Valkenburgh, B., and R.E. Molnar. 2002. "Dinosaurian and Mammalian Predators Compared." *Paleobiology* 28 (4): 527–43. http://dx.doi.org/10.1666/0094-8373(2002)028<0527:DAMPC>2.0.CO;2.

Varricchio, D.J., P.C. Sereno, Z. Xijin, T. Lin, J.A. Wilson, and G.H. Lyon. 2008. "Mud-Trapped Herd Captures Evidence of Distinctive Dinosaur Sociality." *Acta Palaeontologica Polonica* 53 (4): 567–78. http://dx.doi.org/10.4202/app.2008.0402.

Verne, J. 1877. *Voyage au Centre de la Terre*. Paris: Hetzel.

Verne, J. 1965. *Journey to the Centre of the Earth*. Trans. R. Baldick. Harmondsworth: Penguin.

Viveiros de Castro, E. 2012. "Cosmological Perspectivism in Amazonia and Elsewhere." *HAU: Masterclass Series* 1: 45–168.

Ward, P.D. 1992. *On Methuselah's Trail: Living Fossils and the Great Extinctions*. New York: W.H. Freeman and Company.

Warner, M. 1994. *Managing Monsters: Six Myths of Our Times. The Reith Lectures 1994*. London: Vintage.

Whitehead, A.N. 2011. *Science and the Modern World*. Cambridge: Cambridge University Press.

Whyte, J. 1988. "Modern Dreams, Ancient Reality." In *Tyrrell Museum of Palaeontology*, ed. J. Foster and D. Harrison. Special Edition of *Alberta: Studies in the Arts and Sciences* 1 (1): 141–47. Edmonton: University of Alberta Press.

Wiley, E.O., and B. Lieberman. 2011. *Phylogenetics: Theory and Practice of Phylogenetic Systematics*. Hoboken, NJ: Wiley-Blackwell. http://dx.doi.org/10.1002/9781118017883.

Wilford, J.N. 1986. *The Riddle of the Dinosaur*. New York: Alfred A. Knopf.

Wilkinson, D.M., E. Nisbet, and G. Ruxton. 2012. "Could Methane Produced by Sauropod Dinosaurs Have Helped Drive Mesozoic Climate Warmth?" *Current Biology* 22 (9): 292–3. http://dx.doi.org/10.1016/j.cub.2012.03.042.

Williams, R. 1985. *Keywords: A Vocabulary of Culture and Society*. New York: Oxford University Press.

Wilson, R. 1988. "The Struthiomimus's Tale: Discourse in the Tyrrell Museum." *Alberta Studies in the Arts and Sciences* 1 (1): 75–95.

Wing, S.L., H. Sues, et al. 1992. "Mesozoic and Early Cenozoic Terrestrial Ecoystems." In *Terrestrial Ecosystems through Time*, ed. A.K. Behrensmeyer et al., 327–416. Chicago: University of Chicago Press.

Winsor, M.P. 1991. *Reading the Shape of Nature: Comparative Zoology at the Agassiz Museum*. Chicago, London: University of Chicago Press. http://dx.doi.org/10.7208/chicago/9780226902081.001.0001.

Witcomb, A. 2006. "Interactivity: Thinking Beyond." In *A Companion to Museum Studies*, vol. 39: 353–61, ed. S. Macdonald. London: John Wiley & Sons. http://dx.doi.org/10.1002/9780470996836.ch21.

Wittgenstein, L. 1958. *Philosophical Investigations*. New York: Macmillan.

Wittgenstein, L. 1969. *On Certainty*. Oxford: Blackwell.

Wittkower, R. 1942. "Marvels of the East: A Study in the History of Monsters." *Journal of the Warburg Institute* 5: 159–97. http://dx.doi.org/10.2307/750452.

Wonders, K. 1993. *Habitat Dioramas: Illusions of Wilderness in Museums of Natural History*. Figura Nova Series 25. Uppsala: Acta Universitatis Usaliensis.

Young, R.J.C. 1995. *Colonial Desire: Hybridity in Theory, Culture, and Race*. London, New York: Routledge.

Young, R.M. 1992. "Science, Ideology and Donna Haraway." *Science as Culture* 3 (2): 165–207. http://dx.doi.org/10.1080/09505439209526344.

Index